D1433860

Introduction to Stochastic Differential Equations with Applications to Modelling in Biology and Finance

Introduction to Stochastic Differential Equations with Applications to Modelling in Biology and Finance

Carlos A. Braumann
University of Évora
Portugal

Registered Offices
John Wiley & Sons, Inc., 111 River Street, Hoboken, NJ 07030, USA
John Wiley & Sons Ltd, The Atrium, Southern Gate, Chichester, West Sussex, PO19 8SQ, UK

Editorial Office
9600 Garsington Road, Oxford, OX4 2DQ, UK

For details of our global editorial offices, customer services, and more information about Wiley products visit us at www.wiley.com.

Wiley also publishes its books in a variety of electronic formats and by print-on-demand. Some content that appears in standard print versions of this book may not be available in other formats.

Library of Congress Cataloging-in-Publication Data

Names: Braumann, Carlos A., 1951- author.
Title: Introduction to stochastic differential equations with applications to modelling in biology and finance / Carlos A. Braumann (University of Évora, Évora [Portugal]).
Other titles: Stochastic differential equations with applications to modelling in biology and finance
Description: Hoboken, NJ : Wiley, [2019] | Includes bibliographical references and index. |
Identifiers: LCCN 2018060336 (print) | LCCN 2019001885 (ebook) | ISBN 9781119166078 (Adobe PDF) | ISBN 9781119166085 (ePub) | ISBN 9781119166061 (hardcover)
Subjects: LCSH: Stochastic differential equations. | Biology–Mathematical models. | Finance–Mathematical models.
Classification: LCC QA274.23 (ebook) | LCC QA274.23 .B7257 2019 (print) | DDC 519.2/2–dc23
LC record available at https://lccn.loc.gov/2018060336

Cover Design: Wiley
Cover Image: © nikille/Shutterstock

Set in 10/12pt WarnockPro by SPi Global, Chennai, India
Printed in Singapore by C.O.S. Printers Pte Ltd

10 9 8 7 6 5 4 3 2 1

To Manuela

Contents

Preface

This is a beginner's book intended as an introduction to stochastic differential equations (SDEs), covering both theory and applications. SDEs are basically differential equations describing the 'average' dynamical behaviour of some phenomenon with an additional stochastic term describing the effect of random perturbations in environmental conditions (environment taken here in a very broad sense) that influence the phenomenon. They have important and increasing applications in basically all fields of science and technology, and they are ubiquitous in modern finance. I feel that the connection between theory and applications is a very powerful tool in mathematical modelling and makes for a better understanding of the theory and its motivations. Therefore, this book illustrates the concepts and theory with several applications. They are mostly real-life applications coming from the biological, bio-economical, and the financial worlds, based on the research experience (concentrated on biological and bio-economical applications) and teaching experience of the author and his co-workers, but the methodologies used are of interest to readers interested in applications in other areas and even to readers already acquainted with SDEs.

This book wishes to serve both mathematically strong readers and students, academic community members, and practitioners from different areas (mainly from biology and finance) that wish to use SDEs in modelling. It requires basic knowledge of calculus, probability, and statistics. The other required concepts will be provided in the book, emphasizing the intuitive ideas behind the concepts and the way to translate from the phenomena being studied to the mathematical model and to translate back the conclusions for application in the real world. But the book will, at the same time, also give a rigorous treatment, with technical definitions and the most important proofs, including several quite technical definitions and proofs that the less mathematically inclined reader can overlook, using instead the intuitive grasp of what is going on. Since the book is also concerned with effective applicability, it includes a first approach to some of the statistical issues of estimation and prediction, as well as Monte Carlo simulation.

A long-standing issue concerns which stochastic calculus for SDEs, Itô or Stratonovich, is more appropriate in a particular application, an issue that has raised some controversy. For a large class of SDE models we have resolved the controversy by showing that, once the unnoticed semantic confusion traditionally present in the literature is cleared, both calculi can be indifferently used, producing the same results.

I prefer to start with the simplest possible framework, instead of the maximum generality, in order to better carry over the ideas and methodologies, and provide a better intuition to the reader contacting them for the first time, thus avoiding obscuring things with heavy notations and complex technicalities. So the book follows this approach, with extensions to more general frameworks being presented afterwards and, in the more complex cases, referred to other books. There are many interesting subjects (like stochastic stability, optimal control, jump diffusions, further statistical and simulation methodologies, etc.) that are beyond the scope of this book, but I am sure the interested reader will acquire here the knowledge required to later study these subjects should (s)he wish.

The present book was born from a mini-course I gave at the XIII Annual Congress of the Portuguese Statistical Society and the associated extended lecture notes (Braumann, 2005), published in Portuguese and sold out for some years. I am grateful to the Society for that opportunity. The material was revised and considerably enlarged for this book, covering more theoretical issues and a wider range of applications, as well as statistical issues, which are important for real-life applications. The lecture notes have been extensively used in classes of different graduate courses on SDEs and applications and on introduction to financial mathematics, and as accessory material, by me and other colleagues from several institutions, in courses on stochastic processes or mathematical models in biology, both for students with a more mathematical background and students with a background in biology, economics, management, engineering, and other areas. The lecture notes have also served me in structuring many mini-courses I gave at universities in several countries and at international summer schools and conferences. I thank the colleagues and students that have provided me with information on typos and other errors they found, as well as for their suggestions for future improvement. I have tried to incorporate them into this new book.

The teaching and research work that sustains this book was developed over the years at the University of Évora (Portugal) and at its Centro de Investigação em Matemática e Aplicações (CIMA), a research centre that has been funded by Fundação para a Ciência e a Tecnologia, Portugal (FCT), the current FCT funding reference being UID/MAT/04674/2019. I am grateful to the university and to FCT for the continuing support. I also wish to thank my co-workers, particularly the co-authors of several papers; some of the material shown here is the result of joint work with them. I am grateful also to Wiley for the invitation and the opportunity to write this book and for exercising some patience when my predictions on the conclusion date proved to be too optimistic.

I hope the reader, for whom this book was produced, will enjoy it and make good use of its reading.

Carlos A. Braumann

About the companion website

This book is accompanied by a companion website:

www.wiley.com/go/braumann/stochastic-differential-equations

The website includes:

- Solutions to exercises

Scan this QR code to visit the companion website.

1

Introduction

Stochastic differential equations (SDEs) are basically differential equations with an additional stochastic term. The deterministic term, which is common to ordinary differential equations, describes the 'average' dynamical behaviour of the phenomenon under study and the stochastic term describes the 'noise', i.e. the random perturbations that influence the phenomenon. Of course, in the particular case where such random perturbations are absent (deterministic case), the SDE becomes an ordinary differential equation.

As the dynamical behaviour of many natural phenomena can be described by differential equations, SDEs have important applications in basically all fields of science and technology whenever we need to consider random perturbations in the environmental conditions (environment taken here in a very broad sense) that affect such phenomena in a relevant manner.

As far as I know, the first SDE appeared in the literature in Uhlenbeck and Ornstein (1930). It is the *Ornstein–Uhlenbeck model* of *Brownian motion*, the solution of which is known as the *Ornestein–Uhlenbeck process*. Brownian motion is the irregular movement of particles suspended in a fluid, which was named after the botanist Brown, who first observed it at the microscope in the 19th century. The Ornstein–Ulhenbeck model improves Einstein treatment of Brownian motion. Einstein (1905) explained the phenomenon by the collisions of the particle with the molecules of the fluid and provided a model for the particle's position which corresponds to what was later called the *Wiener process*. The Wiener process and its relation with Brownian motion will be discussed on Chapters 3 and 4.

Although the first SDE appeared in 1930, we had to wait till the mid of the 20th century to have a rigorous mathematical theory of SDEs by Itô (1951). Since then the theory has developed considerably and been applied to physics, astronomy, electronics, telecommunications, civil engineering, chemistry, seismology, oceanography, meteorology, biology, fisheries, economics, finance, etc. Using SDEs, one can study phenomena like the dispersion of a pollutant in water or in the air, the effect of noise on the transmission of telecommunication

Introduction to Stochastic Differential Equations with Applications to Modelling in Biology and Finance, First Edition. Carlos A. Braumann.
© 2019 John Wiley & Sons Ltd. Published 2019 by John Wiley & Sons Ltd.
Companion Website: www.wiley.com/go/braumann/stochastic-differential-equations

signals, the trajectory of an artificial satellite, the location of a ship, the thermal noise in an electric circuit, the dynamics of a chemical reaction, the control of an insulin delivery device, the dynamics of one or several populations of living beings when environmental random perturbations affect their growth rates, the optimization of fishing policies for fish populations subject to random environmental fluctuations, the variation of interest rates or of exchange rates, the behaviour of stock prices, the value of a call or put financial option or the risk immunization of investment portfolios or of pension plans, just to mention a few examples.

We will give special attention to the modelling issues, particularly the translation from the physical phenomenon to the SDE model and back. This will be illustrated with several examples, mainly in biological or financial applications. The dynamics of biological phenomena (particularly the dynamics of populations of living beings) and of financial phenomena, besides some clear trends, are frequently influenced by unpredictable components due to the complexity and variability of environmental or market conditions. Such phenomena are therefore particularly prone to benefit from the use of SDE models in their study and so we will prioritize examples of application in these fields. The study of population dynamics is also a field to which the author has dedicated a good deal of his research work. As for financial applications, it has been one of the most active research areas in the last decades, after the pioneering works of Black and Scholes (1973), Merton (1971), and Merton (1973). The 1997 Nobel prize in Economics was given to Merton and Scholes (Black had already died) for their work on what is now called financial mathematics, particularly for their work on the valuation of financial options based on the stochastic calculus this book will introduce you to. In both areas, there is a clear cross-fertilization between theory and applications, with the needs induced by applications having considerably contributed to the development of the theory.

This book is intended to be read by both more mathematically oriented readers and by readers from other areas of science with the usual knowledge of calculus, probability, and statistics, who can skip the more technical parts. Due to the introductory character of this presentation, we will introduce SDEs in the simplest possible context, avoiding clouding the important ideas which we want to convey with heavy technicalities or cumbersome notations, without compromising rigour and directing the reader to more specialized literature when appropriate. In particular, we will only study stochastic differential equations in which the perturbing noise is a continuous-time *white noise*. The use of white noise as a reasonable approximation of real perturbing noises has a great advantage: the cumulative noise (i.e. the integral of the noise) is the *Wiener process*, which has the nice and mathematically convenient property of having independent increments.

The Wiener process, rigorously studied by Wiener and Lévy after 1920 (some literature also calls it the Wiener–Lévy process), is also frequently named

Brownian motion in the literature due to its association with the Einstein's first description of the Brownian motion of a particle suspended in a fluid in 1905. We personally prefer not to use this alternative naming since it identifies the physical phenomenon (the Brownian motion of particles) with its first mathematical model (the Wiener process), ignoring that there is an improved more realistic model (the Ornstein–Uhlenbeck process) of the same phenomenon. The 'invention' of the Wiener process is frequently attributed to Einstein, probably because it was thought he was the first one to use it (although at the time not yet under the name of 'Wiener process'). However, Bachelier (1900) had already used it as a (not very adequate) model for stock prices in the Paris Stock Market.

With the same concern of prioritizing simple contexts in order to more effectively convey the main ideas, we will deal first with unidimensional SDEs. But, of course, if one wishes to study several variables simultaneously (e.g. the value of several financial assets in the stock market or the size of several interacting populations), we need multidimensional SDEs (systems of SDEs). So, we will also present afterwards how to extend the study to the multidimensional case; with the exception of some special issues, the ideas are the same as in the unidimensional case with a slightly heavier matrix notation.

We assume the reader to be knowledgeable of basic probability and statistics as is common in many undergraduate degree studies. Of course, sometimes a few more advanced concepts in probability are required, as well as basic concepts in stochastic processes (random variables that change over time). Chapter 2 intends to refresh the basic probabilistic concepts and present the more advanced concepts in probability that are required, as well as to provide a very brief introduction to basic concepts in stochastic processes. The readers already familiar with these issues may skip it. The other readers should obviously read it, focusing their attention on the main ideas and the intuitive meaning of the concepts, which we will convey without sacrificing rigour.

Throughout the remaining chapters of this book we will have the same concern of conveying the main ideas and intuitive meaning of concepts and results, and advise readers to focus on them. Of course, alongside this we will also present the technical definitions and theorems that translate such ideas and intuitions into a formal mathematical framework (which will be particularly useful for the more mathematically trained readers).

Chapter 3 presents an example of an SDE that can be used to study the growth of a biological population in an environment with abundant resources and random perturbations that affect the population growth rate. The same model is known as the *Black–Scholes model* in the financial literature, where it is used to model the value of a stock in the stock market. This is a nice illustration of the universality of mathematics, but the reason for its presentation is to introduce the reader to the Wiener process and to SDEs in an informal manner.

Chapter 4 studies the more relevant aspects of the Wiener process. Chapter 5 introduces the diffusion processes, which are in a certain way generalizations of the Wiener process and which are going to play a key role in the study of SDEs. Later, we will show that, under certain regularity conditions, diffusion processes and solutions of SDEs are equivalent.

Given an initial condition and an SDE, i.e. given a Cauchy problem, its solution is the solution of the associated stochastic integral equation. In a way, either in the deterministic case or in the case of a stochastic environment, a Cauchy problem is no more than an integral equation in disguise, since the integral equation is the fulcrum of the theoretical treatment. In the stochastic world, it is the integral version of the SDE that truly makes sense since derivatives, as we shall see, do not exist in the current sense (the derivatives of the stochastic processes we deal with here only exist in a generalized sense, i.e. they are not proper stochastic processes). Therefore, for the associated stochastic integral equations to have a meaning, we need to define and study stochastic integrals. That is the object of Chapter 6. Unfortunately, the classical definition of Riemann–Stieltjes integrals (alongside trajectories) is not applicable because the integrator process (which is the Wiener process) is almost certainly of unbounded variation. Different choices of intermediate points in the approximating Riemann–Stieltjes sums lead to different results. There are, thus, several possible definitions of stochastic integrals. Itô's definition is the one with the best probabilistic properties and so it is, as we shall do here, the most commonly adopted. It does not, however, satisfy the usual rules of differential and integral calculus. The *Itô integral* follows different calculus rules, the *Itô calculus*; the key rule of this stochastic calculus is the Itô rule, given by the Itô theorem, which we present in Chapter 6. However, we will mention alternative definitions of the stochastic integral, particularly the *Stratonovich integral*, which does not have the nice probabilistic properties of the Itô integral but does satisfy the ordinary rules of calculus. We will discuss the use of one or the other calculus and present a very useful conversion formula between them. We will also present the generalization of the stochastic integral to several dimensions.

Chapter 7 will deal with the Cauchy problem for SDEs, which is equivalent to the corresponding stochastic integral equation. A main concern is whether the solution exists and is unique, and so we will present the most common existence and uniqueness theorem, as well as study the properties of the solution, particularly that of being a diffusion process under certain regularity conditions. We will also mention other results on existence and uniqueness of the solutions under weaker hypotheses. We end with the generalization to several dimensions. This chapter also takes a first look at how to perform Monte Carlo simulations of trajectories of the solution in order to get a random sample of such trajectories, which is particularly useful in applications.

Chapter 8 will study the Black–Scholes model presented in Chapter 3, obtaining the explicit solution and looking at its properties. Since the solutions under the Itô and the Stratonovich calculi are different (even on relevant qualitative properties), we will discuss the controversy over which calculus, Itô or

Stratonovich, is more appropriate for applications, a long-lasting controversy in the literature. This example serves also as a pretext to present, in Chapter 9, the author's result showing that the controversy makes no sense and is due to a semantic confusion. The resolution of the controversy is explained in the context of the example and then generalized to a wide class of SDEs.

Autonomous SDEs, in which the coefficients of the deterministic and the stochastic parts of the equation are functions of the state of the process (state that varies with time) but not direct functions of time, are particularly important in applications and, under mild regularity conditions, the solutions are homogeneous diffusion processes, also known as *Itô diffusions*.

In Chapter 10 we will talk about the Dynkin and the Feynman–Kac formulas. These formulas relate the expected value of certain functionals (that are important in many applications) of solutions of autonomous SDEs with solutions of certain partial differential equations.

In Chapter 11 we will study the unidimensional Itô diffusions (solutions of unidimensional autonomous SDEs) on issues such as first passage times, classification of the boundaries, and existence of stationary densities (a kind of stochastic equilibrium or stochastic analogue of equilibrium points of ordinary differential equations). These models are commonly used in many applications. For illustration, we will use the Ornstein–Uhlenbeck process, the solution of the first SDE in the literature.

In Chapter 12 we will present several examples of application in finance (the Vasicek model, used, for instance, to model interest rates and exchange rates), in biology (population dynamics model with the study of risk of extinction and distribution of the extinction time), in fisheries (with extinction issues and the study of the fishing policies in order to maximize the fishing yield or the profit), and in the modelling of the dynamics of human mortality rates (which are important in social security, pension plans, and life insurance). Often, SDEs, like ordinary differential equations, have no close form solutions, and so we need to use numerical approximations. In the stochastic case this has to be done for the several realizations or trajectories of the process, i.e. for the several possible histories of the random environmental conditions. Since it is impossible to consider all possible histories, we use Monte Carlo simulation, i.e. we do computer simulations to obtain a random sample of such histories. Like in statistics, sample quantities, like, for example, the sample mean or the sample distribution of quantities of interest, provide estimates of the corresponding mean or distribution of such quantities. We will be taking a look at these issues as they come up, reviewing them in a more organized way in Chapter 12 in the context of some applications.

Chapter 13 studies the problem of changing the probability measure as a way of modifying the SDE drift term (the deterministic part of the equation, which is the average trend of the dynamical behaviour) through the Girsanov theorem. This is a technical issue extremely important in the financial applications covered in the following chapter. The idea in such applications with risky financial assets is to change its drift to that of a riskless asset. This basically amounts to changing the risky asset average rate of return so that it becomes equal to the rate of return r of a riskless asset. Girsanov theorem shows that you can do this by artificially replacing the true probabilities of the different market histories by new probabilities (not the true ones) given by a probability measure called the equivalent martingale measure. In that way, if you discount the risky asset by the discount rate r, it becomes a martingale (a concept akin to a fair game) with respect to the new probability measure. Martingales have nice properties and you can compute easily things concerning the risky and derivative assets that interest you, just being careful to remember that results are with respect to the equivalent martingale measure. So, at the end you should reverse the change of probability measure to obtain the true results (results with respect to the true probability measure).

Chapter 14 assumes that there is no arbitrage in the markets and deals with the theory of option pricing and the derivation of the famous Black–Scholes formula, which are at the foundations of modern financial mathematics. Basically, the simple problem that we start with is to price a European call option on a stock. That option is a contract that gives you the right (but not the obligation) to buy that stock at a future prescribed time at a prescribed price, irrespective of the market price of that stock at the prescribed date. Of course, you only exercise the option if it is advantageous to you, i.e. if such a market price is above the option prescribed price. How much should you fairly pay for such a contract? The Black–Scholes formula gives you the answer and, as a by-product, also determines what can be done by the institution with which you have the contract in order to avoid having a loss. Basically, starting with the money you have paid for the contract and using it in a self-sustained way, the institution should buy and sell certain quantities of the stock and of a riskless asset following a so-called hedging strategy, which ensures that, at the end, it will have exactly what you gain from the option (zero if you do not exercise it or the difference between the market value and the exercise value if you do exercise it). We will use two alternative ways of obtaining the Black–Scholes formula. One uses Girsanov theorem and is quite convenient because it can be applied in other more complex situations for which you do not have an explicit expression; in such a case, we can recur to an approximation, the so-called binomial model, which we will also study. We will also consider European put options and take a quick look at American options. Other types of options and generalizations to more complex situations (like dealing with several risky assets instead of just one) will be considered but without going into details. In fact, this chapter is just intended as an introduction which will enable you to follow more specialized literature should you wish to get involved with more complex situations in mathematical finance.

Chapter 15 presents a summary of the most relevant issues considered in this book in order to give you a synthetic final view in an informal way. Since this will prioritize intuition, reading it right away might be a good idea if we are just interested in a fast intuitive grasp of these matters.

Throughout the book, there are indications on how to implement computing algorithms (e.g. for Monte Carlo simulations) using a spreadsheet or R language codes.

From Chapter 4 onwards there are proposed exercises for the reader. Exercises marked with * are for the more mathematically oriented reader. Solutions to exercises can be found in the Wiley companion website to this book.

2

Revision of probability and stochastic processes

2.1 Revision of probabilistic concepts

Consider a *probability space* $(\Omega, \mathcal{F}, P)$, where $(\Omega, \mathcal{F})$ is a *measurable space* and P is a *probability* defined on it. Usually, it is a model for a real-world phenomenon or an experiment that depends on chance (i.e. is random) and we shall now see what each element of the triplet $(\Omega, \mathcal{F}, P)$ means.

The *universal set* or *sample space* Ω is a non-empty set containing all possible conditions that may influence the outcome of the random phenomenon or experiment.

If we throw two dice simultaneously, say one white and one black, and are interested in the outcome (number of dots on each of the two dice), the space Ω could be the set of all possible 'physical scenarios' describing the throwing of the dice, such as the position of the hands, how strongly and in what direction we throw the dice, the density of the air, and many other factors, some of which we are not even aware. To each such physical scenario there would correspond an outcome in terms of number of dots, but we know little or nothing about the probabilities of the different scenarios or about the correspondence between scenarios and outcomes. Therefore, actually working with this complex space of 'physical scenarios' is not very practical. Fortunately, what really interests us are the actual outcomes determined by the physical scenarios and the probabilities of occurrence of those outcomes. It is therefore legitimate to adopt, as we do, the simplified version of using as Ω the much simpler space of the possible outcomes of the throwing of the dice. So, we will use as our sample space the 36-element set $\Omega = \{1 \circ 1, 1 \circ 2, 1 \circ 3, 1 \circ 4, 1 \circ 5, 1 \circ 6, 2 \circ 1, 2 \circ 2, 2 \circ 3, 2 \circ 4, 2 \circ 5, 2 \circ 6, 3 \circ 1, 3 \circ 2, 3 \circ 3, 3 \circ 4, 3 \circ 5, 3 \circ 6, 4 \circ 1, 4 \circ 2, 4 \circ 3, 4 \circ 4, 4 \circ 5, 4 \circ 6, 5 \circ 1, 5 \circ 2, 5 \circ 3, 5 \circ 4, 5 \circ 5, 5 \circ 6, 6 \circ 1, 6 \circ 2, 6 \circ 3, 6 \circ 4, 6 \circ 5, 6 \circ 6\}$. For instance, the element $\omega = 3 \circ 4$ represents the outcome 'three dots on the white dice and four dots on the black dice'. This outcome is an elementary or simple event, but we may be interested in more complex events, such as having '10 or more dots' on the launching of the two dice,

Introduction to Stochastic Differential Equations with Applications to Modelling in Biology and Finance, First Edition. Carlos A. Braumann.
© 2019 John Wiley & Sons Ltd. Published 2019 by John Wiley & Sons Ltd.
Companion Website: www.wiley.com/go/braumann/stochastic-differential-equations

event that will happen if any of the outcomes 4∘6, 5∘5, 5∘6, 6∘4, 6∘5 or 6∘6 occurs. This event can then be identified with the set of all six individual outcomes that are favourable to its realization, namely the six-element set $A = \{4∘6, 5∘5, 5∘6, 6∘4, 6∘5, 6∘6\}$. For simplicity, an elementary event will be also defined as a set having a single element, for instance the elementary event 'three dots on the white dice and four dots on the black dice' will correspond to the one-element set $C = \{3∘4\}$ and its probability is $P(C) = \frac{1}{36}$, assuming we have fair dice. In such a way, an event, whether elementary or more complex, is always a subset of Ω. But the reverse is not necessarily true and it is up to us to decide, according to our needs and following certain mandatory rules, which subsets of Ω are we going to consider as events. The set of all such events is the class $\mathcal{F}$ referred to above. It is a class, i.e. a higher-order set, because its constituting elements (the events) are sets.

What are the mandatory rules we should obey in choosing the class $\mathcal{F}$ of events? Only one: $\mathcal{F}$ should be a σ-*algebra* of subsets of Ω, which means that all its elements are subsets of Ω and the following three properties are satisfied:

- $\Omega \in \mathcal{F}$, i.e. the universal set must be an event.
- $\mathcal{F}$ is closed to complementation, i.e. if a set A is in $\mathcal{F}$, so is its complement $A^c = \{\omega \in \Omega : \omega \notin A\}$ (the set of elements ω that do not belong to A). **Note:** Since $\Omega \in \mathcal{F}$, also the empty set $\emptyset = \Omega^c \in \mathcal{F}$.
- $\mathcal{F}$ is closed to countable unions of sets. This means that, given any countable collection (i.e. a collection with a finite or a countably infinite number) of sets A_n ($n = 1, 2, \ldots$) that are in $\mathcal{F}$, the union $\bigcup_n A_n$ is also in $\mathcal{F}$. **Note:** To be clear, given an uncountable number of sets in $\mathcal{F}$, we do not require (nor forbid) their union to be in $\mathcal{F}$.

The sets $A \in \mathcal{F}$ are called *events* or *measurable sets* and are the sets for which the probability P is defined. [1] We can loosely interpret $\mathcal{F}$ as the available 'information', in the sense that the events in $\mathcal{F}$ will have its probability defined, while the other sets (those not belonging to $\mathcal{F}$) will not.

The probability P is a function from $\mathcal{F}$ to the $[0, 1]$ interval which is normed and σ-*additive*. By normed we mean that $P(\Omega) = 1$. By σ-additive we mean that, if $A_n \in \mathcal{F}$ ($n = 1, 2, \ldots$) is a countable collection of pairwise disjoint sets,

---

1 One may think that, ideally, we could put in $\mathcal{F}$ all subsets of Ω. Unless other reasons apply (e.g. restrictions on available information), that is indeed the typical choice when Ω is a finite set, like in the example of the dice, or even when Ω is an infinite countable set. However, when Ω is an infinite uncountable set, for example the set of real numbers or an interval of real numbers, this choice is, in most applications, not viable; in fact, such $\mathcal{F}$ would be so huge and would have so many 'strange' subsets of Ω that we could not possibly define the probabilities of their occurrence in a sensible way without running into a contradiction. In such cases, we choose a σ-algebra $\mathcal{F}$ that contains not all subsets of Ω, but rather all subsets of Ω that are really of interest in applications.

then $P\left(\bigcup_n A_n\right) = \sum_n P(A_n)$.[2] For each event $A \in \mathcal{F}$, $P(A)$ is a real number ≥ 0 and ≤ 1 that represents the probability of occurrence of A in our phenomenon or experiment. These properties of probabilities seem quite natural.

In the example of the two dice, assuming they are fair dice, all elementary events (such as, for example, the event $C = \{3 \circ 4\}$ of having 'three dots on the white dice and four dots on the black dice') have probability $\frac{1}{36}$. In this example, we take $\mathcal{F}$ as the class that includes all subsets of Ω (the reader should excuse me for not listing them, but they are too many, exactly $2^{36} = 68719476736$). Since an event with N elements is the union of its disjoint constituent elementary events, its probability is the sum of the probabilities of the elementary events, i.e. $\frac{N}{36}$; for example, the probability of the event $A = \{4 \circ 6, 5 \circ 5, 5 \circ 6, 6 \circ 4, 6 \circ 5, 6 \circ 6\} = \{4 \circ 6\} + \{5 \circ 5\} + \{5 \circ 6\} + \{6 \circ 4\} + \{6 \circ 5\} + \{6 \circ 6\}$ ('having 10 or more dots') is $P(A) = \frac{1}{36} + \frac{1}{36} + \frac{1}{36} + \frac{1}{36} + \frac{1}{36} + \frac{1}{36} = \frac{6}{36}$.

The properties of the probability explain the reason why $\mathcal{F}$ should be closed to complementation and to countable unions. In fact, from the properties of P, if one can compute the probability of an event A, one can also compute the probability of its complement $P(A^c) = 1 - P(A)$ and, if one can compute the probabilities of the events A_n $(n = 1, 2, \ldots)$, one can compute the probability of the event $\bigcup_n A_n$ (it is easy if they are pairwise disjoint, in which case the probability of their union is just the sum of their probabilities, and is a bit more complicated, but it can be done, if they are not pairwise disjoint). Therefore, we can consider A^c and $\bigcup_n A_n$ also as events and it would be silly (and even inconvenient) not to do so.

When studying, for instance, the evolution of the price of a stock of some company, it will be influenced by the 'market scenario' that has occurred during such evolution. By market scenario we may consider a multi-factorial description that includes the evolution along time (past, present, and future) of everything that can affect the price of the stock, such as the sales of the company, the prices of other stocks, the behaviour of relevant national and international economic variables, the political situation, armed conflicts, the psychological reactions of the market stakeholders, etc. Although, through the use of random variables and stochastic processes (to be considered later in this chapter), we will in practice work with a different much simpler space, the space of outcomes, we can conceptually take this complex space of market scenarios as being our sample space Ω, even though we know very little about it. In so doing, we can say that, to each concrete market scenario belonging to Ω there corresponds as an outcome a particular time evolution of the stock price. The same question arises when, for example, we are dealing with the evolution of the size

---

2 A collection of sets is pairwise disjoint when any pair of distinct sets in the collection is disjoint. A pair of sets is disjoint when the two sets have no elements in common. When dealing with pairwise disjoint events, it is customary to talk about the sum of the events as meaning their union. So, for example, we write $A + B$ as an alternative to $A \bigcup B$ when A and B are disjoint. The σ-additive property can therefore be written in the suggestive notation $P\left(\sum_n A_n\right) = \sum_n P(A_n)$.

of a population of living beings, which is influenced by the 'state of nature' (incorporating aspects such as the time evolution of weather, habitat, other interacting populations, etc.); here, too, we may conceptually consider the set of possible states of nature as our sample space Ω, such that, to each particular state in Ω, there corresponds as an outcome a particular time evolution of the population size.

The concrete market scenario [or the state of nature] ω that really occurs is an element of Ω 'chosen at random' according to the probability law P. You may think of the occurring market scenario [state of nature] as the result of throwing a huge dice with many faces, each corresponding to a different possible market scenario [state of nature]; however, such dice will not be fair, i.e. the faces will not have the same probability of occurrence, but rather have probabilities of occurrence equal to the probabilities of occurrence of the corresponding market scenarios [states of nature]. $\mathcal{F}$ is the σ-algebra of the subsets of Ω (events) for which the probability P is defined. The probability P assigns to each event (set of market scenarios [states of nature]) $A \in \mathcal{F}$ the probability $P(A)$ of that event happening, i.e. the probability that the occurring market scenario [state of nature] ω belongs to the set A.

We can assume, without loss of generality, and we will do it from now on, that the probability space $(\Omega, \mathcal{F}, P)$ is *complete*. In fact, when the space is not complete, we can proceed to its completion in a very easy way.[3]

We will now remind you of the definition of a *random variable* (r.v.). Consider the example above of throwing two dice and the random variable $X =$ 'total number of dots on the two dice'. For example, if the outcome of the launching was $\omega = 3\circ4$ ('three dots on the white dice and four dots on the black dice'), the corresponding value of X would be $3 + 4 = 7$. This r.v. X is a function that assigns to each possible outcome $\omega \in \Omega$ a real number, in this case the sum of the number of dots of the two dice. So, if the outcome is $\omega = 3\circ4$, we may write $X(\omega) = 7$, which is often abbreviated to $X = 7$. Taking the outcome ω as representing (being determined by) 'chance', we may say that a random variable is a function of 'chance'. Other examples of random variables are the tomorrow's closing rate of a stock, the dollar–euro exchange rate 90 days from now, the height of a randomly chosen person or the size of a population one year from now.

To give the general definition, we need first to consider a σ-algebra structure on the set of real numbers $\mathbb{R}$, where X takes its values. In fact, we would like to obtain probabilities of the random variable taking certain values. For instance,

---

3 These are more technical issues that we now explain for the so interested readers. The probability space is complete if, given any set $N \in \mathcal{F}$ such that $P(N) = 0$, all subsets Z of N will also belong to $\mathcal{F}$; such sets Z will, of course, also have zero probability. If the probability space is not complete, its completion consists simply in enlarging the class $\mathcal{F}$ in order to include all sets of the form $A \cup Z$ with $A \in \mathcal{F}$ and in extending the probability P to the enlarged σ-algebra by putting $P(A \cup Z) = P(A)$.

in the example we just gave, we may be interested in the probability of having $X \geq 10$. The choice of including in the σ-algebra all subsets of $\mathbb{R}$ will not often work properly (may be too big) and in fact we are only interested in subsets of $\mathbb{R}$ that are intervals of real numbers or can be constructed by countable set operations on such intervals. So we choose in $\mathbb{R}$ the *Borel σ-algebra B*, which is the σ-algebra generated by the intervals of $\mathbb{R}$, i.e. the smallest σ-algebra that includes all intervals of real numbers. Of course, it also includes other sets such as $] - 2, 3[\cup[7, 25[\cup\{100\}$.[4] Interestingly, if we use the open sets of $\mathbb{R}$ instead of the intervals, the σ-algebra generated by them is exactly the same σ-algebra B. B is also generated by the intervals of the form $] - \infty, x]$ (with $x \in \mathbb{R}$). The sets $B \in B$ are called *Borel sets*.

How do we proceed to compute the probability $P[X \in B]$ that a r.v. X takes a value in the Borel set B? It should be the probability of the favourable set of all the $\omega \in \Omega$ for which $X(\omega) \in B$. This event is called the *inverse image* of B by X and can be denoted by $X^{-1}(B)$ or alternatively by $[X \in B]$; formally, $X^{-1}(B) = [X \in B] := \{\omega \in \Omega : X(\omega) \in B\}$. In other words, the inverse image of B is the set of all elements $\omega \in \Omega$ which direct image $X(\omega)$ is in B. For example, in the case of the two dice and the random variable $X =$ 'total number of dots on the two dice', the probability that $X \geq 10$, which is to say $X \in B$ with $B = [10, +\infty[$, will be the probability of the inverse image $A = X^{-1}(B) = \{4\circ6, 5\circ5, 5\circ6, 6\circ4, 6\circ5, 6\circ6\}$ (which is the event 'having 10 or more dots on the two dice'). So, $P[X \geq 10] = P[X \in B] = P(A) = \frac{6}{36}$.[5] Notice the use of parenthesis and square brackets. In fact, we use $P(A)$ to define the probability of an event $A \in \mathcal{F}$; since the probability of $X \in B$ is the same as $P(A)$ with $A = [X \in B]$, one should write $P([X \in B])$, but, to lighten the notation, one simplifies to $P[X \in B]$.

Remember that $P(A)$ is only defined for events $A \in \mathcal{F}$. So, to allow the computation of $P(A)$ with $A = [X \in B]$, it is required that $A \in \mathcal{F}$. This requirement that the inverse images by X of Borel sets should be in $\mathcal{F}$, which is called the *measurability* of X, needs therefore to be included in the formal definition of random variable. Of course, this requirement is automatically satisfied in the example of the two dice because we have taken $\mathcal{F}$ as the class of all subsets of Ω.

Summarizing, we can state the formal definition of *random variable* (r.v.) X, also called *$\mathcal{F}$-measurable function* (usually abbreviated to *measurable function*), defined on the measurable space $(\Omega, \mathcal{F})$. It is a function from Ω to $\mathbb{R}$ such that, given any Borel set $B \in B$, its inverse image $X^{-1}(B) = [X \in B] \in \mathcal{F}$.

---

4 In fact, since $\{100\} = [100,100]$, this is the union of three intervals and we know that σ-algebras are closed to countable unions.

5 Since X only takes integer values between 2 and 12, $X \geq 10$ is also equivalent to $X \in B_1$ with $B_1 = \{10, 11, 12\}$ or to $X \in B_2$ with $B_2 =]9.5, 12.7]$. This is not a problem since the inverse images by X of the Borel sets B_1 and B_2 coincide with the inverse image of B, namely the event $A = \{4\circ6, 5\circ5, 5\circ6, 6\circ4, 6\circ5, 6\circ6\}$.

Given the probability space $(\Omega, \mathcal{F}, P)$ and a r.v. X (on the measurable space $(\Omega, \mathcal{F})$), its *distribution function* (d.f.) will be denoted here by F_X. Let me remind the reader that the d.f. of X is defined for $x \in \mathbb{R}$ by $F_X(x) := P[X \in] - \infty, x]] = P[X \leq x]$. It completely characterizes the probability distribution of X since we can, for any Borel set B, use it to compute $P[X \in B] = \int_B 1 dF_X(y)$.[6]

The class $\sigma(X)$ formed by the inverse images by X of the Borel sets is a sub-σ-algebra of $\mathcal{F}$, called the *σ-algebra generated* by X; it contains the 'information' that is pertinent to determine the behaviour of X. For example, when we throw two dice and Y is the r.v. that takes the value 1 if the two dice have equal number of dots and takes the value 0 otherwise, the σ-algebra generated by Y is the class $\sigma(Y) = (\emptyset, \Omega, D, D^c)$, with $D = [Y = 1] = \{1 \circ 1, 2 \circ 2, 3 \circ 3, 4 \circ 4, 5 \circ 5, 6 \circ 6\}$ (note that $D^c = [Y = 0]$).

In most applications, we will work with random variables that are either discrete or absolutely continuous, although there are random variables that do not fall into either category.

In the example of the two dice, $X =$ 'total number of dots on the two dice' is an example of a *discrete r.v.* A r.v. X is discrete if there is a countable set $S = \{a_1, a_2, \dots\}$ of real numbers such that $P[X \in S] = 1$; we will denote by $p_X(k) = P[X = a_k]$ $(k = 1, 2, \dots)$ its *probability mass function* (pmf). Notice that, since $P[X \in S] = 1$, we have $\sum_k p_X(k) = 1$. The pmf completely characterizes the distribution function since $F_X(x) = \sum_{k:\, a_k \leq x} p_X(k)$. One can compute the probability $P[X \in B]$ of a Borel set B by $P[X \in B] = \sum_{k:\, a_k \in B} p_X(k)$.

If $p_X(k) > 0$, we say that a_k is an *atom* of X. In the example, the atoms are $a_1 = 2$, $a_2 = 3$, $a_3 = 4$, ..., $a_{11} = 12$ and the probability mass function is $p_X(1) = P[X = a_1] = \frac{1}{36}$ (the only favourable case is $1 \circ 1$), $p_X(2) = \frac{2}{36}$ (the favourable cases are $1 \circ 2$ and $2 \circ 1$), $p_X(3) = \frac{3}{36}$ (the favourable cases are $1 \circ 3$, $2 \circ 2$ and $3 \circ 1$), ..., $p_X(11) = \frac{1}{36}$ (the only favourable case is $6 \circ 6$). Figure 2.1 shows the pmf and the d.f. In this example, $P[X \geq 10] = P[X \in B]$ with $B = [10, +\infty[$, and so $P[X \in B] = \sum_{k:\, a_k \in B} p_X(k) = p_X(9) + p_X(10) + p_X(11) = \frac{3}{36} + \frac{2}{36} + \frac{1}{36} = \frac{6}{36}$, since $a_9 = 10$, $a_{10} = 11$ and $a_{11} = 12$ are the only atoms that belong to B.

A r.v. X is said to be *absolutely continuous* (commonly, but not very correctly, one abbreviates to 'continuous r.v.') if there is a non-negative integrable function $f_X(x)$, called the *probability density function* (pdf), such that $F_X(x) = \int_{-\infty}^x f_X(y) dy$. Since $F_X(+\infty) := \lim_{x \to +\infty} F_X(x) = P[X < +\infty] = 1$,

---

6 As a technical curiosity, note that, if P is a probability in $(\Omega, \mathcal{F})$, the function $P_X(B) = P[X \in B] := P(X^{-1}(B))$ is well defined for all Borel sets B and is a probability, so we have a new probability space $(\mathbb{R}, \mathcal{B}, P_X)$. In the example of the two dice and $X =$ 'total number of dots on the two dice', instead of writing $P[X \in B_1] = P[X \in \{10, 11, 12\}] = \frac{6}{36}$, we could write equivalently $P_X(\{10, 11, 12\}) = \frac{6}{36}$.

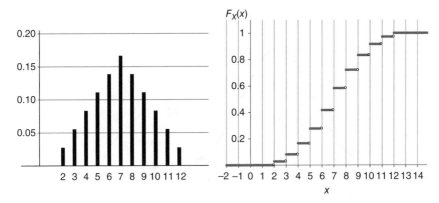

Figure 2.1 Example of the r.v. X = 'total number of dots on the two dice'. Looking at the left figure, it depicts the pmf: the values of the atoms $a_1 = 2, a_2 = 3, \ldots, a_{11} = 12$ appear on the horizontal axis and the corresponding probabilities $p_X(k)$ ($k = 1, 2, \ldots, 11$) are the heights of the corresponding bars. The figure on the right shows the d.f. $F_X(x)$; notice that, at the atoms a_k, this function is continuous on the left but discontinuous on the right.

we have $\int_{-\infty}^{+\infty} f_X(y)dy = 1$. Now, we should not look at the probability of each possible value x the r.v. X can take, since $P[X = x]$ would be zero and we would get no information from it. Instead, we should look at the probability of small neighbourhoods $[x, x + dx]$ (with dx small), which will be approximately given by $f_X(x)dx$, so $f_X(x)$ is not the probability of $X = x$ (that probability is zero) but rather a probability density. Notice that the d.f., being the integral of the pdf, is just the area underneath $f_X(x)$ for x varying in the interval $]-\infty, x]$ (see Figure 2.2). One can see that the pdf completely characterizes the d.f. If X is absolutely continuous, then its d.f. $F_X(x)$ is a continuous function. It is even a differentiable function almost everywhere, i.e. the exceptional set N of real numbers where F_X is not differentiable is a *negligible set*.[7] The derivative of the d.f. is precisely the pdf, i.e. $f_X(x) = dF_X(x)/dx$.[8] Given a Borel set B, we have $P[X \in B] = \int_B f_X(y)dy$, which is the area underneath $f_X(x)$ for x varying in B (see Figure 2.2).

---

7 A negligible set on the real line is a set with zero *Lebesgue measure*, i.e. a set with zero length (the Lebesgue measure extends the concept of length of an interval on the real line to a larger class of sets of real numbers). For example, all sets with a finite or countably infinite number of points like $\{3.14\}$ and $\{1, 2, 3, \ldots\}$, are negligible. An example of a non-negligible set is the interval $[4, 21.2[$, which has Lebesgue measure 17.2.

8 When $N \neq \emptyset$, the derivative is not defined for the exceptional points $x \in N$ but one can, for mathematical convenience, arbitrarily attribute values to the derivative $f_X(x)$ at those points in order for the pdf to be defined, although not uniquely, for all $x \in \mathbb{R}$. This procedure does not affect the computation of the d.f. $F_X(x) = \int_{-\infty}^{x} f_X(y)dy$ nor the probabilities $P[X \in B]$ for Borel sets B. So, basically, one does not care what values are thus attributed to $f_X(x)$ at the exceptional points.

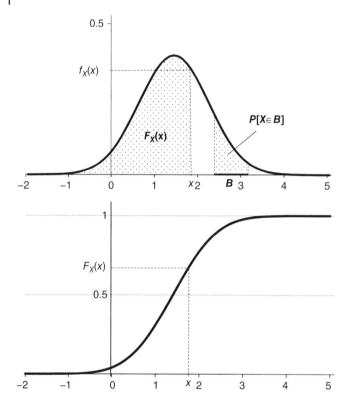

Figure 2.2 Example of a Gaussian random variable X with mean 1.4 and standard deviation 0.8. On top, the pdf $f_X(x)$ is shown; the two shaded areas represent the distribution function $F_X(x) = \int_{-\infty}^{x} f_X(y)dy$ of point x and the probability $P[X \in B] = \int_B f_X(y)dy$ of the Borel set B (which is, in this example, an interval). At the bottom, the d.f. is shown.

Figure 2.2 depicts the particular case of a normal r.v. X with mean 1.4 and standard deviation 0.8 (i.e. with variance $0.8^2 = 0.64$). A *normal* or *Gaussian* or *normally distributed* r.v. X with mean μ and *standard deviation* $\sigma > 0$ (which means that the variance is $\sigma^2 > 0$) is an absolutely continuous r.v. with pdf

$$f_X(x) = \frac{1}{\sigma\sqrt{2\pi}} \exp\left(-\frac{(x - \mu)^2}{2\sigma^2}\right) \quad (-\infty < x < +\infty).$$

We will sometimes use $X \frown \mathcal{N}(\mu, \sigma^2)$ to say that the r.v. X has a normal distribution with mean μ and variance σ^2.[9] The *standard normal distribution* has mean 0 and standard deviation 1; its pdf is commonly denoted by

---

9 Some authors prefer to use instead $X \frown \mathcal{N}(\mu, \sigma)$ to mean that X has a normal distribution with mean μ and standard deviation σ. So, one should check if the second argument stands for the variance or for the standard deviation.

$\phi(x) = \frac{1}{\sqrt{2\pi}} \exp\left(-\frac{x^2}{2}\right)$ $(-\infty < x < +\infty)$ and its d.f. by $\Phi(x) = \int_{-\infty}^{x} \phi(y)dy$. Note that, if $X \frown \mathcal{N}(\mu, \sigma^2)$, one can *standardize* it to a new r.v. $Z = \frac{X-\mu}{\sigma} \frown \mathcal{N}(0,1)$.

Most spreadsheets provide direct computations of the pdf (for absolutely continuous r.v.) or the pmf (for discrete r.v.) and of the d.f. of the most commonly used distributions. Looking at Figure 2.2 depicting the pdf of the normally distributed r.v. $X \frown \mathcal{N}(1.4, 0.64)$, when $x = 1.85$, we can, using a spreadsheet, compute $f_X(1.85) = 0.4257$ and $F_X(1.85) = 0.7131$ (the shadowed area on the left); one can also compute $P[X \in B]$ where $B =]2.35, 3.15]$ using $P[X \in]2.35, 3.15]] = F_X(3.15) - F_X(2.35) = 0.985647 - 0.882485 = 0.1032$ (the shadowed area on the right). Note that, since the r.v. X is absolutely continuous, it does not matter whether the interval B is closed, open or semi-closed on either side, for example $P[X \in [2.35, 3.15]] = P[X \in]2.35, 3.15]]$, since the probability of the single point 2.35 is $P[X = 2.35] = P[X \in [2.35, 2.35]] = \int_{[2.35, 2.35]} f_X(y)dy = \int_{2.35}^{2.35} f_X(y)dy = 0$.

If you use the computer language R, you can use the package 'stats' and compute:

- the pdf at $x = 1.85$ using 'dnorm(1.85, mean=1.4, sd=0.8)'
- the d.f. at $x = 1.85$ using 'pnorm(1.85, mean=1.4, sd=0.8)'
- the probability $P[X \in B]$, with $B =]2.35, 3.15]$, using
  'pnorm(3.15, mean=1.4, sd=0.8)−pnorm(2.35, mean=1.4, sd=0.8)'.

Another very useful distribution is the *uniform distribution* on a finite interval $]a, b]$ $(-\infty < a < b < +\infty)$ of real numbers (it does not matter whether the interval is closed, open or semi-open). A r.v. X with such distribution is absolutely continuous and has pdf (its value at $x = a$ or $x = b$ can be arbitrarily given)

$$f_X(x) = \begin{cases} \dfrac{1}{b-a} & \text{for } a < x < b \\ 0 & \text{otherwise.} \end{cases}$$

We can write $X \frown \mathbf{U}(a, b)$. The d.f. is

$$F_X(x) = \begin{cases} 0 & \text{for } x \leq a \\ \dfrac{x-a}{b-a} & \text{for } a < x < b \\ 1 & \text{for } x \geq b \end{cases}$$

Two r.v. X and Y on the same probability space are said to be *equivalent* or *almost equal* or *equal with probability one* if $P[X = Y] = 1$. This is equivalent to $P[X \neq Y] = 0$, i.e. the two r.v. only differ on a set $N = [X \neq Y] = \{\omega \in \Omega : X(\omega) \neq Y(\omega)\}$ with null probability. When X and Y are almost equal, it is customary to write $X = Y$ w.p. 1 (*with probability one*), $X = Y$ a.s. (*almost surely*),

$X = Y$ a.c. (*almost certainly*) or $X = Y$ a.a. (*almost always*). You can choose what you like best.[10]

We will now remember the concept of *mathematical expectation*, also known as *expectation, expected value, mean value* or simply *mean* of a r.v. X.

Let us look first at the particular case of X being a discrete r.v. with atoms a_k ($k = 1, 2, \ldots$) and pmf $p_X(k) = P[X = a_k]$ ($k = 1, 2, \ldots$). As the term indicates, the expected value or mean value of X is simply a mean or average of the values a_k the r.v. X can effectively take, but, of course, we should give more importance to the values that are more likely to occur. So, we should use a weighted average of the a_k values with weights given by the probabilities $p_X(k)$ of their occurrence. Therefore, the expectation of X is given by $\mathbb{E}[X] = \sum_k a_k p_X(k)$ (there is no need to divide by the sum of the weights since that sum is $\sum_k p_X(k) = 1$). In cases where the number of atoms is countably infinite, the $\sum_k = \sum_{k=1}^{+\infty}$ becomes a series and we only consider the expected value to be properly defined if the series is absolutely convergent, i.e. if $\mathbb{E}[|X|] = \sum_{k=1}^{+\infty} |a_k| p_X(k)$ is finite; when this is infinite, we say that X does not have a mathematical expectation.

The *variance* of a r.v. X, if it exists, is simply the expectation of the r.v. $Y = (X - \mathbb{E}[X])^2$, i.e. $VAR[X] = \mathbb{E}[(X - \mathbb{E}[X])^2]$. It is easy to show that $VAR[X] = \mathbb{E}[X^2] - (\mathbb{E}[X])^2$. The *standard deviation* is the positive square root of the variance and gives an idea of the dispersion of X about its mean value.

In the example of the two dice with $X = $ 'total number of dots on the two dice', we have a finite number of atoms and $\mathbb{E}[X] = \sum_{k=1}^{k=11} a_k p_X(k) = 2 \times \frac{1}{36} + 3 \times \frac{2}{36} + 4 \times \frac{3}{36} + 5 \times \frac{4}{36} + 6 \times \frac{5}{36} + 7 \times \frac{6}{36} + 8 \times \frac{5}{36} + 9 \times \frac{4}{36} + 10 \times \frac{3}{36} + 11 \times \frac{2}{36} + 12 \times \frac{1}{36} = 7$. The variance is $VAR[X] = \sum_{k=1}^{k=11} (a_k - 7)^2 p_X(k) = (2 - 7)^2 \times \frac{1}{36} + (3 - 7)^2 \times \frac{2}{36} + \cdots + (11 - 7)^2 \times \frac{2}{36} + (12 - 7)^2 \times \frac{1}{36} = \frac{210}{36} = 5.833$ and the standard deviation is $SD[X] = \sqrt{VAR[X]} = 2.415$.

When the r.v. X is absolutely continuous with pdf f_X, the expectation is given by $\mathbb{E}[X] = \int_{-\infty}^{+\infty} x f_X(x)dx$, which is also the weighted average of the values X can take, the weights $f_X(x)dx$ now being given by the pdf. This is basically the same as in the discrete case except we have to adjust to the fact that X now takes values x on a continuous set, the real numbers, so the weights should not be the probabilities of the individual x values (those probabilities would be zero) and the role of the sum should now be played by the integral (which is a kind of sum over a continuous set). Again, the mathematical expectation only exists if the integral that defines it is absolutely convergent, i.e. if $\mathbb{E}[|X|] = \int_{-\infty}^{+\infty} |x| f_X(x)dx$ is finite.

---

10 In general, we say that a given property concerning random variables holds with probability one (w.p. 1) [or almost surely (a.s.) or almost certainly (a.c.) or almost always (a.a.)] if the set of values of $\omega \in \Omega$ for which the property is true has probability one, which is equivalent to say that $P(N) = 0$, where N is the set of exceptional values $\omega \in \Omega$ for which the property is not true.

If $X \frown U(a, b)$, since $f_X(x) = \frac{x-a}{b-a}$ for $a < x < b$ and $f_X(x) = 0$ otherwise, the mean is $\mathbb{E}[X] = \int_a^b x \frac{x-a}{b-a} dx = \frac{a+b}{2}$, the variance is $VAR[X] = \int_a^b \left(x - \frac{a+b}{2}\right)^2 \frac{x-a}{b-a} dx = \frac{(b-a)^2}{12}$ and the standard deviation is $SD[X] = \frac{b-a}{\sqrt{12}}$.

If $X \frown \mathcal{N}(\mu, \sigma^2)$, one can show that the mean is

$$\mathbb{E}[X] = \int_{-\infty}^{+\infty} x \frac{1}{\sigma\sqrt{2\pi}} \exp\left(-\frac{(x-\mu)^2}{2\sigma^2}\right) dx = \mu$$

and the variance is

$$VAR[X] = \int_{-\infty}^{+\infty} (x-\mu)^2 \frac{1}{\sigma\sqrt{2\pi}} \exp\left(-\frac{(x-\mu)^2}{2\sigma^2}\right) dx = \sigma^2.$$

The formal general definition of mathematical expectation, which works for all r.v. X, is $\mathbb{E}[X] = \int_{-\infty}^{+\infty} x dF_X(x)$ or, equivalently, using the *Lebesgue integral* with respect to the probability P, $\mathbb{E}[X] = \int_\Omega X(\omega) dP(\omega)$ (one usually abbreviates to $\mathbb{E}[X] = \int_\Omega X dP$).[11] Again, the expectation only exists if the integral is

---

11 For those interested, the Lebesgue integral with respect to a probability P can be constructed by steps in the following way:

- For simple functions $X(\omega) = \sum_{k=1}^n c_k I_{A_k}(\omega)$ (where the sets $A_k \in \mathcal{F}$ are pairwise disjoint with $\bigcup_{k=1}^n A_k = \Omega$, I_{A_k} are their indicator functions and c_k are real numbers), the integral is defined by $\int_\Omega X dP = \sum_{k=1}^n c_k P(A_k)$.

  For the throwing of two dice, the r.v. X = 'total number of dots on both dice' is a simple function. In fact, if we choose for c_k the atoms $a_k = k+1$ ($k = 1, 2, \ldots, 11$) and put $A_k = [X = a_k]$, we have $X(\omega) = \sum_{k=1}^{11} c_k I_{A_k}(\omega)$ and therefore $\int_\Omega X dP = \sum_{k=1}^{11} c_k P(A_k)$ is precisely the way we have calculated above the expectation $\mathbb{E}[X]$ (remember that $P(A_k) = p_X(k)$). Notice, however, that the c_k do not have to be different from one another and so there are different ways of writing X, for example we could choose for the A_k the elementary sets $A_1 = \{1 \bullet 1\}, A_2 = \{1 \bullet 2\}, \ldots, A_7 = \{2 \bullet 1\}, A_8 = \{2 \bullet 2\}, \ldots, A_{36} = \{6 \bullet 6\}$ and for c_k the values X takes at A_k, namely $c_1 = X(1 \bullet 1) = 2, c_2 = X(1 \bullet 2) = 3, \ldots, c_7 = X(2 \bullet 1) = 3, c_8 = X(2 \bullet 2) = 4, \ldots, c_{36} = X(6 \bullet 6) = 12$; still we would have $X(\omega) = \sum_{k=1}^n c_k I_{A_k}(\omega)$ and, since now $P(A_k) = \frac{1}{36}$ for all k, we have $\int_\Omega X dP = \sum_{k=1}^{36} c_k P(A_k) = \sum_{k=1}^{36} c_k \frac{1}{36} = \frac{252}{36} = 7$. Of course, the result is the same, we have just used two different 'bookkeeping' methods.

- For non-negative r.v. X, the integral is defined by $\int_\Omega X dP = \lim_{n\to\infty} \int_\Omega X_n dP$, where X_n is any non-decreasing sequence of non-negative simple functions converging to X with probability one. This procedure is insensitive w.p. 1 to the choice of the approximating sequence of simple functions.

- For arbitrary r.v. X, the integral is defined by $\int_\Omega X dP = \int_\Omega X^+ dP - \int_\Omega X^- dP$, where $X^+(\omega) = X(\omega) I_{[X \geq 0]}(\omega)$ and $X^-(\omega) = -X(\omega) I_{[X < 0]}(\omega)$.

  The Lebesgue integral can be defined similarly with respect to any measure μ; a probability P is just a particular case of a measure that is normed ($P(\Omega) = 1$). Another particular case of a measure μ is the Lebesgue measure (the measure that extends the concept of length to a large class of real number sets); in this case, the Lebesgue integral generalizes the classical Riemann integral, allowing the integration of a larger class of functions. The Riemann integral of a function f is based on approximating the area underneath the graph of f by vertical rectangular slices, which is easy to compute; the Lebesgue integral with respect to the Lebesgue measure uses horizontal instead of vertical slices. For general measures μ, the Lebesgue integral is a generalization of the Riemann–Stieltjes integral $\int f(x) dg(x)$.

absolutely convergent, i.e. if $\mathbb{E}[|X|] = \int_{-\infty}^{+\infty} |X| dF_X(x) = \int_\Omega |X| dP$ is finite; if that happens, we also say that the r.v. X is *integrable*.

If X is discrete or absolutely continuous, these formal definitions simplify to the expressions we have seen above. In the absolutely continuous case, this is quite trivial to show, since we can use the fact that the derivative of $F_X(x)$ is $f_X(x)$ to obtain $\mathbb{E}[X] = \int_{-\infty}^{+\infty} x dF_X(x) = \int_{-\infty}^{+\infty} x \frac{dF_X(x)}{dx} dx = \int_{-\infty}^{+\infty} x f_X(x) dx$.

Note that the *indicator function* I_A (also called the *characteristic function*) I_A of a set $A \in \mathcal{F}$ is a r.v. defined by $I_A(\omega) = 1$ if $\omega \in A$ and $I_A(\omega) = 0$ if $\omega \notin A$. It is a discrete r.v. with two atoms 0 and 1 and we have $\mathbb{E}[I_A] = \int_\Omega I_A(\omega) dP(\omega) = 0 \times P(A^c) + 1 \times P(A) = P(A)$. In conclusion, $P(A) = \mathbb{E}[I_A]$, i.e. the probability of an event $A \in \mathcal{F}$ is just the mathematical expectation of its indicator function.

In general, for $A \in \mathcal{F}$, we define $\int_A X dP = \int_\Omega X I_A dP$. Notice that the average value (weighted by the probability) of X on the set A is $\frac{\int_A X dP}{P(A)}$ (now we need to divide by the sum of the weights $\int_A 1 dP = P(A)$, which may not be equal to one).

Equivalent random variables X and Y differ only on a set of probability zero, irrelevant for our purposes, and have the same d.f. and therefore the same probabilistic properties. In particular, if $X = Y$ a.c., X and Y will have the same mathematical expectation. For these reasons we can safely, in an abuse of language, follow the common practice of simply writing $X = Y$ instead of the more correct notation $X = Y$ a.c. (or the other notations referred to above). In lay terms, we do not distinguish between random variables that are almost equal.[12]

Adopting the common practice of identifying equivalent random variables (i.e. random variables that are almost equal), we can define, for $p \geq 1$, the space $L^p(\Omega, \mathcal{F}, P)$ or (abbreviating) L^p, of the random variables X[13] for which the *moment of order* p, $\mathbb{E}[|X|^p] = \int_\Omega |X|^p dP < +\infty$ exists. Notice that $\mathbb{E}[|X|^p]$ is just the *mathematical expectation* of the r.v. $Y = |X|^p$, i.e. $\mathbb{E}[Y] = \int_\Omega Y dP = \int_{-\infty}^{+\infty} y dF_Y(y) = \int_{-\infty}^{+\infty} |x|^p dF_X(x)$.

A L^p space is a *Banach space* for the L^p *norm* defined by $\|X\|_p = (\mathbb{E}[|X|^p])^{1/p}$. For $p = 2$ it is even a *Hilbert space* with *inner product* $\langle X, Y \rangle = \mathbb{E}[XY]$; the norm L^2 is associated to the inner product by $\|X\|_2 = (\langle X, X \rangle)^{1/2}$.[14]

---

12 This usual identification of r.v. that, although not exactly identical, are equivalent or equal with probability one, is an informal way of saying that we are going to work with equivalence classes of random variables instead of the random variables themselves. So, when speaking about a r.v. X, we take it as a representative of the collection of all random variables that are equivalent to X, i.e. a representative of the equivalence class where X belongs.

13 In reality, the space of the equivalence classes of random variables.

14 Here we will work with the field $\mathbb{R}$ of real numbers. A Banach space is a complete normed vector space over the field $\mathbb{R}$. By complete we mean that the Cauchy sequences (with respect to the norm) converge in the norm. The norm also defines a distance, the distance between random variables X and Y being $\|X - Y\|_p$. Notice that, if $X = Y$ w.p. 1 (in which case we use the common practice of identifying the two random variables), the distance is zero. The reverse is also true, i.e. if $\|X - Y\|_p = 0$, then $X = Y$ w.p. 1. A Hilbert space is a Banach space where the norm is associated to an inner product.

We now review some different concepts of the convergence of a sequence of random variables X_n $(n = 1, 2, \ldots)$ to a r.v. X, all of them on the same probability space $(\Omega, \mathcal{F}, P)$.

One says that X_n *converges* a.s. (*almost surely*), a.c. (*almost certainly*), a.a. (*almost always*) or w.p. 1 (*with probability one*) to X if $X_n \to X$ a.s., i.e. the set $\{\omega \in \Omega : X_n(\omega) \to X(\omega)\}$ has probability one, abbreviated to $P[X_n \to X] = 1$; this is equivalent to saying that the set of exceptional $\omega \in \Omega$ for which the sequence of real numbers $X_n(\omega)$ does not converge to the real number $X(\omega)$ has zero probability. We write $X_n \to X$ a.s. (or a.c. or a.a. or w.p. 1) or $\lim_{n \to +\infty} X_n = X$ a.s. (or a.c. or a.a. or w.p. 1). Sometimes we even abbreviate further and simply write $X_n \to X$ or $\lim_{n \to +\infty} X_n = X$.

We now present a different concept of convergence that neither implies nor is implied by the convergence w.p. 1. If the r.v. X and X_n $(n = 1, 2, \ldots)$ are in L^p $(p \geq 1)$, we say that $X_n \to X$ in p-*mean* if X_n converges to X in the L^p norm, i.e. if $\|X_n - X\|_p \to 0$ as $n \to +\infty$, which is equivalent to $\mathbb{E}[|X_n - X|^p] \to 0$ as $n \to +\infty$. When $p = 2$, we also speak of *convergence in mean square* or *mean square convergence* and write ms-$\lim_{n \to +\infty} X_n = X$ or l.i.m.$_{n \to +\infty} X_n = X$ or $X_n \underset{ms}{\to} X$. If $p \leq q$, convergence in q-mean implies convergence in p-mean.

Let us present a weaker concept of convergence. We say that X_n *converges in probability* or *converges stochastically* to X if, for every $\varepsilon > 0$, $P[|X_n - X| > \varepsilon] \to 0$ as $n \to +\infty$, which means that the probability than X_n is not in an ε-neighbourhood of X is vanishingly small; this is equivalent to $P[|X_n - X| \leq \varepsilon] \to 1$. We write $P-\lim_{n \to +\infty} X_n = X$ or $X_n \underset{P}{\to} X$ or $X_n \to X$ in probability. If X_n converges to X w.p. 1 or in p-mean, it also converges to X in probability, but the reverse may fail.

An even weaker concept is the convergence in distribution. We say that X_n *converges in distribution* to X if the distribution functions of X_n converge to the distribution function of X at every continuity point of the latter, i.e. if $F_{X_n}(x) \to F_X(x)$ when $n \to +\infty$ for all x that are continuity points of $F_X(x)$. We write $X_n \underset{d}{\to} X$ or $X_n \to X$ in distribution. If X_n converges to X w.p. 1 or in p-mean or in probability, it also converges to X in distribution, but the reverse statements may fail.

The concept of r.v. can be generalized to several dimensions. A n-dimensional r.v. or *random vector* $\mathbf{X} = [X_1, X_2, \ldots, X_n]^T$ (T means 'transposed' and so, as is customary, $\mathbf{X}$ is a column vector) is simply a vector with n random variables defined on the same measurable space. We can define its distribution function $F_{\mathbf{X}}(\mathbf{x}) = F_{X_1, X_2, \ldots, X_n}(x_1, x_2, \ldots, x_n) := P[X_1 \leq x_1, X_2 \leq x_2, \ldots, X_n \leq x_n]$ for $\mathbf{x} = [x_1, x_2, \ldots, x_n]^T \in \mathbb{R}^n$, also called the joint d.f. of the r.v. $X_1, X_2, \ldots, X_n$.

The *mathematical expectation of a random vector* $\mathbf{X}$ is the column vector $\mathbb{E}[\mathbf{X}] = [\mathbb{E}[X_1], \mathbb{E}[X_2], \ldots, \mathbb{E}[X_n]]^T$ of the mathematical expectations of the coordinates of $\mathbf{X}$ and exists if such expectations all exist. Besides defining the variance $VAR[X_i] = \mathbb{E}[(X_i - \mathbb{E}[X_i])^2]$ of a coordinate r.v. X_i, we remind that the definition of the *covariance* of any two coordinate random variables X_i and X_j, which is $COV [X_i, X_j] := \mathbb{E}[(X_i - \mathbb{E}[X_i])(X_j - \mathbb{E}[X_j])]$, if this expectation exists; this is equal to $\mathbb{E}[X_iX_j] - \mathbb{E}[X_i]\mathbb{E}[X_j]$. Of course, when $i = j$, the two random variables coincide and the covariance becomes the variance. We also remember the definition of *correlation* between X_i and X_j, which is $CORR [X_i, X_j] := \frac{COV [X_i, X_j]}{SD [X_i] SD [X_j]}$. One can also define the *variance-covariance matrix* $\mathbf{\Sigma}[\mathbf{X}] = \mathbb{E}[(\mathbf{X} - \mathbb{E}[\mathbf{X}])(\mathbf{X} - \mathbb{E}[\mathbf{X}])^T]$, where one collects on the diagonal the variances $\sigma_{ii} = \sigma_i^2 = VAR [X_i]$ $(i = 1, 2, \dots, n)$ and off the diagonal the covariances $\sigma_{ij} = COV [X_i, X_j]$ $(i, j = 1, 2, \dots, n; i \neq j)$. Since $COV [X_i, X_j] = COV [X_j, X_i]$, this matrix is symmetric; it is also non-negative definite.

If there is a countable set of atoms $S = \{\mathbf{a}_1, \mathbf{a}_2, \dots \} \subset \mathbb{R}^n$ such that $P[\mathbf{X} \in S] = 1$, we say that the random vector $\mathbf{X}$ is discrete and the joint probability mass function is given, for any atom $\mathbf{a}_k = [a_{k,1}, a_{k,2}, \dots, a_{k,n}]^T$, by $p_{\mathbf{X}}(k) = P[\mathbf{X} = \mathbf{a}_k]$. If there is a joint pdf $f_{\mathbf{X}}(\mathbf{x}) = f_{X_1, X_2, \dots, X_n}(x_1, x_2, \dots, x_n)$ such that

$$F_{\mathbf{X}}(\mathbf{x}) = \int_{-\infty}^{x_1} \int_{-\infty}^{x_2} \dots \int_{-\infty}^{x_n} f_{X_1, X_2, \dots, X_n}(y_1, y_2, \dots, y_n) dy_n \dots dy_2 dy_1,$$

then $\mathbf{X}$ is absolutely continuous and $f_{\mathbf{X}}(\mathbf{x}) = \frac{\partial^n F_{X_1, X_2, \dots, X_n}(x_1, x_2, \dots, x_n)}{\partial x_1 \partial x_2 \dots \partial x_n}$.[15] For example, a *normal* random vector or *Gaussian* random vector $\mathbf{X} \frown \mathcal{N}(\boldsymbol{\mu}, \mathbf{\Sigma})$ with mean vector $\boldsymbol{\mu}$ and variance-covariance matrix $\mathbf{\Sigma}$, which we assume to be a positive definite matrix, is an absolutely continuous random vector with pdf $f_{\mathbf{X}}(\mathbf{x}) = (2\pi)^{-n/2}(\det(\mathbf{\Sigma}))^{-1/2} \exp\left(-\frac{1}{2}(\mathbf{x} - \boldsymbol{\mu})^T \mathbf{\Sigma}^{-1}(\mathbf{x} - \boldsymbol{\mu})\right)$.

The concepts of L^p space and L^p norm can be generalized to n-dimensional random vectors $\mathbf{X}$ by interpreting $|\mathbf{X}|$ as meaning the euclidean norm $(X_1^2 + X_2^2 + \dots + X_n^2)^{1/2}$. When $p = 2$, the concept of inner product can be generalized by using $\langle \mathbf{X}, \mathbf{Y} \rangle = \mathbb{E}[\mathbf{X}^T \mathbf{Y}]$.

Box 2.1 Summary revision of probabilistic concepts

- Probability space $(\Omega, \mathcal{F}, P)$
  - Ω is the universal set or sample space.
  - $\mathcal{F}$ is a σ-algebra (includes Ω and is closed to complementation and countable unions), the members of which are called events.
  - $(\Omega, \mathcal{F})$ is called a measurable space.
  - P is a probability, i.e. a normed and σ-additive function from $\mathcal{F}$ to $[0, 1]$. Normed means that $P(\Omega) = 1$. σ-additive means that, if $A_n \in \mathcal{F}$ $(n=1,2,\dots)$ are pairwise disjoint, then $P\left(\bigcup_n A_n\right) = \sum_n P(A_n)$.

---

15 As in the one-dimensional case, this derivative may not be defined for exceptional points in $\mathbb{R}^n$ which form a negligible set, i.e. a set with zero n-dimensional Lebesgue measure, and we may, without loss of generality, give arbitrary values to the pdf at those exceptional points. This measure is an extension of the concept of n-dimensional volume (the length if $n = 1$, the area if $n = 2$, the ordinary volume if $n = 3$, hyper-volumes if $n > 3$).

- We will assume $(\Omega, \mathcal{F}, P)$ to be complete, i.e. $\mathcal{F}$ includes all subsets Z of the sets $N \in \mathcal{F}$ satisfying $P(N) = 0$.
- Borel σ-algebra $\mathcal{B}$ in $\mathbb{R}$
  - Is the σ-algebra generated by the intervals of real numbers, i.e. the smallest σ-algebra that includes such intervals.
  - It can also be generated, for example, by the open sets or by the intervals of the form $]-\infty, x]$.
  - Its elements are called Borel sets.
- Random variable X
  - Is a function from Ω to $\mathbb{R}$ such that the inverse images $X^{-1}(B) = [X \in B] = \{\omega \in \Omega : X(\omega) \in B\}$ of Borel sets B are in $\mathcal{F}$.
  - Its distribution function (d.f.) is $F_X(x) = P[X \le x]$ $(-\infty < x < +\infty)$.
  - $P[X \in B] = \int_B 1 dF_X(y)$ for Borel sets B.
  - X is a discrete r.v. if there is a countable number of atoms $a_k \in \mathbb{R}$ $(k = 1, 2, \ldots)$ with probability mass function (pmf) $p_X(k) = P[X = a_k]$ such that $\sum_k p_X(k) = 1$. Its d.f. is given by $F_X(x) = \sum_{k: a_k \le x} p_X(k)$ and, for a Borel set B, $P[X \in B] = \sum_{k: a_k \in B} p_X(k)$.
  - X is an absolutely continuous r.v. (sometimes simply called 'continuous') if there is a pdf $f_X(x)$, i.e. a non-negative function such that $F_X(x) = \int_{-\infty}^x f_X(y) dy$. Then $f_X(x) = dF_X(x)/dx$ (the derivative may not exist for a negligible set of x-values) and, for a Borel set B, $P[X \in B] = \int_{x \in B} f_X(x) dx$.
  - The σ-algebra generated by X (i.e. the σ-algebra generated by the inverse images by X of the Borel sets) is denoted by $\sigma(X)$.
  - $X \frown \mathcal{N}(\mu, \sigma^2)$ (Gaussian or normal r.v. with mean μ and variance $\sigma^2 > 0$) has pdf $f_X(x) = \frac{1}{\sigma\sqrt{2\pi}} \exp\left(-\frac{(x-\mu)^2}{2\sigma^2}\right)$ $(-\infty < x < +\infty)$.
  - $\mathcal{N}(0, 1)$ is the standard normal (or Gaussian) distribution and it is usual to denote its pdf by $\phi(x)$ and its d.f. by $\Phi(x)$.
  - $U \frown U(a, b)$ $(-\infty < a < b < +\infty)$, uniform r.v. on the interval $]a, b[$: has pdf $f_U(x) = \frac{x-a}{b-a}$ for $a < x < b$ and $f_U(x) = 0$ otherwise. Note: it does not matter whether the interval is open, closed or semi-open.
  - $X = Y$ w.p. 1 (or X and Y are equivalent or $X = Y$ a.s. or $X = Y$ a.c. or $X = Y$ a.a.) if $P[X = Y] = 1$. In such a case, it is common to abuse language and identify the two r.v. since they have the same d.f. and probabilistic properties.
- Mathematical expectation (or expectation, expected value, mean value or mean) $\mathbb{E}[X]$ of a r.v. X
  - Is a weighted mean of X given by $\mathbb{E}[X] = \int_{-\infty}^{+\infty} x dF_X(x) = \int_\Omega X dP$ (the two definitions are equivalent) and it exists if $\mathbb{E}[|X|] = \int_{-\infty}^{+\infty} |x| dF_X(x) = \int_\Omega |X| dP < +\infty$.

(Continued)

Box 2.1 (Continued)

- The definition simplifies to $\mathbb{E}[X] = \sum_k a_k p_X(k)$ if X is discrete and to $\mathbb{E}[X] = \int_{-\infty}^{+\infty} x f_X(x) dx$ if X is absolutely continuous.
- $P(A) = \mathbb{E}[I_A]$, where I_A is the indicator function of the set $A \in \mathcal{F}$ defined by $I_A(\omega) = 1$ if $\omega \in A$ and $I_A(\omega) = 0$ if $\omega \notin A$.
- If it exists, the expectation of the r.v. $Y = |X|^p$ is $\mathbb{E}[|X|^p] = \int_\Omega |X|^p dP = \int_{-\infty}^{+\infty} y dF_Y(y) = \int_{-\infty}^{+\infty} |x|^p dF_X(x)$ and is called the moment of order p of X. The centred moment of order 2, if it exists, is called the variance $VAR[X] = \mathbb{E}[(X - \mathbb{E}[X])^2]$ and its positive square root is called the standard deviation $SD[X] = \sqrt{VAR[X]}$.

- $L^p(\Omega, \mathcal{F}, P)$ space (abbrev. L^p space) with $p \geq 1$
  - If we identify random variables on $(\Omega, \mathcal{F}, P)$ that are equivalent, this is the space of the r.v. X for which $\mathbb{E}[|X|^p] < +\infty$, endowed with the L^p-norm $\|X\|_p = (\mathbb{E}[|X|^p])^{1/p}$.
  - It is a Banach space.
  - For $p = 2$, it is a Hilbert space, since there is an inner product $\langle X, Y \rangle = \mathbb{E}[XY]$ associated with the L^2-norm.

- Convergence concepts for a sequence of r.v. X_n ($n = 1, 2, \ldots$) to a r.v. X as $n \to +\infty$
  - Almost sure convergence or convergence with probability one: $X_n \to X$ a.s. or $\lim_{n \to +\infty} X_n = X$ a.s. (or a.c. or a.a. or w.p. 1). Often abbreviated to $X_n \to X$ or $\lim_{n \to +\infty} X_n = X$.
    Means that $P[X_n \to X] = 1$ (the set of $\omega \in \Omega$ for which $X_n(\omega) \to X(\omega)$ has probability one).
  - Convergence in p-mean ($p \geq 1$) for $X_n, X \in L^p$: $X_n \to X$ in p-mean.
    Means that X_n converges to X in the L^p norm, i.e. $\|X_n - X\|_p \to 0$ or equivalently $\mathbb{E}[|X_n - X|^p] \to 0$.
    For $p \leq q$, convergence in q-mean implies convergence in p-mean.
  - Mean square convergence: $\text{ms-}\lim_{n \to +\infty} X_n = \text{l.i.m.}_{n \to +\infty} X_n = X$ or $X_n \xrightarrow{ms} X$.
    Is the particular case of convergence in p-mean with $p = 2$.
  - Convergence in probability or stochastic convergence: $P - \lim_{n \to +\infty} X_n = X$ or $X_n \xrightarrow{P} X$ or $X_n \to X$ in probability.
    Means that, for every $\varepsilon > 0$, $P[|X_n - X| > \varepsilon] \to 0$. Convergence w.p. 1 and convergence in p-mean imply convergence in probability (the reverse may fail).
  - Convergence in distribution: $X_n \xrightarrow{d} X$ or $X_n \to X$ in distribution.
    Means that $F_{X_n}(x) \to F_X(x)$ for all continuity points x of $F_X(x)$. Convergence w.p. 1, convergence in p-mean and convergence in probability imply convergence in distribution (the reverse may fail).

- n-dimensional random vectors: see main text.

2.2   Monte Carlo simulation of random variables

In some situations we may not be able to obtain explicit expressions for certain probabilities and mathematical expectations of random variables of interest to us. Instead of giving up, we can recur to *Monte Carlo simulations*, in which we perform, on the computer, simulations of the real experiment and use the results to approximate the quantities we are interested in studying.

For example, instead of really throwing two dice, we may simulate that on a computer by using appropriate random number generators.

Some computer languages (like R) have generators for several of the most commonly used probability distributions. Others, like some spreadsheets, only consider the uniform distribution on the interval $]0, 1[$ because that is the building block from which we can easily generate random variables with other distributions. We will now see how because, even when you have available random generators for several distributions, you may want to work with a distribution that is not provided, including one that you design yourself.

So, assume your computer can generate a random variable $U \frown U(0, 1)$, i.e. a r.v. U with uniform distribution in the interval $]0, 1[$, which is absolutely continuous with pdf $f_U(u) = 1$ for $0 < u < 1$ and $f_U(u) = 0$ otherwise.[16]

In a spreadsheet, you can generate a $U(0, 1)$ random number on a cell (typically with '=RAND()' or the corresponding term in languages other than English) and then drag it to other cells to generate more numbers.

If you are using the computer language R, you should use the package 'stats' and the command 'runif(1)' to generate one randomly chosen value U from the $U(0, 1)$ distribution. If you need more randomly independent chosen values, say a sequence of 1000 independent random numbers, you use the command 'runif(1000)'. Of course, since these are random numbers, if you repeat the command 'runif(1000)' you will get a different sequence of 1000 random numbers. If, for some reason, you want to use the same sequence of 1000 random numbers at different times, you should, before starting simulating random numbers, define a *seed* and use the same seed at those different times. To define a seed in R you use 'set.seed(m)', where m is a kind of pin number of your choice (a positive integer number) that identifies your seed.

You should be aware that the usual algorithms for random number generation are not really random, but rather pseudo-random. This means that the algorithm is in fact deterministic, but the sequence of numbers it produces (say, for example, a sequence of 1000 numbers) looks random and has statistical properties almost identical to true sequences of 1000 independent random numbers with a $U(0, 1)$ distribution.

---

16 As we have seen, the probabilistic properties are insensitive to whether the interval is closed, open or semi-open and it is convenient here that, as it is often the case, the generator never produces the numbers 0 and 1.

You should remember that the probability $P[U \in [a, b[]$ (with $0 \le a < b \le 1$) of U to fall in an interval $[a, b[$ (it does not matter if it closed, open or semi-open) is $= \int_a^b f_U(u)du = \int_a^b 1du = b - a$, precisely the length of the interval.

To simulate the result of throwing the white dice, one basically divides the interval $]0, 1[$ into six intervals corresponding to the six possible number of dots of the outcome. Since, assuming the dice is fair, each of these outcomes has equal probability $\frac{1}{6}$, one should choose intervals of equal length $\frac{1}{6}$, say the intervals $]0, \frac{1}{6}[$ (corresponding to 1 dot, so that you say the result of throwing the white dice is 1 dot if U falls on this interval), $[\frac{1}{6}, \frac{2}{6}[$ (corresponding to 2 dots), etc., $[\frac{5}{6}, 1[$ (corresponding to 6 dots). Because the intervals are of equal length, we can use a simple R command to make that correspondence and give the number of dots of the white dice: 'trunc(6*runif(1)+1)'. You generate a new $U(0, 1)$ random number and repeat the procedure to simulate the result of the throw for the black dice. So, for each throw of the dice you can compute the value of the r.v. X = 'total number of dots on the two dice' using the command 'trunc(6*runif(1)+1)+trunc(6*runif(1)+1)'. In this case, it is easy to compute the mean value $\mathbb{E}[X] = 7$, but suppose this was not possible to obtain analytically. You could run the above procedure, say a 1000 times, to have a simulated sample of $n = 1000$ values of the number of dots on the throwing of two dice; the sample mean is a good approximation of the expected value $\mathbb{E}[X] = 7$. I did just that by using 'mean(trunc(6*runif(1000)+1)+trunc(6*runif(1000)+1))' and obtained 7.006. Of course, doing it again, I got a different number: 6.977. If you try it, you will probably obtain a slightly different sample mean simply because your random sample is different. The sample mean converges to the true expected value 7 when the number n of simulations (also called 'runs') goes to infinity.

Of course, we could directly simulate the r.v. X without the intermediate step of simulating the throwing of the two dice. This r.v. is discrete and has atoms $a_1 = 2, a_2 = 3, ..., a_{11} = 12$ with probabilities $p_X(1) = 1/36, p_X(2) = 2/36, ..., p_X(11) = 1/36$ (see Section 2.1). So, we divide the interval $]0, 1[$ into 11 intervals corresponding to the 11 atoms a_k, each having a length equal to the corresponding probability $p_X(k)$. The first interval would be $]0, 1/36[$ (length = $p_X(1) = 1/36$) and would correspond to $X = a_1 = 2$, the second would be $[1/36, 3/36[$ (length = $p_X(2) = 2/36$) and would correspond to $X = a_2 = 3$, etc. Then we would generate a random number U with distribution $U(0, 1)$ and choose the value of X according to whether U would fall on the first interval, the second interval, etc.

This would be more complicated to program in R (try it) but this is the general method that works for any discrete r.v.

Let us now see how to simulate an absolutely continuous r.v. with d.f. $F(x)$, assuming that the d.f. is invertible for values in $]0, 1[$, which happens in many applications. Let the inverse function be $u = F^{-1}(x)$. This means that, given a value $u \in]0, 1[$, there is a unique $x \in \mathbb{R}$ such that $F(x) = u$. Then, it can easily be proved that, if U is a uniform r.v. on the interval $]0, 1[$, then the r.v. $X = F^{-1}(U)$ has precisely the d.f. F we want, i.e. $F_X(x) = F(x)$. So, to simulate the r.v. X with d.f. F_X we can simulate values u of a uniform r.v. U on $]0, 1[$ and use the values $x = F_X^{-1}(u)$ as the simulations of the desired r.v. X.[17]

The inverse distribution function is also called the *quantile function* and we say that the quantile of order u $(0 < u < 1)$ or u *-quantile* of X is $x = F_X^{-1}(u)$. Most spreadsheets and R have the inverse distribution function or quantile function of the most used continuous distributions.

For example, in R, to determine the 0.975-quantile of $X \frown \mathcal{N}(1.4, 0.8^2)$, which is the inverse distribution function computed at 0.975, you can use the package 'stats' and the command 'qnorm(0.975, mean=1.4, sd=0.8)', which retrieves the value 2.968 (note that this is $= 1.4 + 0.8 \times 1.96$, where 1.96 is the 0.975-quantile of a standard normal r.v.). To simulate a sample with 1000 independent values of $X \frown \mathcal{N}(1.4, 0.8^2)$ in R, one can use the command 'qnorm(runif(1000), mean=1.4, sd=0.8)'; if we want to use directly the Gaussian random number generator, this command is equivalent to 'rnorm (1000, mean=1.4, sd=0.8)'. To do one simulation of X on a spreadsheet you can use the standard normal d.f. and the command '=1.4+NORMSINV(RAND())*0.8' (or a similar command depending on the spreadsheet and the language) or directly the normal d.f. with mean 1.4 and standard deviation 0.8 using the command '=NORMINV(RAND(), 1.4, 0.8)'; by dragging to neighbour cells, you can produce as many independent values of X as you wish.

Of course, the above procedure relies on the precision of the computations of the inverse distribution function, for which there usually are no explicit expressions. Frequently, the numerical methods used to compute such inverses have a lower precision for values of u very close to 0 or to 1. So, if you need simulations to study what happens for more extreme values of X, it might be worth working with improved precision when computing the inverse distribution functions near 0 or 1 or recur to alternative methods that have been designed for specific distributions. This section is intended for general purpose use of simulations and the reader should consult the specialized literature for simulations with special requirements.

Box 2.2 Review of Monte Carlo simulation of random numbers

- Suppose you want to generate a simulated sequence $x_1, x_2, \ldots, x_n$ of n independent random values having the same d.f. F_X. Some computer languages like R have specific commands to do it for the most used distributions. Alternatively, you can use the random number generator for the uniform $U(0, 1)$ distribution and do some transformations using an appropriate algorithm that

(Continued)

---

17 For the 'exotic' cases in which the d.f. of an absolutely continuous r.v. is not invertible, you can replace in the above procedure the inverse distribution function by the generalized inverse defined by $F_X^{-1}(u) = \inf\{x : F_X(x) = u\}$.

Box 2.2 (Continued)

we now summarize (if you use spreadsheets, the same ideas work with appropriate adjustments).

- Generation of a random sequence of size n for the $U(0, 1)$ distribution.
  Say $n = 1000$. In R, use the package 'stats' and the command 'runif(1000)'. In spreadsheets, one typically writes '=RAND()' on a cell (spreadsheets in a language other than English may use a translation of 'RAND' instead), and then drags that cell to the next 999 cells.
- Generation of a random sequence for discrete r.v. X.
  Let a_k ($k = 1, 2, \ldots, kmax$) be the atoms and $p_X(k)$ ($k = 1, 2, \ldots, kmax$) the pmf ($kmax =$ number of atoms; if they are countably infinite, put a large number so that the generated uniform random number u will fall in the first '$kmax$' intervals with very large probability).
  Algorithm:
  for (i in 1:n)
  {
  1. Generate a random number u from a uniform $U(0, 1)$ distribution (see above).
  2. Divide the interval $(0, 1)$ into subintervals of lengths $p_X(1)$, $p_X(2)$, $p_X(3)$, ... ($p_X(1)$, $p_X(1) + p_X(2)$, $p_X(1) + p_X(2) + p_X(3)$, ... will be the separation points) and check in which subinterval u falls; if u falls in the subinterval number j, then the simulated value of X_i will be $x_i = a_j$. If $kmax$ is small, this can be easily implemented with a sequence of 'if...else' commands. If $kmax$ is large, you can use instead:
  (a) Initialization: set $k = 0$, $s = 0$ (the separation point), $j = 0$.
  (b) For k from 1 to $kmax$ do the following:
    – Update the separation point $s = s + p_X(k)$
    – If $j = 0$ and $u < s$, put $j = k$; else leave j unchanged.
  (c) The ith simulated value of X is $x_i = a_j$.
  }
- Generation of a random sequence for absolutely continuous r.v. X.
  We assume that the d.f. F_X is invertible for values in $]0, 1[$.
  Algorithm:
  for (i in 1:n)
  {
  1. Generate a random number u from a uniform $U(0, 1)$ distribution (see above).
  2. Determine the ith simulated value of X, $x_i = F_X^{-1}(u)$, where F_X^{-1} is the inverse distribution function.
  }
- If you need to repeat the same sequence of random numbers on different occasions, define a seed and use the same seed on each occasion.

2.3 Conditional expectations, conditional probabilities, and independence

This section can be skipped by readers who already have an informal idea of conditional probability, conditional expectation, and independence, and are less concerned with a more theoretical solid ground approach.

Given a r.v. $X \in L^1(\Omega, \mathcal{F}, P)$ and a sub-σ-algebra $\mathcal{H} \subset \mathcal{F}$, [18] there is a r.v. $Z = \mathbb{E}[X \mid \mathcal{H}]$, called the *conditional expectation* of X *given* $\mathcal{H}$, which is $\mathcal{H}$-mesaurable and such that $\int_H X \, dP = \int_H Z dP$ for all events $H \in \mathcal{H}$. [19]

Note that both X and Z are $\mathcal{F}$-measurable (i.e. inverse images of Borel sets are in the σ-algebra $\mathcal{F}$), but, in what concerns Z, we require the stronger property of being also $\mathcal{H}$-measurable (i.e. inverse images of Borel sets should be in the smaller σ-algebra $\mathcal{H}$). Furthermore, X and Z have the same average behaviour on the sets in $\mathcal{H}$. In fact, for a set $H \in \mathcal{H}$, we have $\int_H X \, dP = \int_H Z dP$ and, therefore, the mean values of X and Z on H, respectively $\frac{\int_H X \, dP}{P(H)}$ and $\frac{\int_H Z \, dP}{P(H)}$, [20] are equal.

Basically, Z is a 'lower resolution' version of X when, instead of having the full 'information' $\mathcal{F}$, we have a more limited 'information' $\mathcal{H}$. Let us give an example. When throwing two dice, consider the r.v. $X = $ 'total number of dots on both dice'. Suppose now that the information is restricted and you only have information on whether the number of dots in the white dice is even or odd. This information is given by the sub-σ-algebra $\mathcal{H} = \{\emptyset, \Omega, G, G^c\}$, where $G = \{2\circ1, 2\circ2, 2\circ3, \dots, 2\circ6, 4\circ1, 4\circ2, \dots, 4\circ6, 6\circ1, 6\circ2, \dots, 6\circ6\}$ is the event 'even number of dots on the white dice' and G^c is the event 'odd number of dots in the white dice'. For Borel sets B, the inverse images by Z can only be $\emptyset$, Ω, G or G^c, i.e. with the available information, when working with Z one cannot distinguish the different ω in G nor the different ω in G^c, so $Z(\omega) = z_1$ for all $\omega \in G$ and $Z(\omega) = z_2$ for all $\omega \in G^c$. To obtain z_1 and z_2, we recur to the other property, $\int_H X dP = \int_H Z dP$ for sets $H \in \mathcal{H}$. Notice that $XI_G = 0 \times I_{G^c} + 3 \times I_{\{2\circ1\}} + 4 \times I_{\{2\circ2\}} + 5 \times I_{\{2\circ3\}} + \cdots + 8 \times I_{\{2\circ6\}} + 5 \times I_{\{4\circ1\}} + 6 \times I_{\{4\circ2\}} + \cdots + 10 \times I_{\{4\circ6\}} + 7 \times I_{\{6\circ1\}} + 8 \times I_{\{6\circ2\}} + \cdots + 12 \times I_{\{6\circ6\}}$

---

18 The symbol $\subset$ of set inclusion is taken in the wide sense, i.e. we allow the left-hand side to be equal to the right-hand side. Unlike us, some authors use $\subset$ only for a proper inclusion (left-hand side is contained in the right-hand side but cannot be equal to it) and use the symbol $\subseteq$ when they refer to the wide sense inclusion.

19 The Radon–Nikodym theorem ensures that a r.v. Z with such characteristics not only exists but is even a.s. unique, i.e. if Z and Z^* have these characteristics, then $Z = Z^*$ a.s. and we can, with the usual abuse of language, identify them.

20 The mean value of a r.v. X on a set H, which is the weighted average of the values X takes on H with weights given by the probability distribution, is indeed given by $\frac{\int_H X \, dP}{P(H)}$. In fact, the integral 'sums' on H the values of X multiplied by the weights, and we have to divide the result by the 'sum' of the weights, which is $\int_H dP = P(H)$, to have a weighted average. That last division step is not necessary when we work with the expected value $\int_\Omega X dP$ (which is the mean value of X on Ω) because the sum of the weights in this case is simply $P(\Omega) = 1$.

(is different from X) and so $\int_G X dP = \int_\Omega X I_G dP = 0 \times \frac{18}{36} + 3 \times \frac{1}{36} + 4 \times \frac{1}{36} + 5 \times \frac{1}{36} + \cdots + 8 \times \frac{1}{36} + 5 \times \frac{1}{36} + 6 \times \frac{1}{36} + \cdots + 10 \times \frac{1}{36} + 7 \times \frac{1}{36} + 8 \times \frac{1}{36} + \cdots + 12 \times \frac{1}{36} = \frac{135}{36}$. As for $Z I_G = 0 \times I_{G^c} + z_1 \times I_G$, since $P[I_G = 0] = P(G^c) = \frac{18}{36}$ and $P[I_G = 1] = P(G) = \frac{18}{36}$, we have $\int_G Z dP = \int_\Omega Z I_G P = 0 \times \frac{18}{36} + z_1 \times \frac{18}{36}$. Since we must have $\int_G X dP = \int_G Z dP$, we conclude that $z_1 = 7.5$. Notice that z_1 is, as it should be, just be the mean value (weighted by the probabilities) of $X(\omega)$ when considering only the values of $\omega \in G$ and therefore could have been computed directly by $z_1 = \frac{\int_G X dP}{P(G)} = \frac{135/36}{18/36} = 7.5$. Similar reasoning with the r.v. $X I_{G^c}$ and $Z I_{G^c}$ gives the mean value X takes in G^c as $z_2 = \frac{\int_{G^c} X dP}{P(G^c)} = \frac{117/36}{18/36} = 6.5$.

In conclusion, the conditional expectation in this example is the r.v. $\mathbb{E}[X \mid \mathcal{H}](\omega) = Z(\omega) = 7.5$ for all $\omega \in G$ and $\mathbb{E}[X \mid \mathcal{H}](\omega) = Z(\omega) = 6.5$ for all $\omega \in G^c$. It is clear that the conditional expectation $Z = \mathbb{E}[X \mid \mathcal{H}]$ is a 'lower resolution' version of the r.v. X. While X gives a precise information on the number of dots of the two dice for each element $\omega \in \Omega$, Z can only distinguish the elements in G from the elements in G^c, for example Z treats all elements in G alike and gives us the average number of dots of the two dice on the set G whatever element $\omega \in G$ we are considering. An analogy would be to think of X as a photo with 36 pixels (the elements in Ω), each pixel having one of 11 possible shades of grey (notice that the number of dots varies between 2 and 12); then Z would be the photo that results from X when you fuse the original 36 pixels into just two large pixels (G and G^c), each of the two fused pixels having a shade of grey equal to the average of its original pixels.

Notice that the (unconditional) expectation $\mathbb{E}[X]$ does not depend on ω (is deterministic) and takes the fixed numerical value 7. The conditional expectation $\mathbb{E}[X \mid \mathcal{H}](\omega)$, which can be abbreviated as $\mathbb{E}[X \mid \mathcal{H}]$, however, is a r.v. Z, and we may determine its expectation $\mathbb{E}[Z] = 7.5 \times P[Z = 7.5] + 6.5 \times P[Z = 6.5] = 7.5 \times P(G) + 6.5 \times P(G^c) = 7.5 \times \frac{18}{36} + 6.5 \times \frac{18}{36} = 7 = \mathbb{E}[X]$. This procedure gives, always using the proper weights on the averages, the average of the averages that X takes on the sets G and G^c; this should obviously give the same result as averaging directly X over $\Omega = G + G^c$. This is not a coincidence or a special property for this example, but rather a general property of weighted averages.

Therefore, given any r.v. $X \in L^1(\Omega, \mathcal{F}, P)$ and a sub-σ-algebra $\mathcal{H}$ of $\mathcal{F}$, we have the *law of total expectation*

$$\mathbb{E}[\mathbb{E}[X \mid \mathcal{H}]] = \mathbb{E}[X].$$

The proof is very simple. Putting $Z = \mathbb{E}[X \mid \mathcal{H}]$, go to the defining property $\int_H Z dP = \int_H X dP$ (valid for all events $H \in \mathcal{H}$) and use $H = \Omega$ to obtain $\int_\Omega Z dP = \int_\Omega X dP$. This proves the result since the left-hand side is $\mathbb{E}[Z] = \mathbb{E}[\mathbb{E}[X \mid \mathcal{H}]]$ and the right-hand side is $\mathbb{E}[X]$.

We now cite the other most important properties of conditional expectations, under the assumption that all random variables involved are in $L^1(\Omega, \mathcal{F}, P)$ and that $\mathcal{G}$ and $\mathcal{H}$ are σ-algebras contained in $\mathcal{F}$:

X is $\mathcal{H}$-measurable $\Rightarrow \mathbb{E}[X \mid \mathcal{H}] = X$

X is $\mathcal{H}$-measurable $\Rightarrow \mathbb{E}[XY \mid \mathcal{H}] = X\,\mathbb{E}[Y \mid \mathcal{H}]$

$\mathcal{G} \subset \mathcal{H} \Rightarrow \mathbb{E}[\mathbb{E}[X \mid \mathcal{H}] \mid \mathcal{G}] = \mathbb{E}[\mathbb{E}[X \mid \mathcal{G}] \mid \mathcal{H}] = \mathbb{E}[X \mid \mathcal{G}].$

For $A \in \mathcal{F}$, we can define the *conditional probability* $P(A \mid \mathcal{H}) := \mathbb{E}[I_A \mid \mathcal{H}]$. This concept is an extension of the classical concept $P(A \mid C) = P(A \cap C)/P(C)$ defined for $A, C \in \mathcal{F}$ such that $P(C) > 0$. In fact, if you put $\mathcal{H} = \{C, C^c, \emptyset, \Omega\}$ (which is the σ-algebra generated by C), $P(A \mid C)$ is nothing more than the common value that $\mathbb{E}[I_A \mid \mathcal{H}](\omega)$ takes for any $\omega \in C$.

When $\mathcal{H} = \sigma(Y)$ is the σ-algebra generated by a r.v. Y, we define $\mathbb{E}[X \mid Y] := \mathbb{E}[X \mid \mathcal{H}]$ and $P[X \in B \mid Y] := P(X^{-1}(B) \mid \mathcal{H})$ for Borel sets B. The law of total expectation now becomes

$$\mathbb{E}[\mathbb{E}[X \mid Y]] = \mathbb{E}[X].$$

Notice that $\mathbb{E}[X \mid Y]$ and $P[X \in B \mid Y]$ are r.v., i.e. they depend on chance ω; more specifically, they depend on the value of $Y(\omega)$. So, for $y \in \mathbb{R}$, we can define $\mathbb{E}[X \mid Y = y]$ as the value (which is unique w.p. 1) of $\mathbb{E}[X \mid Y]$ when $Y = y$. We can define $P[X \in B \mid Y = y]$ similarly and interpret it as the probability of X having a value in B given that (or 'when' or 'if' or 'conditional on') $Y = y$. The *conditional distribution function* of X given $Y = y$ is $F_{X|Y=y}(x) := P[X \leq x | Y = y]$.

If X and Y are discrete random variables, then the conditional pmf is, for atoms (x, y) of the joint distribution, given by $p_{X|Y=y}(x) = \frac{p_{X,Y}(x,y)}{p_Y(y)}$ when the marginal pmf $p_Y(y) > 0$. In the example of two dice throwing with $X = $ 'total number of dots on the two dice' and $Y = 0$ or 1 according to whether the number of dots on the white dice is even or odd, notice that $\sigma(Y) = \{G, G^c, \emptyset, \Omega\}$ is precisely the σ-algebra $\mathcal{H}$ considered in the example above. Therefore, $\mathbb{E}[X \mid Y = 0] = 7.5$ (expected value of X under the condition $Y = 0$) and $\mathbb{E}[X \mid Y = 1] = 6.5$ (expected value of X given $Y = 1$). Since $p_Y(0) = P[Y = 0] = P(G) = 18/36 > 0$ we have that the probability of $X = 5$ given that $Y = 0$ is $P[X = 5 \mid Y = 0] = p_{X|Y=0}(5) = \frac{p_{X,Y}(5,0)}{p_Y(0)} = \frac{P[X=5, Y=0]}{P[Y=0]} = \frac{P(\{2 \circ 3, 4 \circ 1\})}{18/36} = \frac{2/36}{18/36} = 2/18.$

If X and Y are absolutely continuous random variables then, for $Y = y$, the conditional pdf is given by $f_{X|Y=y}(x) = \frac{f_{X,Y}(x,y)}{f_Y(y)}$ when the marginal pdf $f_Y(y) > 0$.

As an example, consider the Gaussian random vector $\mathbf{X} = \begin{bmatrix} X_1 \\ X_2 \end{bmatrix} \frown \mathcal{N}(\boldsymbol{\mu}, \boldsymbol{\Sigma})$, with mean vector $\boldsymbol{\mu} = \begin{bmatrix} \mu_1 \\ \mu_2 \end{bmatrix}$ and variance-covariace matrix $\boldsymbol{\Sigma} = \begin{bmatrix} \sigma_1^2 & \sigma_{12} \\ \sigma_{12} & \sigma_2^2 \end{bmatrix}$. The conditional pdf of X_1 given $X_2 = x_2$ is, with $\mathbf{x} = [x_1, x_2]^T$, $f_{X_1|X_2=x_2}(x_1) = $

$\dfrac{f_{X_1,X_2}(x_1,x_2)}{f_{X_2}(x_2)} = \dfrac{1}{\sigma_{X_1|X_2=x_2}\sqrt{2\pi}}\exp\left(-\dfrac{(x_1-\mu_{X_1|X_2=x_2})^2}{2\sigma^2_{X_1|X_2=x_2}}\right)$, which is the pdf of a univariate

Gaussian distribution with mean $\mu_{X_1|X_2=x_2} = \mu_1 + \rho\dfrac{\sigma_1}{\sigma_2}(x_2-\mu_2)$ and variance $\sigma^2_{X_1|X_2=x_2} = (1-\rho^2)\sigma^2_1$, where $\rho = \dfrac{\sigma_{12}}{\sigma_1\sigma_2}$ is the correlation coefficient.

Independence is an important probabilistic concept. Given a probability space $(\Omega,\mathcal{F},P)$, the events $A_1,A_2,\ldots,A_n \in \mathcal{F}$ are *independent* if, for any number $k \in \{1,2,\ldots,n\}$ of indices and any choice of k indices $1 \le i_1 < i_2 < \cdots < i_k \le n$, we have $P(A_{i_1} \cap A_{i_2} \cap \cdots \cap A_{i_k}) = P(A_{i_1})P(A_{i_2})\ldots P(A_{i_k})$; since this is obviously true for $k = 1$, we need just to verify it for $k = 2,3,\ldots,n$. Of course, in the case $n = 2$ of two events A_1 and A_2, they are independent if $P(A_1 \cap A_2) = P(A_1)P(A_2)$; if that happens, then $P(A_2\,|\,A_1) = P(A_2)$ and $P(A_1\,|\,A_2) = P(A_1)$, [21] so two independent events satisfy the intuitive idea of independence, namely that the occurrence of one event does not influence the probability of the other. For three events A_1,A_2,A_3 to be independent we need to verify that $P(A_1 \cap A_2) = P(A_1)P(A_2)$, $P(A_1 \cap A_3) = P(A_1)P(A_3)$, $P(A_2 \cap A_3) = P(A_2)P(A_3)$, and $P(A_1 \cap A_2 \cap A_3) = P(A_1)P(A_2)P(A_3)$.

In the example of throwing two dice, the event G of having an even number of dots on the white dice is independent of the event $E = \{1\circ1, 2\circ1, 3\circ1, 4\circ1, 5\circ1, 6\circ1\}$ of having one dot on the black dice. This is intuitively built in the concept of fairness of the dice, which leads us to admit that the result we get for the white dice does not influence the result for the black dice. Mathematically, one sees that the independence property $P(G \cap E) = P(G)P(E)$ is indeed verified since $P(G)P(E) = \dfrac{18}{36} \times \dfrac{6}{36} = \dfrac{3}{36}$ and $P(G \cap E) = P(\{2\circ1, 4\circ1, 6\circ1\}) = \dfrac{3}{36}$. The events G and $F = \{5\circ6, 6\circ5, 6\circ6\}$ (having a number of dots on the two dice larger or equal to 11) are not independent.

Given σ-algebras $\mathcal{F}_1, \mathcal{F}_2, \ldots, \mathcal{F}_n$ contained in $\mathcal{F}$, they are said to be *independent* if the events $A_1, A_2, \ldots, A_n$ are independent for all possible choices of $A_1 \in \mathcal{F}_1$, $A_2 \in \mathcal{F}_2, \ldots, A_n \in \mathcal{F}_n$.

The random variables $X_1, X_2, \ldots, X_n$ are *independent* if the σ-algebras $\sigma(X_1)$, $\sigma(X_2)$, ..., $\sigma(X_n)$ generated by them are independent. This is equivalent to say that, given any Borel sets $B_1, B_2, \ldots, B_n$, the events $[X_1 \in B_1], [X_2 \in B_2], \ldots, [X_n \in B_n]$ are independent. Since the Borel σ-algebra can be generated from Borel sets of the particular form $\,]-\infty, x]$, it is enough to show the property for Borel sets of this particular form. Since the d.f. gives the probability of such sets, one can show that the independence of the random variables $X_1, X_2, \ldots, X_n$ is equivalent to having $F_{X_1,X_2,\ldots,X_n}(x_1,x_2,\ldots,x_n) = F_{X_1}(x_1)F_{X_2}(x_2)\ldots F_{X_n}(x_n)$ for all possible choices of

---

21 In fact, $P(A_2\,|\,A_1) = \dfrac{P(A_1 \cap A_2)}{P(A_1)} = \dfrac{P(A_1)P(A_2)}{P(A_1)} = P(A_2)$ and the relation $P(A_1\,|\,A_2) = P(A_1)$ can be shown similarly.

real numbers $x_1, x_2, \ldots, x_n$.[22] For absolutely continuous random variables (or for discrete) r.v., a similar property relating the joint pdf (or the joint pmf) with the product of the individual pdf (or the individual pmf) is also equivalent.

In the particular case of random variables X and Y, if they are independent, then, for any Borel sets B and C, we have $P[X \in B, Y \in C] = P[X \in B] P[Y \in C]$ and also $P[X \in B \mid Y \in C] = P[X \in B], P[Y \in C \mid X \in B] = P[Y \in C]$. We have also $F_{X|Y=y}(x) = F_X(x)$ and $F_{Y|X=x}(y) = F_Y(y)$ for any real numbers x and y. The same holds for the conditional pdf or for the conditional pmf if the r.v. are, respectively, absolutely continuous or discrete. For mathematical expectations, if X and Y are L^2 independent r.v. and the expectations exist, we have $\mathbb{E}[XY] = \mathbb{E}[X]\mathbb{E}[Y]$, and also $\mathbb{E}[Y \mid X] = \mathbb{E}[Y], \mathbb{E}[X \mid Y] = \mathbb{E}[X]$.

So, if $X, Y \in L^2$ are independent, its covariance $COV[X, Y] = \mathbb{E}[XY] - \mathbb{E}[X]\mathbb{E}[Y] = \mathbb{E}[X]\mathbb{E}[Y] - \mathbb{E}[X]\mathbb{E}[Y] = 0$ and correlation $CORR[X, Y] = 0$ are both zero. So, we conclude that two independent random variables are uncorrelated. The reciprocal, however, although true for some cases like the case of X and Y having a Gaussian distribution, is not true in general.

In the example of throwing two dice, if X_1 is the number of dots on the white dice and X_2 is the number of dots on the black dice, these two r.v. are independent.

If, for example, X and Y represent, respectively, the height and the weight of the same randomly chosen individual, these r.v. are not independent and are positively correlated; in fact, there is a tendency (just a tendency since there are many exceptions) for tall individuals to be heavier than shorter individuals. If you pick an individual at random and determine its height X and repeat the experiment by again piking an individual at random and determining its weight Z, then the random variables X and Z are independent.

The concept of independence can be extended to an infinite collection of events, σ-algebras or random variables. The full collection is said to be independent if any finite sub-collection is independent.

---

22 One would think at first that this would not be enough and that the product property for the d.f. should be verified for any number $k \in \{1, 2, \ldots, n\}$ and any sub-collection $X_{i_1}, X_{i_2}, \ldots, X_{i_k}$ of the given random variables. However, due to the special properties of distribution functions, if the product property is valid for the full collection of the n given random variables, it is also automatically valid for the sub-collections, so there is no need to perform such verifications. In fact, if you take $F_{X_1, X_2, \ldots, X_n}(x_1, x_2, \ldots, x_n) = F_{X_1}(x_1) F_{X_2}(x_2) \ldots F_{X_n}(x_n)$ and, for the indices j that are not in the sub-collection, you let $x_j \to +\infty$, this will make $F_{X_j}(x_j) \to 1$ and, since the left-hand side will converge to $F_{X_{i_1}, X_{i_2}, \ldots, X_{i_k}}(x_{i_1}, x_{i_2}, \ldots, x_{i_k})$, you will get the product property valid for the sub-collection.

Box 2.3 Review on conditioning and independence

- Conditional expectations
  - $\mathbb{E}[X \mid \mathcal{H}]$ denotes the conditional expectation of X given $\mathcal{H}$, where $X \in L^1(\Omega, \mathcal{F}, P)$ and $\mathcal{H} \subset \mathcal{F}$ is a σ-algebra. It is by definition a $\mathcal{H}$-measurable r.v. Z such that $\int_H X dP = \int_H Z dP$ for all $H \in \mathcal{H}$. It is a.s. unique.
  - Properties (assuming all r.v. involved are in $L^1(\Omega, \mathcal{F}, P)$ and $\mathcal{G}$ and $\mathcal{H}$ are σ-algebras contained in $\mathcal{F}$):

    $$\mathbb{E}[\mathbb{E}[X \mid \mathcal{H}]] = \mathbb{E}[X] \text{ (law of total expectation)}$$
    $$X \text{ is } \mathcal{H}\text{-measurable} \Rightarrow \mathbb{E}[X \mid \mathcal{H}] = X$$
    $$X \text{ is } \mathcal{H}\text{-measurable} \Rightarrow \mathbb{E}[XY \mid \mathcal{H}] = X \mathbb{E}[Y \mid \mathcal{H}]$$
    $$\mathcal{G} \subset \mathcal{H} \Rightarrow \mathbb{E}[\mathbb{E}[X \mid \mathcal{H}] \mid \mathcal{G}] = \mathbb{E}[\mathbb{E}[X \mid \mathcal{G}] \mid \mathcal{H}] = \mathbb{E}[X \mid \mathcal{G}].$$

  - $\mathbb{E}[X \mid Y]$ denotes the conditional expectation of the r.v. X given the r.v. Y. It is the r.v. $\mathbb{E}[X \mid \mathcal{H}]$ with $\mathcal{H} = \sigma(Y)$. The law of total expectation now reads $\mathbb{E}[\mathbb{E}[X \mid Y]] = \mathbb{E}[X]$. The value that the r.v. $\mathbb{E}[X \mid Y]$ takes when $Y = y$ can be denoted by $\mathbb{E}[X \mid Y = y]$.
- Conditional probabilities
  - $P(A \mid C) = P(A \cap C)/P(C)$ for $A, C \in \mathcal{F}$ and $P(C) > 0$.
  - $P(A \mid \mathcal{H}) = \mathbb{E}[I_A \mid \mathcal{H}]$ for $A \in \mathcal{F}$ and a σ-algebra $\mathcal{H} \subset \mathcal{F}$.
  - $P[X \in B \mid Y] = \mathbb{E}[I_{[X \in B]} \mid Y]$ and $P[X \in B \mid Y = y] = \mathbb{E}[I_{[X \in B]} \mid Y = y]$ for Borel sets B.

    The conditional d.f. of X given $Y=y$ is $F_{X \mid Y=y}(x) = P[X \le x \mid Y=y]$.

    For discrete (absolutely continuous) r.v., the conditional pmf (pdf) is $p_{X \mid Y=y}(x) = \frac{p_{X,Y}(x,y)}{p_Y(y)}$ $\left[f_{X \mid Y=y}(x) = \frac{f_{X,Y}(x,y)}{f_Y(y)}\right]$, when the denominator is positive.
- Independence
  - Given a probability space $(\Omega, \mathcal{F}, P)$, the events $A_1, A_2, \dots, A_n \in \mathcal{F}$ are independent if, for any number $k \in \{1, 2, \dots, n\}$ of indices and any choice of k indices $1 \le i_1 < i_2 < \cdots < i_k \le n$, we have $P(A_{i_1} \cap A_{i_2} \cap \cdots \cap A_{i_k}) = P(A_{i_1}) P(A_{i_2}) \dots P(A_{i_k})$.
  - The σ-algebras $\mathcal{F}_1, \mathcal{F}_2, \dots, \mathcal{F}_n \subset \mathcal{F}$ are independent if the events $A_1, A_2, \dots, A_n$ are independent for all possible choices of $A_1 \in \mathcal{F}_1$, $A_2 \in \mathcal{F}_2, \dots, A_n \in \mathcal{F}_n$.
  - The r.v. $X_1, X_2, \dots, X_n$ are independent if the σ-algebras $\sigma(X_1), \sigma(X_2), \dots, \sigma(X_n)$ are independent.

    This is equivalent to $[X_1 \in B_1], [X_2 \in B_2], \dots, [X_n \in B_n]$ being independent events for any Borel sets $B_1, B_2, \dots, B_n$.

    It is also equivalent to $F_{X_1, X_2, \dots, X_n}(x_1, x_2, \dots, x_n) = F_{X_1}(x_1) F_{X_2}(x_2) \dots F_{X_n}(x_n)$ for all real numbers $x_1, x_2, \dots, x_n$; for discrete (absolutely continuous) r.v., this is equivalent to having a similar product property for the pmf (pdf).
  - If X and Y are independent, then, when the involved quantities are defined, we have (with B, C Borel sets and x, y real numbers):

$P[X \in B, \ Y \in C] = P[X \in B] \, P[Y \in C]$
$P[X \in B \mid Y \in C] = P[X \in B]$
$P[Y \in C \mid X \in B] = P[Y \in C]$;
$F_{X|Y=y}(x) = F_X(x)$ and $F_{Y|X=x}(y) = F_Y(y)$; a similar property holds for pmf (pdf) when the r.v. are discrete (absolutely continuous);
$\mathbb{E}[XY] = \mathbb{E}[X] \, \mathbb{E}[Y]$
$\mathbb{E}[Y \mid X] = \mathbb{E}[Y]$
$\mathbb{E}[X \mid Y] = \mathbb{E}[X]$;
$COV \, [X, Y] = 0$ and $CORR \, [X, Y] = 0$ (uncorrelated)
The reverse may fail, i.e. there are uncorrelated r.v. that are not independent. However, if X and Y are Gaussian and are uncorrelated, they are also independent.

2.4 A brief review of stochastic processes

A *stochastic process* or *random process* on a probability space $(\Omega, \mathcal{F}, P)$ is simply an indexed collection $\{X_t\}_{t \in I}$ of random variables. In our case, t will be interpreted as time and the *index set* I will usually be a time interval of the form $[0, +\infty[, \ ] - \infty, +\infty[$ or $[a, b]$ (stochastic processes in continuous time). However, in other situations, I can be the set of the integers or of the non-negative integers (stochastic processes in discrete time), an interval in $\mathbb{R}^d$ (spatial processes) or any convenient set.

Since each random variable $X_t = X_t(\omega)$ is a function of 'chance' $\omega \in \Omega$, a stochastic process can be considered as a function of two variables, time $t \in I$ and 'chance' $\omega \in \Omega$; like we do for random variables it is customary to abbreviate the notation and simply write X_t instead of $X_t(\omega)$, but we should keep in mind that the stochastic process depends on 'chance' ω even when such dependence is not explicitly written. This function of t and ω must, of course, when we fix the time t, satisfy the property of being a r.v. (i.e. a measurable function of ω). If we fix the 'chance' ω, we obtain a function of time alone, which is called a *trajectory* or *sample path* or *realization* of the stochastic process. So, a stochastic process can also be considered as a collection of trajectories, one trajectory for each state of the 'chance' ω.

The price X_t (abbreviation of $X_t(\omega)$) of a stock at time t for $t \in I = [0, +\infty[$ is an example of a stochastic process. Now ω represents the market scenario, which we may think of as the evolution along time (past, present, and future) of everything that can affect the price of the stock. Obviously, different market scenarios would lead to different price evolutions. For a $t \in I$ fixed, $X_t(\omega)$ is a r.v., therefore a function of 'chance' that associates to each market scenario $\omega \in \Omega$ the corresponding price $X_t(\omega)$ of the stock at time t. For a fixed market scenario $\omega \in \Omega$, the corresponding trajectory $X_t(\omega)$ is a function of time that associates to each time instant $t \in I$ the corresponding price of the stock under that market scenario. When we observe the variation over time of a stock price and draw the corresponding graph, we are indeed drawing a trajectory, the one corresponding to the market scenario ω that, by 'chance', has effectively occurred. Figure 2.3 shows an example of an observed trajectory of a stock price (in log scale) in Euronext Paris (data obtained from www.euronext.com). Figure 2.4 shows two trajectories for two other hypothetical market scenarios (obtained by simulation using the geometric Brownian motion model, to be studied later on this book).

Sometimes it is more convenient to use the alternative notation $X(t, \omega)$ (or the abbreviated form $X(t)$) instead of $X_t(\omega)$ (or X_t). We will use the two

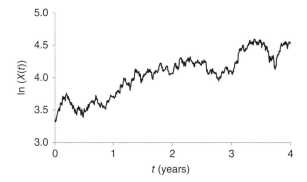

Figure 2.3 Observed trajectory of the stochastic process $\ln X(t)$, where $X(t)$ is the price in euros of the Renault stock during the period 2 January 2012 ($t = 0$) to 31 December 2015 in Euronext Paris. It corresponds to the ω (market scenario) that has effectively occurred.

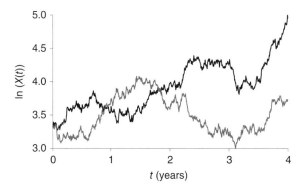

Figure 2.4 Two hypothetical trajectories obtained by Monte Carlo simulation of the logarithm of $X(t)$, where $X(t)$ is geometric Brownian motion with parameters $r = R - \sigma^2/2 = 0.303$/year and $\sigma^2 = 0.149$/year (see Chapter 8). This process was used as a model of the price of the Renault stock during the period 2 January 2012 ($t = 0$) to 31 December 2015 in Euronext Paris (prices are in euros). If the model is correct, these trajectories correspond to two hypothetical randomly chosen market scenarios.

notations indifferently. The *state space* is usually the range of values (also called counter-domain) the function $X(t, \omega)$ can take, but sometimes, for convenience, one uses as state space a larger set containing the range.

The *finite-dimensional distributions* of a stochastic process $\{X_t\}_{t \in I}$ are the probability distributions of a finite number of their random variables $X_{t_1}, X_{t_2}, \ldots, X_{t_n}$ ($t_1, t_2, \ldots, t_n \in I$); they are characterized by the corresponding joint distribution functions

$$F_{t_1, t_2, \ldots, t_n}(x_1, x_2, \ldots, x_n) := P[X_{t_1} \leq x_1, X_{t_2} \leq x_2, \ldots, X_{t_n} \leq x_n],$$

which we may call *finite-dimensional distribution functions*. The family of all finite-dimensional distributions (i.e. defined for all $n = 1, 2, \ldots$ and all the $t_1, t_2, \ldots, t_n \in I$) determines the probabilistic properties of the stochastic process (but not necessarily all its properties). This family obviously satisfies, for all $n = 1, 2, \ldots$ and all $t_1, t_2, \ldots, t_n \in I$ and $x_1, x_2, \ldots, x_n \in \mathbb{R}$, the following properties: [23]

1) $F_{t_{\sigma(1)}, t_{\sigma(2)}, \ldots, t_{\sigma(n)}}(x_{\sigma(1)}, x_{\sigma(2)}, \ldots, x_{\sigma(n)}) = F_{t_1, t_2, \ldots, t_n}(x_1, x_2, \ldots, x_n)$ for all permutations σ of $\{1, 2, \ldots, n\}$ (*symmetry*, i.e. the order of the random variables is not relevant).

2) For $k = 2, \ldots, n - 1$, $F_{t_1, t_2, \ldots, t_k, t_{k+1}, \ldots, t_n}(x_1, x_2, \ldots, x_k, +\infty, \ldots, +\infty) = F_{t_1, t_2, \ldots, t_k}(x_1, x_2, \ldots, x_k)$ (*compatibility of finite-dimensional distributions*).

Two stochastic processes $\{X_t\}_{t \in I}$ and $\{Y_t\}_{t \in I}$ on the same probability space are said to be *equivalent* (it is also said that each one is a *version* of the other) if, for each $t \in I$, one has $X_t = Y_t$ w.p. 1. Equivalent processes have the same finite-dimensional distributions and, therefore, the same probabilistic properties, but may have different analytical properties, as we shall soon see.

In the remainder of this section we will assume that I is an interval of the form $[a, b]$ or $]-\infty, b]$ or $[a, +\infty[$ or $]-\infty, +\infty[$.

---

23 For those more interested in the mathematical details, we remind you of the *Kolmogorov extension theorem*. It says that, given a family of distribution functions $F_{t_1, t_2, \ldots, t_n}$ (with domain $\mathbb{R}^n$), defined for all $n = 1, 2, \ldots$ and all $t_1, t_2, \ldots, t_n \in I$ and satisfying the properties of symmetry and compatibility, there is at least a probability space and a stochastic process on that space such that the family of its finite-dimensional distribution functions is precisely the given family. The proof is constructive. Kolmogorov chose as sample space Ω the set $\mathbb{R}^I$ of the real functions $\omega = \omega(\cdot)$ defined in I. In that space, he considered the class of cylindrical sets of the form $A = \{\omega \in \mathbb{R}^I : (\omega(t_1), \omega(t_2), \ldots, \omega(t_n)) \in B\}$ ($n = 1, 2, \ldots$), which have as bases B the $\mathbb{R}^n$ intervals (Cartesian products of n one-dimensional real number intervals). For the σ-algebra of events, Kolmogorov chose the Borel σ-algebra in $\mathbb{R}^I$, B^I, which is the σ-algebra generated by that class of cylindrical sets. Kolmogorov shows that the probability P on the measurable space $(\mathbb{R}^I, B^I)$ is uniquely characterized (is the unique 'extension') if one knows the probabilities of the above cylindrical sets A; Kolmogorov defined such probabilities by $P(A) = \int_B dF_{t_1, t_2, \ldots, t_n}$. In this probability space $(\mathbb{R}^I, B^I, P)$, Kolmogorov considered the stochastic process $X_t(\omega) = \omega(t)$ and showed (which is quite simple to prove) that its family of finite-dimensional distribution functions is the given family. Basically, he used as elementary events ω the trajectories of the stochastic process themselves.

For the more curious readers, it may happen that, given two equivalent stochastic processes, one version has trajectories that are all continuous functions of time, while the other version has trajectories that are all non-continuous functions. For an example see footnote. [24]

A stronger concept than equivalence of stochastic processes is that of processes with identical trajectories with probability one or (abbreviating) *processes with identical trajectories*, which means $P[X_t = Y_t$ for all $t \in I] = 1$.[25] When the two processes are both continuous, the two concepts coincide. By a *continuous process* we mean a process for which trajectories are continuous with probability one (that is to say, the set of trajectories that are not continuous has probability zero).

To avoid some technical problems (like the one referred above), it is advantageous to work with *separable processes*, which are characterized, for continuity purposes, by the values they take on a countable set dense in I. [26] If we have two equivalent separable processes and one is continuous, so is the other.

Since all stochastic processes have a *separable version*, we do not loose generality if we assume that our processes are separable. So, from now on we will adopt the convention of always working with separable versions of stochastic processes.

The *Kolmogorov criterion* is very useful. It says that, if there are positive constants C, α and β such that $\mathbb{E}[|X_t - X_s|^\alpha] \leq C|t - s|^{1+\beta}$ for all $s, t \in I$, then there

---

24 Let us give an example. Let $\Omega = [0, 1]$, $I = [0, 1]$, $\mathcal{F} = \mathcal{B}_{[0,1]}$ (where by $\mathcal{B}_{[0,1]}$ we denote the Borel σ-algebra of the interval $[0, 1]$, which is generated by the open sets contained in $[0, 1]$) and P is the uniform distribution on $[0, 1]$ (the probability of an interval contained in $[0, 1]$ is equal to the interval's length). Consider the stochastic processes

$$X_t(\omega) = \begin{cases} 0 & \text{if } \omega \neq t \\ 1 & \text{if } \omega = t, \end{cases}$$

(which trajectories are all non-continuous) and $Y_t(\omega) \equiv 0$ (which trajectories are all continuous). Note that the exceptional set $N_t = \{t\}$ of ω values for which $X_t(\omega) \neq Y_t(\omega)$ has zero probability. So $P[X_t = Y_t] = 1$ for any $t \in [0, 1]$ and the two processes are equivalent. However, the exceptional set for which the trajectories of the two processes differ (at least in one time instant) is $N = \bigcup_{t \in I} N_t = [0, 1]$ and has not zero probability, so $P[X_t = Y_t$ for all $t \in I] < 1$ and there is no reason why the two processes should have the same continuity properties. Do not be surprised, since, while it is obvious that a countable union of sets of zero probability has zero probability, this property may fail (and in this case it does fail) for uncountable unions such as $\bigcup_{t \in I} N_t$.
25 In this case $P(N) = 0$, where $N \subset \Omega$ is the exceptional set of trajectories where the processes show some difference.
26 More precisely, for the more interested reader, $\{X_t\}_{t \in I}$ is *separable* if there is a countable set J dense in I such that, for every open subinterval $]a, b[\subset I$ and every closed set $A \subset \mathbb{R}$, the difference between the sets $\{\omega : X_t(\omega) \in A$ for all $t \in]a, b[\cap J\}$ and $\{\omega : X_t(\omega) \in A$ for all $t \in]a, b[\}$ has zero probability (we are assuming our probability space is complete so that these two sets both belong to $\mathcal{F}$). Separable processes do not have the problem of the above example since we just need to work with countable unions of exceptional sets. In the example above, X_t is not separable but Y_t is a separable version of $X(t)$.

exists a separable version of X_t which is continuous. [27] With the convention adopted above of always working with separable versions of stochastic processes, a process that satisfies Kolmogorov criterion is continuous.

A continuous process is also a *measurable process*, [28] which means that it is measurable as a joint function of the two variables t and ω (it is not enough to be separately measurable with respect to each variable t and ω). Notice that all stochastic process are, by definition, measurable with respect to ω.

We can define an *n-dimensional stochastic process* $\{\mathbf{X}(t)\}_{t \in I}$ either as an indexed collection of n-dimensional random vectors or as an n-dimensional vector $\mathbf{X(t)} = [X_1(t), X_2(t), \ldots, X_n(t)]^T$ in which the coordinates $X_i(t)$ $(i = 1, 2, \ldots, n)$ are stochastic processes with index set I defined on the same probability space.

Box 2.4 Brief review of stochastic processes

- Stochastic process
  - Definition: A stochastic process $X_t = X_t(\omega)$ or $X(t) = X(t, \omega)$ with $t \in I$ (I=index set) is a collection of random variables indexed by t on the same probability space $(\Omega, \mathcal{F}, P)$. Here, we will interpret t as time and use for I an interval on the real line.
  - Trajectory or sample path: the function of t obtained by fixing 'chance' ω in $X(t, \omega)$.
  - State space: range of values taken by $X(t, \omega)$ or a superset of it, according to convenience.
- Finite-dimensional distributions
  Finite-dimensional distribution functions $F_{t_1, t_2, \ldots, t_n}(x_1, x_2, \ldots, x_n) :=$ $P[X_{t_1} \leq x_1, X_{t_2} \leq x_2, \ldots, X_{t_n} \leq x_n]$, defined for all $n = 1, 2, \ldots$ and all $t_1, t_2, \ldots, t_n \in I$, characterize probabilistically the stochastic process. These d.f. satisfy the symmetry and compatibility properties.
- Some concepts ($\{X_t\}_{t \in I}$ and $\{Y_t\}_{t \in I}$ are stochastic processes on the same probability space $(\Omega, \mathcal{F}, P)$):
  - Equivalent processes: $X_t = Y_t$ w.p. 1 for each $t \in I$. They have the same finite-dimensional distributions.
  - Processes with identical trajectories: $P[X_t = Y_t$ for all $t \in I] = 1$.

(Continued)

---

27 That does not mean that the original process X_t is continuous, as the example mentioned above shows. The process X_t in that example is not continuous despite satisfying the Kolmogorov criterion. It has, however, a continuous separable version Y_t.

28 For the more interested reader, $X(t, \omega)$ is *measurable* if, considering the *product measurable space* $(I \times \Omega, \mathcal{B}_I \times \mathcal{F})$, the joint inverse image $\{(t, \omega) \in I \times \Omega : X(t, \omega) \in B\}$ of any Borel set $B \in \mathcal{B}$ is in the σ-algebra $\mathcal{B}_I \times \mathcal{F}$. We have denoted by $\mathcal{B}_I$ the Borel σ-algebra on the interval I. The product σ-algebra $\mathcal{B}_I \times \mathcal{F}$ is the σ-algebra generated by the sets of the form $G \times A$ with $G \in \mathcal{B}_I$ and $A \in \mathcal{F}$.

Box 2.4 (Continued)

- Measurable processes: $\{X(t, \omega)\}_{t \in I}$ is measurable if it is jointly measurable in t and ω (see text for details). Note that every stochastic process is measurable with respect to ω.
- Convention: We assume that we work with a separable version, which always exists, of a stochastic process (see text for the concept of separable process).
- Continuous process: $\{X_t\}_{t \in I}$ is continuous if the trajectories are continuous with probability one. Continuous processes are measurable.
- Kolmogorov criterion: if there are positive constants C, α and β such that $\mathbb{E}[|X_t - X_s|^\alpha] \leq C|t - s|^{1+\beta}$ for all $s, t \in I$, then X_t has a continuous separable version. Adopting the Convention, we can say that it is continuous.
- Multidimensional stochastic processes: vectors of one-dimensional stochastic processes on the same probability space and with the same index set.

2.5 A brief review of stationary processes

Let $\{X(t)\}_{t \in I}$ be a stochastic process where the index set I is a real line interval.

A stochastic process is *strictly stationary* if its finite-dimensional distributions are invariant with respect to time translations, i.e.

$$F_{t_1, t_2, \dots, t_n}(x_1, x_2, \dots, x_n) = F_{t_1 + \tau, t_2 + \tau, \dots, t_n + \tau}(x_1, x_2, \dots, x_n)$$

for all $n = 1, 2, \dots$, $x_1, x_2, \dots, x_n \in \mathbb{R}$, $t_1, t_2, \dots, t_n \in I$, $\tau \in \mathbb{R}$ such that $t_1 + \tau$, $t_2 + \tau, \dots, t_n + \tau \in I$.

A stochastic process $\{X(t)\}_{t \in I}$ is said to be a *second-order process* if $X(t) \in L^2$ for all $t \in I$.

Given a second-order process $X(t)$, if it is strictly stationary, then it is also *second-order stationary*, also called *wide-sense stationary*, *weakly stationary* or, more briefly, *stationary*. This means that the first- and second-order moments are invariant to time translations. Therefore, a second-order stationary process $X(t)$ has constant mean $\mathbb{E}[X(t)] \equiv m$ and autocovariances depending only on the time differences, i.e. $COV[X(s), X(t)] = C(t - s)$; the function $C(\cdot)$ is called the *autocovariance function*. The reverse is not necessarily true; there are second-order processes that are second-order stationary but not strictly stationary. However, for *Gaussian processes*, which are processes with Gaussian finite-dimensional distributions, therefore completely characterized by the first- and second-order moments, the two concepts of strict stationarity and second-order stationarity are equivalent.

A second-order process $\{X(t)\}_{t \in I}$ is *mean-square continuous* (or *m.s. continuous* or *continuous in mean-square*) in I if $X(t)$ is continuous in I with respect to the L^2 norm, that is, for any $t_0 \in I$, $\|X(t) - X(t_0)\|_2 \to 0$ when $t \to t_0$ (with $t \in I$);

notice that $\mathbb{E}[(X(t) - X(t_0))^2] = (\|X(t) - X(t_0)\|_2)^2$ and so the above property can also be written as $\mathbb{E}[(X(t) - X(t_0))^2] \to 0$ when $t \to t_0$.

A note for the more interested reader: if the stochastic process $X(t)$ is second-order stationary and m.s. continuous, it has a *spectral distribution function $F(\lambda)$* (which indicates how frequencies of harmonic oscillations of $X(t)$ are distributed) and the autocovariance is given by $C(t) = \int_{-\infty}^{+\infty} \exp(it\lambda)dF(\lambda)$. If, in particular, F has a density f, the *spectral density*, and C is integrable, then the spectral density is the Fourier transform of the autocovariance function, i.e. $f(\lambda) = \frac{1}{2\pi} \int_{-\infty}^{+\infty} \exp(-it\lambda)C(t)dt$.

Box 2.5 Brief review of stationary processes

- $\{X(t)\}_{t\in I}$ is strictly stationary if the finite-dimensional distributions are invariant with respect to time translations.
- $\{X(t)\}_{t\in I}$ is a second-order process if $X(t) \in L^2$ for all $t \in I$.
- A second-order process $\{X(t)\}_{t\in I}$ is second-order stationary (wide-sense stationary, weakly stationary or stationary) if it has first- and second-order moments invariant with respect to time translations, i.e. $\mathbb{E}[X(t)] \equiv m$ (m constant) and $COV[X(s), X(t)] = C(t - s)$ (C is called the autocovariance function).
- A second-order strictly stationary process is second-order stationary; the reciprocal may fail, but it is true for Gaussian processes (processes with Gaussian finite-dimensional distributions).
- A second-order process $\{X(t)\}_{t\in I}$ is m.s. continuous (mean-square continuous, continuous in mean-square) in I if it is continuous w.r.t. the L^2 norm, i.e. for all $t_0 \in I$, $\|X(t) - X(t_0)\|_2 \to 0$ (which is equivalent to $\mathbb{E}[(X(t) - X(t_0))^2] \to 0$) when $t \to t_0$ (with $t \in I$).
- If $\{X(t)\}_{t\in I}$ is second-order stationary and m.s. continuous, then $C(t) = \int_{-\infty}^{+\infty} \exp(it\lambda)dF(\lambda)$, where $F(\lambda)$ is the spectral distribution function. If F has a spectral density f and C is integrable, then f is the Fourier transform of C, i.e. $f(\lambda) = \frac{1}{2\pi} \int_{-\infty}^{+\infty} \exp(-it\lambda)C(t)dt$.

2.6 Filtrations, martingales, and Markov times

Consider a time interval $I = [0, T]$ with $0 \le T \le +\infty$; when $T = +\infty$, we interpret it as $I = [0, +\infty[$. Labelling our starting time as being time 0, as we do, is just convenient, but nothing prevents one giving it a different label, say t_0, and working in a time interval of the form $[t_0, T]$.

Consider a probability space $(\Omega, \mathcal{F}, P)$ and a stochastic process $\{X_t\}_{t\in I}$. Let $\{\mathcal{F}_t\}_{t\in I}$ be a family of sub-σ-algebras of $\mathcal{F}$ such that $\mathcal{F}_s \subset \mathcal{F}_t$ when $s \le t$. Such a family is called a *filtration*.

Basically, we wish to use, at each time instant t, not the whole 'information' F, but a filtered shorter version F_t carrying the present (time t) and past (times $s < t$) 'information'. Of course, as time goes by, we accumulate the previous information with the new-coming information, which is expressed by the property $F_s \subset F_t$ when $s \le t$.

If we are interested in the stochastic process $X(t)$, we often choose the *natural filtration* $F_t = \sigma(X_s; 0 \le s \le t)$ of the σ-algebras generated by the stochastic process up to (and including) time t, which contains the past and present 'information' of the process. Sometimes, however, other choices may be required if one needs the filtration to include other information, such as the information contained in the initial condition of a stochastic differential equation. In this case, we will normally need the stochastic process to still be adapted to the filtration, i.e. the filtration should still include the present and past information carried by the stochastic process. This leads to the formal definition:

We say that the stochastic process $\{X_t\}_{t \in I}$ is *adapted to the filtration* $\{F_t\}_{t \in I}$ or is F_t *-adapted* if, for each $t \in I$, X_t is F_t-measurable. This implies that the inverse images by X_t of Borel sets, which are obviously in F by definition of stochastic process, satisfy the stronger property of being in F_t.

Of course, a stochastic process $\{X_t\}_{t \in I}$ is always adapted to its natural filtration, which is actually the smallest filtration to which it is adapted. But it is also adapted to filtrations carrying additional information, i.e. filtrations which σ-algebras contain, for each $t \in I$, the corresponding σ-algebras $\sigma(X_s; 0 \le s \le t)$ of the natural filtration.

Another requirement that we often need filtrations to fulfil is being non-anticipative, i.e. not being able to use 'information' that only future observations could reveal. This concept of 'no clairvoyance' will be dealt with later on this book.

Given a stochastic process $\{X_t\}_{t \in I}$ on the probability space (Ω, F, P) and a filtration $\{F_t\}_{t \in I}$ in that space, we say that the stochastic process is a F_t-*martingale* if it is adapted to the filtration, $X_t \in L^1$ (i.e. $\mathbb{E}[|X_t|] < +\infty$) for all $t \in I$, and

$$\mathbb{E}[X_t \mid F_s] = X_s \text{ a.s. for any } s \le t. \tag{2.1}$$

Using the convention of identifying a.s. equal random variables, the property (2.1) can be abbreviated to $\mathbb{E}[X_t \mid F_s] = X_s$ for any $s \le t$. In the case of the natural filtration, it can be further simplified to $\mathbb{E}[X_t \mid X_u, 0 \le u \le s] = X_s$ for any $s \le t$. When it is clear from the context which is the filtration we are working with, it is customary to abbreviate and speak of a *martingale* instead of a F_t-martingale. It is also customary to simply say *martingale* when no filtration is mentioned in the context, in which case it is understood we are referring to the natural filtration.

The concept of martingale is linked to the concept of fair game. In fact, if $X(s)$ are one's accumulated gains at the present time s and the game is fair, then, given the information available up to s, one expects on average to maintain the present gains $X(s)$ at any future time $t > s$.

If we replace = by ≤ (respectively by ≥) in (2.1), we have a *supermartingale* (respectively a *submartingale*).

Here we are only interested in stochastic processes in continuous time, but all the above concepts (filtration, natural filtration, adapted process, martingale, submartingale, supermartingale) have analogous definitions for processes in discrete time, except for the fact that the index set is of the form $I = \{0, 1, 2, ..., T\}$ with $0 \leq T \leq +\infty$ (when $T = +\infty$, we interpret it as $I = \mathbb{N}_0 = \{0, 1, 2, ...\}$).

If X_t is a martingale (resp. supermartingale, submartingale), then $\mathbb{E}[X_t]$ is a constant (resp. non-increasing, non-decreasing) function of t. If $X_t \in L^p$ with $p \geq 1$ is a martingale, then, for finite intervals $[a, b] \subset I$ and for $c > 0$, we have the *martingale maximal inequalities*:

$$P\left[\sup_{t \in [a,b]} |X_t| \geq c\right] \leq \frac{|X_b|^p}{c^p} \tag{2.2}$$

$$\mathbb{E}\left[\sup_{t \in [a,b]} |X_t|^p\right] \leq \left(\frac{p}{p-1}\right)^p \mathbb{E}[|X_b|^p] \quad \text{for } p > 1. \tag{2.3}$$

Let us now define the concept of Markov time or stopping time for a stochastic processes in continuous time $\{X_t\}_{t \in [0,+\infty[}$ with index set $I = [0, +\infty[$. Keep in mind, however, that the same concept can be defined in a similar manner for stochastic processes in discrete time with index set $I = \mathbb{N}_0$.

Consider a probability space $(\Omega, \mathcal{F}, P)$ with a filtration $\{\mathcal{F}_t\}_{t \in I}$ and let $I = [0, +\infty[$. Let Θ be an *extended random variable* (also called *improper random variable*), i.e. a function from Ω to $\overline{\mathbb{R}} := \mathbb{R} + \{+\infty\}$ such that the inverse images of Borel sets of $\overline{\mathbb{R}}$ [29] are in $\mathcal{F}$. We assume that Θ takes values in $[0, +\infty]$, i.e. its range is contained in that interval. We say that Θ is a $\mathcal{F}_t$-*Markov time* or $\mathcal{F}_t$-*stopping time* if, given any fixed time instant t, the event $[\Theta \leq t] = \{\omega \in \Omega : \Theta(\omega) \leq t\} \in \mathcal{F}_t$.

This definition serves an important purpose: to know whether the event $\Theta \leq t$ has or has not occurred, we only need to use the 'information' on $\mathcal{F}_t$, that is the 'information' available up to (and including) time t. So, we can determine the Markov time d.f. $F_\Theta(t) = P[\Theta \leq t]$ using just such information. If, as is often the case, the filtration is the natural filtration of a stochastic process X_t, then determining whether $\Theta \leq t$ has or has not occurred requires only information on the past and present values of the stochastic process. Of course, sometimes we may need to incorporate additional information not contained in the stochastic process, in which case we have to use a filtration larger than the natural filtration, taking care that the stochastic process is adapted to the filtration.

---

29 The Borel sets of $\overline{\mathbb{R}}$ are the sets in the σ-algebra generated by the intervals of $\overline{\mathbb{R}}$, which are the intervals of $\mathbb{R}$ and the intervals of the form $[a, +\infty]$ (with $a \in \mathbb{R}$).

Since Θ is an extended r.v., its d.f. F_Θ can be an *improper d.f.*, so it may happen that $F_\Theta(+\infty) := \lim_{t \to +\infty} F_\Theta(t) < 1$, which cannot happen for a proper d.f. However, this happens if and only if $P[\Theta = +\infty] > 0$ and we have $F_\Theta(+\infty) + P[\Theta = +\infty] = 1$.

When it is clear from context which is the filtration we are working with, it is customary to abbreviate and simply speak of a 'Markov time' or a 'stopping time', dropping the reference to the filtration. The same abbreviation is used if no filtration is mentioned in the context, in which case one assumes the natural filtration is in use.

The most typical example of a Markov time is the *first passage time* T_a of a stochastic process X_t by some threshold value $a \in \mathbb{R}$ (see Figure 2.5), which is by definition

$$T_a = \inf\{t \geq 0 : X_t = a\}.$$

We are reminded that, by convention, the infimum of an empty set of values is equal to $+\infty$. So, given an ω, we have $T_a(\omega) = +\infty$ if and only if the trajectory $X_t(\omega)$ never passes by a (i.e. if there is no $t \geq 0$ such that $X_t(\omega) = a$). It is easy to recognize that T_a is a Markov time (for the natural filtration of X_t). In fact, in order to know whether the event $[T_a \leq t]$ has or has not occurred, we only need to know the trajectory of X_t up to (and including) time t. The same cannot be said for the last passage time by a, which is clearly not a Markov time.

Informally, having a filtration is a way of indicating, for each fixed time t, the available 'information' $\mathcal{F}_t$ accumulated up to that time. The more curious reader may ask the question: can we do the same, i.e. define $\mathcal{F}_\Theta$ for a Markov time Θ, which is not a fixed value of time but rather a random time (e.g. the first passage time of a process by some threshold value)? The answer is yes. One considers the smallest σ-algebra $\mathcal{F}_{+\infty}$ that contains all σ-algebras $\mathcal{F}_t$ for $t \geq 0$ and defines $\mathcal{F}_\Theta$ as the σ-algebra formed by all the sets $A \in \mathcal{F}_{+\infty}$ for which $A \cap [\Theta \leq t] \in \mathcal{F}_t$.

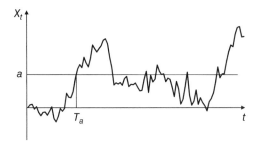

Figure 2.5 First passage time T_a of the stochastic process X_t by a. The figure shows a particular trajectory $X_t(\omega)$ of the stochastic process and the corresponding value $T_a(\omega)$ of the first passage time.

Box 2.6 Filtrations, martingales, and Markov times

Consider a probability space $(\Omega, \mathcal{F}, P)$ and a stochastic process $\{X_t\}_{t \in I}$ on it, with index set $I = [0, T]$ $(0 \leq T \leq +\infty)$; if $T = +\infty$, we interpret as $I = [0, +\infty[$.

- Filtration is a family $\{\mathcal{F}_t\}_{t \in I}$ of sub-σ-algebras of $\mathcal{F}$ such that $\mathcal{F}_s \subset \mathcal{F}_t$ when $s \leq t$. Informally, $\mathcal{F}_t$ contains the 'information' available up to (and including) time t.
- Natural filtration of $\{X(t)\}_{t \in I}$ is defined by $\mathcal{F}_t = \sigma(X_s; 0 \leq s \leq t)$.
- Adapted processes. The stochastic process X_t is adapted to the filtration $\{\mathcal{F}_t\}_{t \in I}$ or is $\mathcal{F}_t$-*adapted* if, for each $t \in I$, X_t is $\mathcal{F}_t$-measurable.
- A process is adapted to its natural filtration (which is the smallest filtration for which that happens) or to a larger filtration.
- Martingale (concept associated to the idea of fair game): $\{X_t\}_{t \in I}$ is a $\mathcal{F}_t$-martingale (abbreviated to 'martingale' when the filtration is clear from the context or is the natural filtration) if it is $\mathcal{F}_t$-adapted, $X_t \in L^1$ for all $t \in I$ and $\mathbb{E}[X_t \mid \mathcal{F}_s] = X_s$ for any $s \leq t$. For the natural filtration the last item simplifies to $\mathbb{E}[X_t \mid X_u, 0 \leq u \leq s] = X_s$ for any $s \leq t$.
- Submartingale (supermatingale): replace = by $\geq$ ($\leq$).
- If X_t is a martingale (supermartingale, submartingale), then $\mathbb{E}[X_t]$ is a constant (non-increasing, non-decreasing) function of t.
- Martingale maximal inequalities (for martingales $X_t \in L^p$ with $p \geq 1$, finite intervals $[a, b] \subset I$ and $c > 0$):

$$P[\sup_{t \in [a,b]} |X_t| \geq c] \leq \frac{|X_b|^p}{c^p}$$
$$\mathbb{E}[\sup_{t \in [a,b]} |X_t|^p] \leq \left(\frac{p}{p-1}\right)^p \mathbb{E}[|X_b|^p] \text{ if } p > 1.$$

- Markov time: An extended r.v. Θ taking values in $[0, +\infty]$ is a $\mathcal{F}_t$-Markov time or $\mathcal{F}_t$-stopping time if, given any fixed t, the event $[\Theta \leq t] = \{\omega \in \Omega : \Theta(\omega) \leq t\}$ $\in \mathcal{F}_t$. One abbreviates to 'Markov time' or 'stopping time' when the filtration is clear from the context or is the natural filtration. The d.f. satisfies the property $F_\Theta(+\infty) + P[\Theta = +\infty] = 1$ and may be improper.
- The first passage time T_a of X_t by a threshold value a is $T_a = \inf\{t \geq 0 : X_t = a\}$. It is a Markov time.

2.7 Markov processes

In common terms, a *Markov process* is a memoryless process, i.e. a stochastic process for which, *given its present value*, future values are independent of past values. Informally, the past is completely 'forgotten' and does not influence future behaviour. In other words, if someone knows *exactly* the present value of the process, knowing or not knowing how the process has evolved in the past to reach the present value is irrelevant to determine the probabilities of future events. The word 'exactly' is essential, since if we only have an imprecise or approximate knowledge about the present value, the independence we have mentioned is not granted.

As in Section 2.6, consider a time interval $I = [0, T]$ with $0 \leq T \leq +\infty$ [30] and a stochastic process $\{X_t\}_{t \in I}$ on the probability space $(\Omega, \mathcal{F}, P)$. The process is a *Markov process in continuous time* [31] if, for any $s, t \in I$, $s \leq t$ and any Borel set $B \in \mathcal{B}$, we have [32]

$$P[X_t \in B \mid X_u, 0 \leq u \leq s] = P[X_t \in B \mid X_s]. \tag{2.4}$$

In the expression (2.4), s plays the role of the present time, t of the future time, and u (when smaller than s) of the past times. Note that $P[X_t \in B | X_u, 0 \leq u \leq s] = P[X_t \in B \mid \mathcal{F}_s]$, where $\mathcal{F}_s$ is the σ-algebra generated by the r.v. $X_u, 0 \leq u \leq s$ (so we are using the natural filtration of the stochastic process).

The *Markov property* (2.4) just mentioned is equivalent to

$$P[X_t \in B \mid X_{t_1} = x_1, \dots, X_{t_{n-1}} = x_{n-1}, X_{t_n} = x_n]$$
$$= P[X_t \in B \mid X_{t_n} = x_n] \tag{2.5}$$

for any $n = 1, 2, \dots$; $t_1 \leq \cdots \leq t_{n-1} \leq t_n \leq t$; $x_1, \dots, x_{n-1}, x_n \in \mathbb{R}$ and for any Borel set B; here t_n plays the role of the present, t of the future, and $t_1, t_2, \dots, t_{n-1}$ of the past. We do not even need to verify this property for all Borel sets, but just for the sets of a generating class of $\mathcal{B}$, as, for example, the class of intervals of the form $] - \infty, x]$. Therefore, the Markov property is equivalent to the following equality on conditional d.f.:

$$F_{X_t | X_{t_1} = x_1, \dots, X_{t_{n-1}} = x_{n-1}, X_{t_n} = x_n}(x) = F_{X_t | X_{t_n} = x_n}(x), \tag{2.6}$$

for any $n = 1, 2, \dots$; $t_1 \leq \cdots \leq t_{n-1} \leq t_n \leq t$ and $x_1, \dots, x_{n-1}, x_n, x \in \mathbb{R}$.

Another property equivalent to the Markov property is having, for any $\mathcal{F}_t$-measurable and integrable r.v. Y and any $s \in I$, $s \leq t$:

$$\mathbb{E}[Y \mid X_u, 0 \leq u \leq s] = \mathbb{E}[Y \mid X_s]. \tag{2.7}$$

Let us consider Borel sets $B, B_1, \dots, B_n$ and $x \in \mathbb{R}$. We will denote the (unconditional) probability distribution at time t by $P_t(B) := P[X_t \in B]$; in the case where it has a pdf, we will denote it by $p(t, y) := f_{X_t}(y) = dF_{X_t}(y)/dy$. The finite-dimensional distributions will be denoted by $P_{t_1, \dots, t_n}(B_1, \dots, B_n) := P[X_{t_1} \in B_1, \dots, X_{t_n} \in B_n]$.

---

30 The starting time could take some other value t_0 but, for simplicity, we will label our starting time as being time 0.

31 Here, we are only interested in Markov processes in continuous time. However, the definition of Markov processes in discrete time is analogous, just the index set, $I = \{0, 1, 2, \dots, T\}$ with integer $0 \leq T \leq +\infty$, is different.

32 Although not explicitly mentioned, the equality of the conditional probabilities is just an equality w.p. 1 and we adopt the usual convention of identifying equivalent r.v.

We can also define the transition probability distributions, which are conditional distributions, by the *transition probabilities* $P(t, B|s, x) :=$ $P[X_t \in B \mid X_s = x]$ for $s \leq t$; $P(t, B|s, x)$ is the probability that the process, being at state x at time s, will make a transition (or will move, if you prefer) to a state in B at time t. In the case where the transition distributions have a pdf, they will be denoted by $p(t, y|s, x) := f_{X_t|X_s=x}(y) = \frac{\partial}{\partial y} F_{X_t|X_s=x}(y)$ and called *transition densities*.

Of course, when $s = t$, we have $P(s, B|s, x) = I_B(x)$ $(= 1$ or $= 0$ according to whether $x \in B$ or $x \notin B$), and therefore, in this particular case, the transition d.f. $F_{X_s|X_s=x}(y) = P[X_s \leq y \mid X_s = x] = H(y - x)$, where $H(z)$ is the Heaviside function $(H(z) = 0$ if $z < 0$ and $H(z) = 1$ if $z \geq 0$). Even when the transition densities exist for all $s < t$, the transition density does not exist when $s = t$, at least in the ordinary sense of a real-valued non-negative function with integral over the real line equal to 1. However, if we allow generalized functions instead of just real-valued functions, we can say that the transition density also exists for $s = t$ and would be $p(s, y|s, x) = \delta(y - x)$, where $\delta(z)$ is the *Dirac delta function*. The Dirac delta function is a generalized function (actually the generalized function derivative of the Heaviside function), which is $= 0$ for all $z \neq 0$, is $= +\infty$ for $z = 0$ and satisfies $\int_{-\infty}^{+\infty} \delta(z)dz = 1$. Sometimes, we will make this generalization.

Using the total probability theorem and one of the forms of the Markov property, one obtains, for $0 < t_1 < t_2 < \cdots < t_{n-1} < t_n$,

$$P_{t_1,\ldots,t_n}(B_1, \ldots, B_n) = \int_{\mathbb{R}} \int_{B_1} \cdots \int_{B_{n-1}} P(t_n, B_n|t_{n-1}, x_{n-1}) \\ P(t_{n-1}, dx_{n-1}|t_{n-2}, x_{n-2}) \ldots P(t_1, dx_1|0, x_0)P_0(dx_0). \tag{2.8}$$

Therefore, to characterize a Markov process from the probabilistic point of view, we only need to know the initial probability distribution P_0 of X_0 and the transition distributions, since from them we can obtain, using the previous expression, all the finite-dimensional distributions.

Again using the total probability theorem and the Markov property, one can easily obtain

$$P_t(B) = \int_{\mathbb{R}} P(t, B|s, x)P_s(dx) \quad (s \leq t) \tag{2.9}$$

and the *Chapman–Kolmogorov equations*

$$P(t, B|s, x) = \int_{\mathbb{R}} P(t, B|u, z)P(u, dz|s, x) \quad (s < u < t). \tag{2.10}$$

In the case where all the pdfs exist, the above properties have a pdf version:

$$p(t, y) = \int_{\mathbb{R}} p(t, y|s, x)p(s, x)dx \quad (s < t) \\ p(t, y|s, x) = \int_{\mathbb{R}} p(t, y|u, z)p(u, z|s, x)dz \quad (s < u < t). \tag{2.11}$$

The Chapman–Kolmogorov equations are easy to prove but it is probably more illuminating to see their intuitive meaning/justification, which depends heavily on the memoryless nature of Markov processes (so we cannot use the equations with other stochastic processes). The idea behind the equations is to break the movement between times s and t into two steps (or more steps if we wish to apply the equations recursively). The first step is the movement between time s and an intermediate time u (during which the process moves from state x to some state z) and the second step is the movement between time u and time t (during which the process goes from z to x). At the moment the second step starts (time u), what has happened in the first step is past and what will happen in the second step is future. Therefore, given the present state z of the process, the memoryless character of Markov processes (the Markov property) says that the two steps are independent. So, given z, the probability of moving from x to y between times s and t is just the product of the two transition probabilities of the steps (product which value depends on the value of z). But we want the total probability of moving from x to y between times s and t no matter what z is, so, by the total probability theorem, we should 'add' (in this case, since z varies continuously, integrate) the products corresponding to the different z. Doing that, we obtain the Chapman–Kolmogorov equations

A *homogeneous Markov process* [33] is a Markov process with stationary transition probabilities, i.e. the transition probabilities satisfy the property

$$P(t + \tau, B|s + \tau, x) = P(t, B|s, x),$$

i.e. they only depend on x, B and $t - s$ (the duration of the time interval), not where the time interval starts or ends. Therefore, we may denote them by $P(t - s, B|x) := P(t, B|s, x)$ or, changing variables to $\tau = t - s$, $P(\tau, B|x) := P(s + \tau, B|s, x) = P(\tau, B|0, x)$. This is the probability of going from a state x to a state in B in a time interval of duration τ (it does not matter what the time interval is, it can be from 2 to $2 + \tau$ or from 5.4 to $5.4 + \tau$ or from 0 to τ, what matters is that its duration is τ). If the transition densities of the homogeneous Markov process exist, we can write $p(\tau, y|x) := p(s + \tau, y|s, x) = p(\tau, y|0, x)$. It is important to distinguish a homogeneous Markov process from a stationary Markov process; the latter has stationary finite-dimensional d.f., while the former has stationary transition probabilities.

Note that, for a homogeneous Markov process with $I = [0, T]$, denoting by $\mathcal{F}_s$ the natural filtration (i.e. the σ-algebra generated by the r.v. $X(u)$ ($0 \le u \le s$)), the Markov property (2.4) can be written in the form

$$P[X_{s+\tau} \in B \mid \mathcal{F}_s] = P[X_\tau \in B \mid X_0]$$

---

33 By 'homogeneous' we mean 'homogeneous in time', not homogeneous in the state space.

for any $\tau \geq 0$ with $s, s + \tau \in I$ and any Borel set B, because $P[X_{s+\tau} \in B \mid X_s] = P[X_\tau \in B \mid X_0]$. The property is equivalent to $\mathbb{E}[h(X_{s+\tau}) \mid \mathcal{F}_s] = \mathbb{E}[h(X_\tau) \mid X_0]$ for all Borel-measurable bounded functions h. [34]

A homogeneous Markov process $\{X_t\}_{t \geq 0}$ is said to be a *strong Markov process* (we can also say that it satisfies the *strong Markov property*) if, for any $\tau \geq 0$ and Borel set B, we have

$$P[X_{S+\tau} \in B \mid \mathcal{F}_S] = P[X_\tau \in B \mid X_0] \tag{2.12}$$

for all Markov times S (with respect to the natural filtration of X_t). This property is equivalent to

$$\mathbb{E}[h(X_{S+\tau}) \mid \mathcal{F}_S] = \mathbb{E}[h(X_\tau) \mid X_0] \tag{2.13}$$

for all Borel-measurable bounded functions h.

The concepts in this section can be easily generalized to n-dimensional stochastic processes. One just needs to replace $\mathbb{R}$ by $\mathbb{R}^n$ and consider now Borel sets of $\mathbb{R}^n$.

Box 2.7 Markov processes

Consider a probability space $(\Omega, \mathcal{F}, P)$ and a stochastic process $\{X_t\}_{t \in I}$ on it, with index set $I = [0, T]$ $(0 \leq T \leq +\infty)$; if $T = +\infty$, we interpret as $I = [0, +\infty[$. Let $\{\mathcal{F}_t\}_{t \in I}$ be the natural filtration of X_t.

- X_t is a Markov process if it satisfies any one of the following equivalent properties:
  - $P[X_t \in B \mid X_u, 0 \leq u \leq s] = P[X_t \in B \mid X_s]$ for any $s, t \in I$, $s \leq t$ and $B \in \mathcal{B}$; note that $P[X_t \in B \mid X_u, 0 \leq u \leq s] = P[X_t \in B \mid \mathcal{F}_s]$;
  - $P[X_t \in B \mid X_{t_1} = x_1, \dots, X_{t_{n-1}} = x_{n-1}, X_{t_n} = x_n] = P[X_t \in B \mid X_{t_n} = x_n]$ for any $n = 1, 2, \dots$; $t_1 \leq \dots \leq t_{n-1} \leq t_n \leq t$; $x_1, \dots, x_{n-1}, x_n \in \mathbb{R}$, and $B \in \mathcal{B}$;
  - $F_{X_t \mid X_{t_1} = x_1, \dots, X_{t_{n-1}} = x_{n-1}, X_{t_n} = x_n}(x) = F_{X_t \mid X_{t_n} = x_n}(x)$ for any $n = 1, 2, \dots$; $t_1 \leq \dots \leq t_{n-1} \leq t_n \leq t$ and $x_1, \dots, x_{n-1}, x_n, x \in \mathbb{R}$;
  - $\mathbb{E}[Y \mid X_u, 0 \leq u \leq s] = \mathbb{E}[Y \mid X_s]$ for any $\mathcal{F}_t$-measurable and integrable r.v. Y and any $s \in I$, $s \leq t$.
- Distribution at time t: $P_t(B) := P[X_t \in B]$ $(B \in \mathcal{B})$
  Initial distribution: $P_0(B)$
  pdf at time t (if it exists): $p(t, y) := f_{X_t}(y) = dF_{X_t}(y)/dy$
- Finite-dimensional distributions: $P_{t_1, \dots, t_n}(B_1, \dots, B_n) := P[X_{t_1} \in B_1, \dots, X_{t_n} \in B_n]$ $(B_1, \dots, B_n \in \mathcal{B})$

(Continued)

---

34 Note that $h(x)$ is not a r.v. but a deterministic function from $\mathbb{R}$ to $\mathbb{R}$. To say that $h(x)$ is Borel-measurable means that we are endowing the domain of the function with the Borel σ-algebra $\mathcal{B}$ and the inverse images by h of Borel sets are in that σ-algebra. Of course, the composite function $h(X_t) = h(X_t(\omega))$ is a stochastic process.

Box 2.7 (Continued)

- Transition probabilities: $P(t, B|s, x) := P[X_t \in B \mid X_s = x]$ $(s \leq t)$
  Transition density (if it exists): $p(t, y|s, x) := f_{X_t|X_s=x}(y) = \frac{\partial}{\partial y} F_{X_t|X_s=x}(y)$ $(s < t)$

- $P_{t_1,\ldots,t_n}(B_1, \ldots, B_n) = \int_{\mathbb{R}} \int_{B_1} \cdots \int_{B_{n-1}} P(t_n, B_n|t_{n-1}, x_{n-1})$
  $P(t_{n-1}, dx_{n-1}|t_{n-2}, x_{n-2}) \ldots P(t_1, dx_1|0, x_0)P_0(dx_0)$
  $$(0 < t_1 < t_2 < \cdots < t_{n-1} < t_n)$$

- $P_t(B) = \int_{\mathbb{R}} P(t, B|s, x) P_s(dx)$ $(s \leq t)$
  If the densities exist, $p(t, y) = \int_{\mathbb{R}} p(t, y|s, x) p(s, x) dx$ $(s < t)$

- Chapman–Kolmogorov equations:
  $P(t, B|s, x) = \int_{\mathbb{R}} P(t, B|u, z) P(u, dz|s, x)$ $(s < u < t)$
  If the densities exist, $p(t, y|s, x) = \int_{\mathbb{R}} p(t, y|u, z) p(u, z|s, x) dz$ $(s < u < t)$

- Homogeneous Markov process: transition probabilities are stationary $P(t + \tau, B|s + \tau, x) = P(t, B|s, x)$, so we can use the notation $P(\tau, B|x) := P(s + \tau, B|s, x) = P(\tau, B|0, x)$.
  If the densities exist, $p(\tau, y|x) := p(s + \tau, y|s, x) = p(\tau|0, x)$.

- For homogeneous Markov processes, the Markov property is equivalent to any of the following properties:
  - $P[X_{s+\tau} \in B \mid \mathcal{F}_s] = P[X_\tau \in B \mid X_0]$ for any $\tau \geq 0$ with $s, s + \tau \in I$ and any $B \in \mathcal{B}$;
  - $\mathbb{E}[h(X_{s+\tau}) \mid \mathcal{F}_s] = \mathbb{E}[h(X_\tau) \mid X_0]$ for all Borel-measurable bounded functions h and any $\tau \geq 0$ with $s, s + \tau \in I$.

- A homogeneous Markov processes is a strong Markov process if it satisfies anyone of the two equivalent properties:
  - $P[X_{S+\tau} \in B \mid \mathcal{F}_S] = P[X_\tau \in B \mid X_0]$ for any $\tau \geq 0$, any $B \in \mathcal{B}$ and all Markov times S;
  - $\mathbb{E}[h(X_{S+\tau}) \mid \mathcal{F}_S] = \mathbb{E}[h(X_\tau) \mid X_0]$ for all Borel-measurable bounded functions h, any $\tau \geq 0$ and all Markov times S.

3

An informal introduction to stochastic differential equations

Let $X = X(t)$ be the size of a wildlife population at time $t \geq 0$ and let $X(0) = x_0$ be the initial population size. Assume there are no food, territorial or other limitations to growth, so that the dynamics of the population size can be described by the *Malthusian model*

$$\frac{dX}{dt} = RX. \tag{3.1}$$

This model simply says that that the (instantaneous) growth rate of the population is proportional to population size, R being the constant of proportionality. This is equivalent to saying that the (instantaneous) *per capita* growth rate $\frac{1}{X}\frac{dX}{dt}$ is constant and equal to R. Remember that R is the difference between the (instantaneous *per capita*) birth and death rates, so it represents the net rate at which the average individual contributes to next generations. The Malthusian model just says that, in the absence of limitations to growth, this rate is constant, therefore does not depend on the size of the population. The situation and the model would be different if there were resource limitations, in which case the amount of resources available for individual survival and reproduction will become scarcer for larger population sizes. Coming back to the Malthusian model, which describes a multiplicative type growth, the solution of the ordinary differential equation (ODE) (3.1) is the Malthusian law or exponential growth law

$$X(t) = x_0 \exp(Rt). \tag{3.2}$$

It is due to Malthus the observation that 'Population, when unchecked, increases in a geometrical ratio' (see Chapter 1 in Malthus (1798)), although he was referring to human rather than to wildlife populations; growing geometrically, when dealing with continuous-time models, is equivalent to exponential growth.

The very same model can be applied if $X(t)$ is the value of a bond with fixed (instantaneous) return rate R or the amount of a bank deposit with fixed

Introduction to Stochastic Differential Equations with Applications to Modelling in Biology and Finance,
First Edition. Carlos A. Braumann.
© 2019 John Wiley & Sons Ltd. Published 2019 by John Wiley & Sons Ltd.
Companion Website: www.wiley.com/go/braumann/stochastic-differential-equations

(instantaneous) interest rate R, or even the price of some good or resource subject to inflation with (instantaneous) fixed rate R.

The ODE (3.1), which can also be written in the form $dX = (Rdt)X$, can be obtained as the limit as $\Delta t \to 0$ of the discrete time model (difference equation) $\Delta X(t) = (R\Delta t)X(t)$, where $\Delta X(t) = X(t + \Delta t) - X(t)$ and where $R\Delta t$ is approximately, when $\Delta t > 0$ is small, the growth/return/interest rate on the time interval $]t, t + \Delta t]$.

Nature and markets, however, are not so steady. The environment where the population grows has random fluctuations that affect the growth rate. Many unpredictable factors in the markets affect the growth rates in the price of goods and resources and the return rates of many financial assets. For example, if $X(t)$ is the price of a stock in the stock market, we do not expect its return rate to be constant, but rather that it fluctuates randomly. Let $B(t)$ (we can also use the alternative notation B_t) be the cumulative effect between time 0 and time t of the environment (resp. market) fluctuations on the growth rate (resp. return rate) of the population (resp. stock or some other financial asset subjected to market fluctuations).

Obviously, $B(t)$ depends on chance ω (i.e. the state of nature or the market scenario), where ω varies in Ω (set of all possible states or scenarios), and we have a probability space structure $(\Omega, \mathcal{F}, P)$, where P gives us the probability of occurrence of the different measurable sets of states or scenarios. To be explicit, we should write $B(t, \omega)$ but we simplify notation by adopting the usual convention of not explicitly mention the dependence on ω. Since $B(t)$ also depends on time, it is a stochastic process. Of course, $B(0) = 0$ and the growth/return rate in a small time interval $]t, t + \Delta t]$ is approximately $R\Delta t + \Delta B(t)$, where $\Delta B(t) = B(t + \Delta t) - B(t)$ is the *increment* of $B(t)$ on the interval $]t, t + \Delta t]$. Thus we obtain as an approximate model the *stochastic difference equation* $\Delta X(t) = (R\Delta t + \Delta B(t))X(t)$. By taking the limit as $\Delta t \to 0$, we obtain $dX(t) = (Rdt + dB(t))X(t)$ or

$$dX(t) = RX(t)dt + X(t)dB(t). \tag{3.3}$$

Since $B(t)$ is a stochastic process, this is a *stochastic differential equation* (SDE), well known in the financial literature as the *Black–Scholes model* for stock prices (and also used for some other financial assets). As we have seen, the same model also describes the Malthusian growth of a population in a randomly varying environment and so we may also call it *stochastic Malthusian model*. Its solution $X = X(t)$ also depends on chance ω (market scenario/state of nature) and is therefore a stochastic process $X(t, \omega)$. The term $RX(t)dt$ describes the mean trend of the dynamical behaviour of this variable, and so R is therefore the (instantaneous) 'average' return/growth rate. The term $X(t)dB(t)$ describes the perturbations around the trend.

What properties should we require the stochastic process $B(t)$ to possess in order to be a reasonable model of its behaviour?

We may assume there are very many sources of random fluctuations in the environment/market that affect the growth/return rate and so, invoking the central limit theorem, we can expect that their joint cumulative effect on such rates, $B(t)$, will have approximately a Gaussian distribution. We will therefore assume that the increment $\Delta B(t)$ on the interval $]t, t + \Delta t]$ is a Gaussian r.v. with mean zero (if the mean were not zero, we should absorb it on the parameter R so that the interpretation of this parameter as the 'average' rate would make sense).

Given two non-overlapping time intervals, it is reasonable to assume that the sources of the random fluctuations on one interval (which affect the increment of B on that interval) are approximately independent from the sources of the random fluctuations on the other interval. We will therefore assume, as a reasonable approximation, that the increments of B on non-overlapping time intervals are independent random variables.

As for the variance of the increment $\Delta B(t)$, it should be the sum of the variances of the many sources of the random fluctuations occurring in the time interval $]t, t + \Delta t]$ (assuming such sources to be approximately independent). As the number of such sources should be approximately proportional to the duration Δt of the interval, the same should happen to the variance of the increment. We will therefore assume that the variance of $\Delta B(t)$ is $\sigma^2 \Delta t$, where σ is a parameter measuring the strength of the environmental fluctuations. This parameter is known in the financial literature as the *volatility*. Let $W(t) = B(t)/\sigma$. Then $\Delta W(t) \frown \mathcal{N}(0, \Delta t)$ (normal distribution with mean zero and variance Δt). The increments $\Delta B(t)$ and $\Delta W(t)$ are stationary (their distributions do not depend on the specific time interval but only on their duration Δt).

A stochastic process $W(t)$ with the properties mentioned above is known as *standard Wiener process* and will be studied in Chapter 4. It is also known as *Brownian motion* because it was the model used by Einstein (1905) in its golden year to describe the Brownian motion of a particle suspended in a fluid. More exactly, if the movement occurs in a plane surface, it is described by $(x_0 + \sigma W_1(t), y_0 + \sigma W_2(t))$, where (x_0, y_0) is the initial position of the particle, $W_1(t)$ and $W_2(t)$ are two independent standard Wiener processes and σ is a coefficient (σ^2 is called the diffusion coefficient). This two dimensional model of Brownian motion is illustrated on Figure 3.1 and it can be extended to three dimensions. Note that in 1905 the process $W(t)$ was not yet known. Brownian motion of a particle was initially observed by the botanist Brown in 1827 when, looking through a microscope at a pollen particle suspended in a liquid, he observed the particle moving in a very irregular way. Einstein attributed the phenomenon to the collision of the particle with molecules of the liquid, which are constantly moving around. Looking to the position of the particle in one of the coordinate axes (say the x-axis), we obtain a Wiener process $x_0 + \sigma W(t)$. The justification is that, on each small time interval of duration Δt, the particle will suffer numerous random independent collisions resulting in an approximately normally distributed cumulative effect on the particle's position; furthermore, the independence of the increments and its

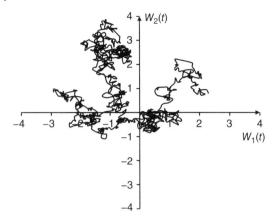

Figure 3.1 Simulation of Brownian motion using the Einstein model with initial position $x_0 = y_0 = 0$ and $\sigma = 1$ (i.e. simulation of $(W_1(t, \omega), W_2(t, \omega))$, where $W_1(t)$ and $W_2(t)$ are independent Wiener processes) for $t \in [0, 10]$ and for a fixed ω chosen randomly. Time was discretized in steps of 0.01 units and the points $(W_1(0.01k), W_2(0.01k))$ ($k = 0, 1, 2, \ldots, 1000$) were joined by straight lines.

variance being proportional to the duration of the time interval are explained by similar reasons to the ones we have used above for population growth or stock prices. The diffusion coefficient can be written as a function of the temperature (higher temperatures result in more agitation of the liquid molecules and so in a larger number of collisions with the particle per unit time), the mass of the suspended particle, and the viscosity of the fluid; its expression, obtained by Einstein, involves the Avogadro number (see Einstein (1905, 1956)). In an experiment, an estimate of $\sigma^2 \Delta t$ that can be obtained with high precision is the sample variance of the increments in particle position (along an axis) on a sample of time intervals of duration Δt. From this estimate, since temperature, particle mass, and fluid viscosity are known, one can obtain a quite precise estimate of the Avogadro number. Although the Einstein model of Brownian motion uses $W(t)$ to reasonably describe the particle position with good precision, I personally do not like to call Brownian motion to $W(t)$, even though many authors use that terminology. The reason is that other more realistic models of the Brownian particle's behaviour were developed later.

Although the 'paternity' of the Wiener process is frequently attributed to Einstein, the development of financial applications of stochastic differential equations brought attention to the pioneer work of Bachelier (1900), which uses precisely the same model $x_0 + \sigma W(t)$ to describe the behaviour of stock prices in the Paris stock market. This is not a very realistic model of stock prices, since, due to the fact that $W(t)$ has a zero expected value, it predicts a constant expected value x_0 for the price.

We may prefer to write (3.3) in the form

$$dX(t) = RX(t)dt + \sigma X(t)dW(t) \quad \text{or} \quad \frac{dX(t)}{dt} = RX(t) + \sigma X(t)\varepsilon(t), \quad (3.4)$$

with $\varepsilon(t) = \varepsilon(t, \omega) = dW(t)/dt$. It so happens that the trajectories of $W(t)$, although a.s. continuous, are extremely irregular (that is the reason they accurately describe the Brownian motion of particles) and, as we shall see in Chapter 4, they are a.s. non-differentiable. Therefore, the derivative $dW(t)/dt$ does not exist in the ordinary sense and only in the sense of generalized functions. Consequently, $\varepsilon(t)$, called *standard white noise*, is not a stochastic process but rather a generalized stochastic process, in the sense that its trajectories are generalized functions of time. Since $W(t)$ has independent and stationary increments, $\varepsilon(t)$ is a generalized stochastic process with independent values at different times, no matter how close these times are to each other. For discrete time processes, the independence of the values at different times poses no difficulties since such times cannot be arbitrarily close and indeed Gaussian white noise in discrete time (which is simply a sequence of independent identically distributed Gaussian random variables) is commonly used to model time series in discrete time. But, in continuous time, the independence of white noise at different times requires the use of generalized stochastic processes. $\varepsilon(t)$ is a Gaussian generalized stochastic process with autocovariance function $C(t) = 0$ for every $t \neq 0$. Curiously, $\Delta W(t)/\Delta t$ has variance $\Delta t/(\Delta t)^2 = 1/\Delta t \to +\infty$ when $\Delta t \to 0$ and, therefore, $C(0) = VAR\,[\varepsilon(t)] = +\infty$. Thus, $C(t) = \delta(t)$ is the *Dirac delta function*, which is a generalized function characterized by $\delta(x) = 0$ for $x \neq 0$, $\delta(x) = +\infty$ for $x = 0$ and $\int_{-\infty}^{+\infty} \delta(x)dx = 1$. Its Fourier transform is a constant function and so $\varepsilon(t)$ has a constant spectral density (all frequencies have equal intensity, a behaviour that is reminiscent of white light, so the name of white noise). One expects natural noises $\tilde{\varepsilon}(t)$ to be coloured, at least slightly, with a non-constant spectral density, small non-zero autocorrelations at neighbour times, and a proper (not generalized) auto-covariance function with a finite peak at the origin. If this peak is sharp, the auto-covariance function can be reasonably approximated by a Dirac delta function (which has an infinite peak at the origin) and the corresponding *coloured noise* can be approximated by a white noise. Contrary to white noise, the integral of such a coloured noise will not be a Wiener process but rather a stochastic process $\tilde{W}(t)$ with 'slightly' dependent increments and smoother trajectories. However, the Wiener process and white noise are mathematically much more tractable (with a very simple $C(t)$, contrary to coloured noises) and are often a very good approximation. For those reasons, we are going to use them in our models.

The equation (3.4) can also be written in the form

$$\frac{dX(t)}{dt} = (R + \sigma\varepsilon(t))X(t). \quad (3.5)$$

This has the nice interpretation of being the model (3.1) with the (instantaneous) growth/return rate R perturbed by a white noise induced by environmental/market random fluctuations. Usually textbooks avoid using the white noise notation and prefer to write the equation using the notation $dX(t) = RX(t)dt + \sigma X(t)dW(t)$, or the abbreviated form $dX = RXdt + \sigma XdW$. Although we have adopted the usual convention of not showing the existing dependence on 'chance' ω, we should keep in mind that $W(t) = W(t, \omega)$ is a stochastic process and so is the solution $X(t) = X(t, \omega)$ of the stochastic differential equation.

4

The Wiener process

4.1 Definition

The Wiener process plays an essential role in the stochastic differential equations we are going to study. It translates the cumulative effect (i.e. is the integral) of the underlying random perturbations affecting the dynamics of the phenomenon under study, so we are assuming that the perturbing noise is continuous-time white noise. In practice, it is enough that this is a good approximation. As mentioned in Chapters 1 and 3, Bachelier used the Wiener process in 1900 to model the price of stocks in the Paris stock market and Einstein used it to model the Brownian motion of a particle suspended on a fluid. However, only after 1920 was this process rigorously studied by Wiener and by Lévy. We will examine here the main properties. We will denote the standard Wiener process indifferently by $W(t)$ (abbreviation of $W(t, \omega)$) or by W_t (abbreviation of $W_t(\omega)$). Let us give the formal definition of the standard Wiener process.

Definition 4.1 (Wiener process) *Given a probability space $(\Omega, \mathcal{F}, P)$, a stochastic process $\{W_t\}_{t \in [0, +\infty[}$ defined on that space is a standard Wiener process, also called Brownian motion, if it satisfies the following properties:*[1]

- $W(0) = 0$ *a.s.*

---

1 An important question is that of existence, i.e. of whether there are stochastic processes satisfying the properties of this definition, for otherwise its content would be void. Property 4.4 of Section 4.2 defines the finite-dimensional distributions of Wiener processes in terms of the pdfs, but these can be integrated to give the finite-dimensional distribution functions, which are effectively proper Gaussian distribution functions. As can be easily seen, the finite-dimensional distributions satisfy the symmetry and compatibility conditions. One can then use the Kolmogorov extension theorem, which insures that there exists (and gives a constructive method to obtain) a probability space and a stochastic process on that space with such finite-dimensional distributions. Thus, Wiener processes do indeed exist.

Introduction to Stochastic Differential Equations with Applications to Modelling in Biology and Finance,
First Edition. Carlos A. Braumann.
© 2019 John Wiley & Sons Ltd. Published 2019 by John Wiley & Sons Ltd.
Companion Website: www.wiley.com/go/braumann/stochastic-differential-equations

- *The increments $W(t) - W(s)$ $(s < t)$ have a Gaussian distribution with mean 0 and variance $t - s$.*
- *Has independent increments, i.e. increments $W(t_i) - W(s_i)$ $(i = 1, \ldots, n)$ on non-overlapping time intervals $]s_i, t_i]$ $(i = 1, 2, \ldots, n)$ are independent r.v.*

To simulate a trajectory of $W(t)$ for a grid of time instants $t_k = k\Delta t$ $(k = 0, 1, 2, \ldots, N$, with $\Delta t > 0)$, one can just simulate the increments $\Delta W(t_k) = W(t_{k+1}) - W(t_k)$ $(k = 0, 1, \ldots, N - 1)$, which are independent r.v. with a Gaussian distribution $\mathcal{N}(0, \Delta t)$, and then obtain the successive grid values starting with $W(t_0) = W(0) = 0$ and adding the successive increments to obtain the desired values $W(t_1) = W(t_0) + \Delta W(t_0)$, $W(t_2) = W(t_1) + \Delta W(t_1)$, $\ldots$, $W(t_N) = W(t_{N-1}) + \Delta W(t_{N-1})$.

For instance, suppose you wish to simulate $W(t)$ in the time interval $[0, 10]$ with a discretization step of $\Delta t = 0.01$ (corresponding to $N = 1000$). In R this can be done using the command 't<-0.01*seq(0,1000)' to produce the vector of instants 0, 0.01, 0.02, ..., 9.99, 10.00 and the command 'W<-c(0,cumsum(rnorm(1000, mean=0, sd=sqrt(0.01))))' to produce the corresponding vector of $W(t)$ simulated values; notice that you start with $W(0) = 0$ and then do the cumulative sums. Repeating the last command produces a different trajectory. You can then plot the graph of a simulated trajectory using 'plot(t,W,type="l")' (plots the $W(t)$ simulated trajectory as a function of t) or use more sophisticated commands and options of the R graphics package to design your axes, figure labels or even plot several simulated trajectories.

On a spreadsheet, you can use three columns, say columns A, B, and C, respectively, for the t_k values $(k = 0, 1, \ldots, 1000)$, the increments $\Delta W(t_k)$ $(k = 0, 1, \ldots, 999)$, and the Wiener process simulations $W(t_k)$. You can use the first line for labels (A1 can be 't', B1 can be 'Delta W(t)' and C1 can be 'W(t)') and the following lines 2 to 1002 for the numerical values at the different 1001 time instants. The cell A2 will be '=0.00' (the $t_0 = 0$ value), the cell B2 will be '=NORMINV(RAND(), 0, SQRT(0.01))' (or a similar command depending on the spreadsheet and the language, corresponding to $\Delta W(t_0)$), and the cell C2 will be '=0' (which is $W(0)$). The cell A3 will be '=A2+0.01' (corresponding to t_1), the cell B3 will again be '=NORMINV(RAND(), 0, SQRT(0.01))' (corresponding to $\Delta W(t_1)$), and the cell C3 will be '=C2+B2' (which is $W(t_0) + \Delta W(t_0) = W(t_1)$). Then, you just need to select the three cells of line 3 (A3, B3, and C3) and drag till line number 1002. If you plot columns A and C, you will get a graph of the simulated sample path of $W(t)$ as a function of t.

Since the distribution of the increment $W(t) - W(s)$ of the Wiener process on a time interval $]s, t]$ (with $s < t$) only depends on its duration $t - s$, the Wiener process has stationary increments. However, the Wiener process is not a stationary process, as you can see from the fact that the variance $VAR\,[W(t)]$ $= VAR\,[W(t) - W(0)] = t - 0 = t$ is not constant. From the definition the Wiener process has independent increments, but the Wiener process itself does not have independent values at different time instants, as can be seen from Property 4.3 of Section 4.2.

A Wiener process need not be a standard one, and a more general Wiener process takes the form $a + \sigma W(t)$, with a and σ constant. However, if there is no risk of confusion, we will simply say 'Wiener process' when we are referring to the standard Wiener process.

4.2 Main properties

We will now study the main properties of the Wiener process $W(t)$.

Property 4.1 *The Wiener process has a separable version which is continuous (i.e. has continuous trajectories a.s.) and we will assume, from now on, that we are working with the continuous version.*

Note: *With this convention there is no need to include the continuity property in the definition of $W(t)$, as some authors prefer to do.*

Proof: This result can be obtained by applying Kolmogorov's criterion (see Section 2.4), since, from the properties of the normal distribution, we have $\mathbb{E}[|W_t - W_s|^4] = 3|t - s|^2$. ∎

Property 4.2 *$W(t)$ has a normal distribution with mean 0 and variance t, abbreviately $W(t) \frown \mathcal{N}(0, t)$.*

Proof: Just notice that $W(t)$ is equal to the increment $W(t) - W(0)$. ∎

Property 4.3 *$COV\,[W(s), W(t)] = \mathbb{E}[W(s)W(t)] = \min(s, t)$.*

Proof: For $s = t$, this is immediate from Property 4.2. For $s < t$, we have $\mathbb{E}[W(s)W(t)] = \mathbb{E}[W^2(s)] + \mathbb{E}[(W(t) - W(s))(W(s) - W(0))] = s + 0$; in fact, $\mathbb{E}[W^2(s)] = s$ and, since $]0, s]$ and $]s, t]$ do not overlap and the corresponding increments are independent, we have $\mathbb{E}[(W(t) - W(s))(W(s) - W(0))] = \mathbb{E}[W(t) - W(s)]\mathbb{E}[W(s) - W(0)] = 0 \times 0 = 0$. The case $s > t$ is similar. ∎

Property 4.4 *$W(t)$ is a Gaussian process and, for $0 = t_0 < t_1 < \cdots < t_n$, the joint pdf $f_{t_1,\ldots,t_n}(x_1, \ldots, x_n)$ of $W(t_1), \ldots, W(t_n)$ is given by (using $x_0 = 0$)*

$$f_{t_1,\ldots,t_n}(x_1, \ldots, x_n) = \prod_{i=1}^{n} \frac{1}{\sqrt{2\pi(t_i - t_{i-1})}} \exp\left(-\frac{(x_i - x_{i-1})^2}{2(t_i - t_{i-1})}\right). \tag{4.1}$$

Proof: The increments $W(t_1) - W(0)$, $W(t_2) - W(t_1)$, $\ldots$, $W(t_n) - W(t_{n-1})$ are independent and so their joint pdf is the product of the Gaussian pdf of its factors, which have mean 0 and variances $t_1 - t_0$, $t_2 - t_1$, $\ldots, t_n - t_{n-1}$. This joint pdf gives the expression on the right-hand side of (4.1). If we make a transformation (change of variables) from the increments to the variables $W(t_1)$, $W(t_2)$, $\ldots$, $W(t_n)$, the joint pdf of $W(t_1)$, $W(t_2)$, $\ldots$, $W(t_n)$ is given by the same expression multiplied by the Jacobian J of the transformation, which in this case is $J = 1$. ∎

Property 4.5 *$W(t)$ is a homogeneous Markov process with transition densities*

$$p(\tau, y | x) = (2\pi\tau)^{-1/2} \exp\left(-\frac{(y-x)^2}{2\tau}\right) \qquad (\tau > 0), \qquad (4.2)$$

i.e. the conditional distribution of $W(s + \tau)$ (with $\tau > 0$) given that $W(s) = x$, is normal with mean x and variance τ:

$$(W_{s+\tau} \mid W_s = x) \frown \mathcal{N}(x, \tau). \qquad (4.3)$$

Proof: From the independence of the increments, we see that the conditional mean and variance are $\mathbb{E}[W_{s+\tau} \mid W_s = x] = \mathbb{E}[W_{s+\tau} - W_s \mid W_s - W_0 = x] + \mathbb{E}[W_s \mid W_s = x] = \mathbb{E}[W_{s+\tau} - W_s] + x = 0 + x = x$ and $VAR[W_{s+\tau} | W_s = x] = \mathbb{E}[(W_{s+\tau} - x)^2 \mid W_s = x] = \mathbb{E}[(W_{s+\tau} - W_s)^2 \mid W_s - W_0 = x] = \mathbb{E}[(W_{s+\tau} - W_s)^2] = (s + \tau - s) = \tau$. The homogeneity comes from the fact that this distribution only depends on the duration τ of the time interval. The Markov property can be proved by showing that (2.6) holds and this is a result of the independence of increments, as can be seen from:

$$F_{W_t | W_{t_1} = x_1, \ldots, W_{t_{n-1}} = x_{n-1}, W_{t_n} = x_n}(x)$$
$$= P[W_t \le x \mid W_{t_1} = x_1, W_{t_2} = x_2, \ldots, W_{t_{n-1}} = x_{n-1}, W_{t_n} = x_n]$$
$$= P[W_t - W_{t_n} \le x - x_n \mid W_{t_1} - W_0 = x_1, W_{t_2} - W_{t_1} = x_2 - x_1, \ldots,$$
$$W_{t_{n-1}} - W_{t_{n-2}} = x_{n-1} - x_{n-2}, W_{t_n} - W_{t_{n-1}} = x_n - x_{n-1}]$$
$$= P[W_t - W_{t_n} \le x - x_n] = P[W_t - W_{t_n} \le x - x_n \mid W_{t_n} - W_0 = x_n]$$
$$= P[W_t \le x \mid W_{t_n} - W_0 = x_n] = F_{W_t | W_{t_n} = x_n}(x). \qquad ∎$$

Property 4.6 *$W(t)$ is a martingale. By default, this means it is a martingale for its natural filtration $\{\mathcal{M}_s\}_{s \ge 0}$, where $\mathcal{M}_s = \sigma(W(u) : 0 \le u \le s)$. Therefore, for $s \le t$, we have $\mathbb{E}[W(t) \mid \mathcal{M}_s] := \mathbb{E}[W(t) \mid W(u), 0 \le u \le s] = W(s)$.*

Proof: Since $W(t)$ is a Markov process, from (2.4) we have $P[W(t) \in B \mid \mathcal{M}_s] = P[W(t) \in B \mid W(s)]$ for any Borel set B. Therefore, $\mathbb{E}[W(t) \mid W(u), 0 \le u \le s] = \mathbb{E}[W(t) \mid W(s)]$. But $\mathbb{E}[W(t) \mid W(s)] = W(s)$ since, due to (4.3), we have $\mathbb{E}[W(t) \mid W(s) = x] = x$ for all x. ∎

Exercise 4.1 *Taking advantage of the properties of the Wiener process, compute or determine:*

(a) $P[W(2.7) > 1.5]$
(b) $P[-1.5 < W(2.7) < 1.5]$
(c) $P[W(2.7) < 1.5 \mid W(1.8) = 1]$
(d) $P[-1.5 < W(2.7) < 1.5 \mid W(1.8) = 1]$
(e) $\mathbb{E}[W(t) \mid W(s), W(u)]$ *with* $0 < u < s < t$
(f) $VAR\,[W(t) \mid W(s), W(u)]$ *with* $0 < u < s < t$
(g) $P[W(2.7) > 1.5 \mid W(1.8) = 1, W(0.5) = -2]$
(h) $\mathbb{E}[W(2.7) \mid W(1.8) = 1, W(0.5) = -2]$
(i) $P[W(1.8) < 1 \mid W(2.7) = 1.5]$. *Be aware that this is not a transition probability since this is not a transition from past to future, but rather the reverse. So, you cannot apply Property 4.5 because it only works for proper transitions, i.e. for* $\tau > 0$ *and here, since* $s = 2.7$ *and* $s + \tau = 1.8$, *we have* $\tau = -0.8 < 0$.
(j) $P[W(1.8) = 1 \mid W(2.7) < 1.5]$
(k) $P[W(2.7) = 1.5, W(1.8) > 1]$
(l) $P[W(2.7) < 1.5, W(1.8) = 1]$
(m) $P[-1 < W(2.7) - W(1.8) < 1.4 \text{ and } 0.5 < W(1.6) - W(0.9) < 1.5]$
(n) $P[-1 < W(2.7) - W(1.8) < 1.4 \mid W(1.6) - W(0.9) = 1.5]$
(o) $\mathbb{E}[W(2.7) \mid \mathcal{M}_{1.8}]$.

The Wiener process can also be considered the continuous version of a simple random walk. This is a Markov process in discrete-time ($t = 0$, Δt, $2\Delta t$, ...) with discrete state space $S = \{\dots, -2\Delta x, -\Delta x, 0, \Delta x, 2\Delta x, \dots\}$ (with $\Delta t > 0$ and $\Delta x > 0$) such that, in each time step, the process moves to the left or the right neighbour state with equal probabilities. We can think of the movement of a particle subject to collisions. If the random walk starts at the 0 position at time 0, when we put $\Delta x = (\Delta t)^{1/2}$ and let $\Delta t \to 0$, the resulting process is the standard Wiener process $W(t)$. To obtain a non-standard Wiener process $x_0 + \sigma W(t)$, we just use a random walk starting at position x_0 at time 0, put $\Delta x = \sigma (\Delta t)^{1/2}$ and let $\Delta t \to 0$.

Exercise 4.2 *Show that:*

(a) *For fixed* $s \geq 0$, *the process* $Y(t) = W(s + t) - W(s)$ $(t \geq 0)$ *is also a standard Wiener process.*
(b) *For constant* $c > 0$, *the process* $Z(t) = \frac{1}{c} W(c^2 t)$ $(t \geq 0)$ *is also a standard Wiener process.*
(c) *The process* $H(t) = W^2(t) - t$ *is a* $\mathcal{M}_t$-*martingale.*

This last property characterizes a Wiener process. In fact, it can be proved (see, for instance, Karatzas and Shreve (1991)) that, given a stochastic process

$X(t)$ $(t \geq 0)$ adapted to a filtration $\mathcal{F}_t \subset \mathcal{F}$, having a.s. continuous trajectories, verifying $X(0) = 0$ a.s., being a $\mathcal{F}_t$-martingale and such that $X^2(t) - t$ is also a $\mathcal{F}_t$-martingale, then $X(t)$ is a Wiener process. This property is known as *Lévy's characterization*.

4.3 Some analytical properties

Although the trajectories of the Wiener process are, as we have seen, a.s. continuous, they are a.s. not differentiable. A proof can be seen in McKean (1969). This is the reason for the strange behaviour we have seen in Chapter 3 of the difference quotient $\frac{\Delta W(t)}{\Delta t} = \frac{W(t+\Delta t) - W(t)}{\Delta t}$, which is Gaussian with mean 0 and variance $\frac{1}{\Delta t} \to +\infty$ as $\Delta t \to 0$. Although the derivative does not exist in the ordinary sense, it can be defined in the generalized function sense, resulting in a generalized process, the standard white noise $\varepsilon(t)$.

For the more interested reader, we now show that the trajectories of the Wiener process are even a.s. functions of *unbounded variation* on any finite time interval $[a, b]$ $(b > a)$. To recognize that, partition that interval into 2^n subintervals of equal duration $\delta_n = (b - a)/2^n$ (which is the mesh of the partition) using the partition points $t_{n,k} = a + k\delta_n$ $(k = 0, 1, \dots, 2^n)$. Note that, as $n \to +\infty$, the mesh $\delta_n \to 0$ and $\sum_{k=1}^{2^n} \delta_n = b - a < +\infty$. Let $V_n = \sum_{k=1}^{2^n} (W(t_{n,k}) - W(t_{n,k-1}))^2$.

Exercise 4.3 *With $V_{n,k} = (W(t_{n,k}) - W(t_{n,k-1}))^2 - \delta_n$, show that*

$$\mathbb{E}[V_n] = \sum_k \delta_n = (b - a)$$

$$VAR[V_n] = \mathbb{E}\left[\left(\sum_k V_{n,k}\right)^2\right] = 2\sum_k (t_{n,k} - t_{n,k-1})^2 \leq 2(b - a)\delta_n \to 0.$$

Suggestion: Note that $\left(\sum_k V_{n,k}\right)^2 = \sum_k V_{n,k}^2 + 2\sum_k \sum_{j>k} V_{n,k} V_{n,j}$ and, due to the independence of the increments of the Wiener process, we get, for $k \neq j$, $\mathbb{E}[V_{n,k} V_{n,j}] = \mathbb{E}[V_{n,k}]\,\mathbb{E}[V_{n,j}]$.

The result of the previous exercise shows that V_n *converges in mean square* (i.e. with respect to the norm-L^2) to $b - a$. Since $\sum_n VAR[V_n] < +\infty$, this implies (a property that can be seen in a good book on probability) that V_n *converges almost surely* (a.s.) to $b - a$, which means that $P[V_n \to b - a \text{ when } n \to +\infty] = 1$. Since

$$V_n \leq (\max_k |W(t_{n,k}) - W(t_{n,k-1})|) \sum_k |W(t_{n,k}) - W(t_{n,k-1})|$$

and since $\max_k |W(t_{n,k}) - W(t_{n,k-1})| \to 0$ a.s. (because $W(t)$ is a.s. continuous on the close interval $[a, b]$ and so is a.s. uniformly continuous on $[a, b]$),

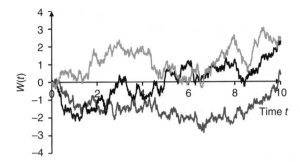

Figure 4.1 Three simulated trajectories (three different values of ω) of the standard Wiener process on the time interval $[0, 10]$. Time was discretized in steps of $\Delta t = 0.01$ time units and simulated values were joined by straight lines.

we conclude that $\sum_k |W(t_{n,k}) - W(t_{n,k-1})| \to +\infty$ a.s.. This shows that the trajectories of the Wiener process are a.s. of unbounded variation on $[a, b]$.

So, although the graph of a trajectory can be drawn without lifting the pencil from the paper (due to its continuity), it will have no smooth pieces and will rather have corners everywhere (due to non-differentiability). Furthermore, the length of a piece of a trajectory (on a time interval $[a, b]$, however small) is infinite (due to unbounded variation), so one would need a pencil of infinite length to draw it.

Figure 4.1 shows three simulated trajectories of the Wiener process on the time interval $[0, 10]$ and gives you an idea of how irregular they are.[2] If one amplifies with a lens a little piece of a trajectory, it would look as irregular as the whole figure.

The mean value of $W(t)$ stays constant with zero value, but the variance of $W(t)$ is t, i.e. it grows when $t \to +\infty$. What would be the asymptotic behaviour of the trajectories of $W(t)$ when $t \to +\infty$? It turns out that $W(t)/t \to 0$ a.s.[3] A more precise idea of the asymptotic behaviour is given by the *law of the iterated logarithm*

$$\lim \sup_{t \to +\infty} \frac{W(t)}{\sqrt{2t \ln \ln t}} = 1 \quad \text{a.s.}$$

$$\lim \inf_{t \to +\infty} \frac{W(t)}{\sqrt{2t \ln \ln t}} = -1 \quad \text{a.s.}$$

(4.4)

---

2 Notice that we could not simulate trajectories of $W(t)$ for all values of t, so we have done it for $t = 0, 0.01, 0.02, \ldots, 10$ and, to help you visualize, we have joined the points in the graph by straight lines, but in reality the values in between the mesh points are not in a straight line.
3 This is a consequence of the *strong law of large numbers*. The proof is simple in the case $t = n$ ($n = 1, 2, \ldots$) is integer. One just needs to notice that the r.v. $W(n) - W(n-1)$ ($n = 1, 2, \ldots$) are independent identically distributed (they are even Gaussian) with common mean $= 0$ and common finite variance $= 1$ and so, by the strong law of large numbers, its average $W(n)/n$ converges a.s. to the common expected mean 0 as $n \to +\infty$. For the general case, one could use the fact that $W(t)/t$ is not 'far' from its value at the integer part of t, but one can also use the law of the iterated logarithm, from which this property can be trivially deduced.

The proof is complex and can be seen, for instance, in Freedman (1983). From the law of the iterated logarithm, one sees that, given a trajectory $W(t, \omega)$ and an arbitrary $\varepsilon > 0$, we have a.s.

$$-(1 + \varepsilon)\sqrt{2t \ln \ln t} \leq W(t, \omega) \leq (1 + \varepsilon)\sqrt{2t \ln \ln t}$$

for sufficiently large t (i.e. there is a $t_0(\omega)$ such that the property holds for $t \geq t_0(\omega)$). This provides us with upper and lower bound curves for the trajectories when t is large.

Exercise 4.4 *(*)* *Show that the process*

$$X(t) := \begin{cases} tW(1/t) & \text{if } t > 0 \\ 0 & \text{if } t = 0 \end{cases}$$

is a standard Wiener process.

In Chapter 5 we will see that the Wiener process is a particular case of a diffusion process.

4.4 First passage times

We have seen that the Wiener process is a homogeneous Markov process. It is even a *strong Markov process*. So, given a Markov time S, since $W(0) = 0$, we can obtain the process $Y(u) := W(S + u) - W(S)$ $(u \geq 0)$ and, using (2.12), recognize that it has the same distribution as $W(u)$ and also its increments have the same distribution as the increments of $W(u)$. Therefore, the increments are uncorrelated and, since the process is Gaussian, they are also independent. We conclude that $Y(u)$ is a Wiener process independent of $\mathcal{F}_S$. This is exactly the property described in Exercise 4.2a of Section 4.2, except that the exercise considered a deterministic time s instead of a random time S. To see proofs of the above facts, the reader can refer to, for example, Gihman and Skorohod (1972) or Karatzas and Shreve (1991). Since $W(0) = 0$ is deterministic, from (2.13), we conclude that

$$\mathbb{E}[h(W(S + u)) \mid \mathcal{F}_S] = \mathbb{E}[h(W(u))] \tag{4.5}$$

for Borel-measurable bounded functions h and $u \geq 0$. The fact that $Y(u)$ is a Wiener process independent of $\mathcal{F}_S$ can be translated, borrowing the terminology of Itô and McKean, by saying that the Wiener process 'starts afresh' at Markov times S. That is to say, if we make a parallel shift of the coordinate axes t and x so that the origin of the new axes is located at the point with original coordinates $(S, W(S))$, we obtain the trajectories of $Y(u)$.

Consider the *first passage time* $T_a = \inf\{t \geq 0 : W(t) = a\}$ of a Wiener process by a constant threshold $a \neq 0$ (the case $a = 0$ is trivial). This is a Markov time. So, to know whether the event $[T \leq t]$ has or has not occurred, we only need to know the trajectory of the Wiener process up to (and including) time t. Let us determine the distribution of T_a using an argument of symmetry and the strong Markov property. We will assume $a > 0$ but similar treatment can be applied to the case $a < 0$.

Since the trajectories are continuous and $W(0) = 0 < a$, if $T_a > t$ the trajectory at time t must be below a (otherwise, the process would have previously crossed a and we would not have $T_a > t$). Therefore, $P[W(t) \geq a \mid T_a > t] = 0$ and so

$$P[W(t) \geq a]$$
$$= P[W(t) \geq a \mid T_a \leq t] \, P[T_a \leq t] + P[W(t) \geq a \mid T_a > t] \, P[T_a > t]$$
$$= P[W(t) \geq a \mid T_a \leq t] \, P[T_a \leq t].$$

We know that $Y(u) = W(T_a + u) - W(T_a) = W(T_a + u) - a$ is a Wiener process. Therefore, for $t \geq T_a$, $P[W(t) \geq a \mid T_a \leq t] = P[Y(t - T_a) \geq 0]$; the last probability can only be equal to $P[Y(t - T_a) \leq 0]$ due to the symmetry of the distribution of the Wiener process, so both probabilities must be equal to $1/2$ (because the probability of the intersection is $P[Y(t - T_a) = 0] = 0$). From here, we obtain

$$P[T_a \leq t] = 2P[W(t) \geq a] = 2 \int_a^{+\infty} \frac{1}{\sqrt{2\pi t}} \exp\left(-\frac{x^2}{2t}\right) dx \quad (a > 0)$$

and, after an appropriate change of variables in the integral, we obtain the d.f. of T_a:

$$F_{T_a}(t) = 2 \int_{|a|/\sqrt{t}}^{+\infty} \frac{1}{\sqrt{2\pi}} e^{-y^2/2} dy = 2\left(1 - \Phi\left(\frac{|a|}{\sqrt{t}}\right)\right) \quad (t > 0). \qquad (4.6)$$

We had written $|a|$ instead of a so that this expression is also valid for the case $a \leq 0$ (the reader can check it as an exercise). The pdf of T_a is just the derivative, and so:

$$f_{T_a}(t) = |a|(2\pi t^3)^{-1/2} \exp\left(-\frac{a^2}{2t}\right) \quad (t > 0). \qquad (4.7)$$

Exercise 4.5

(a) Determine the probability that $W(t)$ reaches for the first time the value -2 after time 1.5.
(b) Despite the fact that $T_a < +\infty$ a.s. (i.e. $P[T_a = +\infty] = 0$), the mean value of T_a is $\mathbb{E}[T_a] = +\infty$.[4] Derive this result.

---

4 Here, there is a slight abuse of language since the integral $\int_\Omega T_a dP$ is infinite and so formally the expected value is not defined and we are saying that it is infinite.

(c) *Since $W(t)$ has a.s. continuous trajectories, there is a.s. the* maximum $X(t) =$ $\max_{0 \leq u \leq t} W(u)$. *Determine the d.f. of* $X(t)$.
Suggestion: Using the continuity of the trajectories, find a relationship between the event $[X(t) \geq a]$ (with $a \geq 0$) and an event concerning T_a.

(d) (*) *Let* $0 < t_1 < t_2$. *Show that the probability of* $W(t)$ *not taking a* 0 *value on the time interval* $[t_1, t_2]$ *is* $(2/\pi) \arcsin \sqrt{t_1/t_2}$ (arcsine law).
Suggestion: Consider the event A that consists in $W(t)$ taking the value 0 at least once on the time interval $[t_1, t_2]$. To determine $P(A)$, condition with respect to $W(t_1) = x$ and use the law of total probability. Note that $P(A \mid W(t_1) = x) = P[T_{|x|} \leq t_2 - t_1]$ and use (4.6). Then, you just need an appropriate change of variables in the resulting double integral to get the desired probability.

4.5 Multidimensional Wiener processes

We can easily defined a *standard m-dimensional Wiener process*

$$\mathbf{W}(t) = [W_1(t),\ W_2(t),\ \ldots,\ W_m(t)]^T.$$

It is just a column vector in which coordinates are m independent unidimensional Wiener processes. If $\mathbf{a}$ is a m-dimensional constant column vector and $\mathbf{C}$ is a $m \times m$ constant definite positive matrix, then $\mathbf{a} + \mathbf{C}\mathbf{W}(t)$ is also called a m-dimensional Wiener process, but now its coordinates may no longer be independent.

5

Diffusion processes

5.1 Definition

To simplify notation, we will use the abbreviation $\mathbb{E}_{s,x}[\ldots]$ for the conditional expectation $\mathbb{E}[\ldots \mid X_s = x]$, where ... represents some r.v. and X_t is a stochastic process. Similarly, for the variance, we use $VAR_{s,x}[\ldots]$ as an abbreviation of $VAR\,[\ldots \mid X(s) = x]$. A similar abbreviation $P_{s,x}[\ldots]$ will be occasionally used for the conditional probability $P[\ldots \mid X_s = x]$, where ... is some event. We will also indifferently use the notations X_t or $X(t)$ for the same stochastic process.

There are in the literature several non-equivalent definitions of what is meant by a diffusion process. We will use here a definition slightly stronger than is usual, but which is sufficient for our purposes.

Definition 5.1 *Let $\{X_t\}_{t \in [0,T]}$ be a stochastic process in a complete probability space $(\Omega, \mathcal{F}, P)$. We say that it is a diffusion process if it is a Markov process with a.s. continuous trajectories such that $X_t \in L^2$ ($t \in [0, T]$) and, for all $x \in \mathbb{R}$ and $s \in [0, T[$, we have, with the convergences being uniform with respect to $s \in [0, T[$,*

$$\lim_{\Delta \to 0^+} \frac{1}{\Delta} P_{s,x}[|X_{s+\Delta} - x| > \varepsilon] = 0 \; \text{for all} \; \varepsilon > 0 \tag{5.1}$$

$$\lim_{\Delta \to 0^+} \mathbb{E}_{s,x}\left[\frac{X_{s+\Delta} - x}{\Delta}\right] = a(s,x) \tag{5.2}$$

$$\lim_{\Delta \to 0^+} \mathbb{E}_{s,x}\left[\frac{(X_{s+\Delta} - x)^2}{\Delta}\right] = b(s,x). \tag{5.3}$$

Observation 5.1 *The most usual definition is weaker, since it does not require the uniform convergence nor that $X_t \in L^2$ ($t \in [0, T]$). This is not enough to guarantee that the moments used (5.2) and (5.3) exist and so such moments are*

Introduction to Stochastic Differential Equations with Applications to Modelling in Biology and Finance,
First Edition. Carlos A. Braumann.
© 2019 John Wiley & Sons Ltd. Published 2019 by John Wiley & Sons Ltd.
Companion Website: www.wiley.com/go/braumann/stochastic-differential-equations

replaced by the always existing truncated moments:

$$a(s, x) := \lim_{\Delta \to 0^+} \mathbb{E}_{s,x} \left[\frac{1}{\Delta} (X_{s+\Delta} - x) I_{|X_{s+\Delta} - x| \le \varepsilon} \right]$$

$$b(s, x) := \lim_{\Delta \to 0^+} \mathbb{E}_{s,x} \left[\frac{1}{\Delta} (X_{s+\Delta} - x)^2 I_{|X_{s+\Delta} - x| \le \varepsilon} \right],$$

with I the indicator function and $\varepsilon > 0$ arbitrary. Property (5.1) suffers no change.

Observation 5.2 *Property (5.1) is often hard to prove, but it is sufficient to prove instead the stronger property* $\lim_{\Delta \to 0^+} \mathbb{E}_{s,x} \left[\frac{|X_{s+\Delta} - x|^{2+\delta}}{\Delta} \right] = 0$ *for some $\delta > 0$.*

The infinitesimal first-order moment $a(s, x)$ is called the *infinitesimal mean, drift coefficient* or simply *drift* and is the speed of change of the mean of the process at time s when $X_s = x$. The infinitesimal second-order moment $b(s, x)$ is called the *infinitesimal variance* or *diffusion coefficient* and is the speed of change of the variance of the process at time s when $X_s = x$ (measures the intensity of fluctuations). From (5.2) and (5.3) we get, when $\Delta \to 0^+$,

$$\mathbb{E}_{s,x}[X_{s+\Delta} - X_s] = a(s, x)\Delta + o(\Delta)$$

$$VAR_{s,x}[X_{s+\Delta} - X_s] = b(s, x)\Delta + o(\Delta).$$

Therefore, $X_{s+\Delta} - X_s \approx a(s, x)\Delta + \sqrt{b(s, x)}Z$, where Z is a r.v. with mean zero and standard deviation $\sqrt{\Delta}$, which can be approximated by $W_{s+\Delta} - W_s$. Letting $\Delta \to 0^+$ and putting $g(t, x) = \sqrt{b(t, x)}$, one obtains $dX_s = a(s, X(s))dt + g(s, X(s))dW_s$, which is the general form of a stochastic differential equation (SDE); the Black–Scholes model mentioned in Chapter 3 is just an example for which $a(s, x) = Rx$ and $g(s, x) = \sigma x$. So, we should not be surprised when later we conclude that, under certain conditions, the solutions of SDEs are diffusion processes.

Since (5.1) can be written in the form $P[|X_{s+\Delta} - X_s| > \varepsilon \mid X_s = x] = o(\Delta)$ when $\Delta \to 0^+$, it is a way of saying that large changes in the process in a small time Δ are not likely to occur.

Definition 5.2 *A homogeneous diffusion process is a diffusion process with drift and diffusion coefficients that do not depend on t. We will denote them by $a(x)$ and $b(x)$, respectively.*

Exercise 5.1

(a) *Show that the Wiener process W_t is a homogeneous diffusion process with zero drift coefficient and a unit diffusion coefficient.*

(b) *Show that the (non-standard) Wiener process $X_t = x_0 + \sigma W_t$, with constant x_0 and σ, is a homogeneous diffusion process with zero drift coefficient and a diffusion coefficient equal to σ^2.*

(c) *Show that $Z_t = x_0 + \mu t + \sigma W_t$, with constant x_0, μ and σ, is a homogeneous diffusion process with drift coefficient μ and diffusion coefficient σ^2. It is called* Brownian motion with drift.

5.2 Kolmogorov equations

Given a diffusion process $\{X_t\}_{t\in[0,T]}$, we can define the *diffusion operator*

$$D = a(s,x)\frac{\partial}{\partial x} + \frac{1}{2}b(s,x)\frac{\partial^2}{\partial x^2}. \tag{5.4}$$

Suppose that $a(s,x)$ and $b(s,x)$ are continuous functions. Let $h(x)$ be a continuous bounded function and, for fixed t with $t > s$, define

$$u^*(s,x) = \mathbb{E}_{s,x}[h(X_t)]. \tag{5.5}$$

If u^* is continuous bounded with first and second derivatives in x also continuous bounded, then u^* is differentiable in s and satisfies the *backward Kolmogorov equation* (BKE)

$$\frac{\partial u^*}{\partial s} + Du^* = 0 \tag{5.6}$$

with the terminal condition

$$\lim_{s\uparrow t} u^*(s,x) = h(x). \tag{5.7}$$

A rigorous proof can be seen, for instance, in Gikhman and Skorohod (1969) and uses a Taylor expansion of u^* about (s,x).[1] In theory, the BKE allows the determination of the transition probabilities since they are uniquely defined

---

1 To give you an idea of the technique used, suppose, to simplify the notation, that the transition density exists. We have $u^*(s,x) = \mathbb{E}_{s,x}[h(X_t)] = \int_{\mathbb{R}} h(y)p(t,y|s,x)dy$. Using the Chapman–Kolmogorov equations to split the transition from time s to time t into the transition from s to $s + \Delta$ (with $\Delta > 0$ such that $s + \Delta < t$) and the transition from $s + \Delta$ to t, one obtains $u^*(s,x) = \int h(y) \int p(t,y|s+\Delta,z)p(s+\Delta,z|s,x)dzdy$. Perform a Taylor expansion of first order on s and second order on x, $p(t,y|s+\Delta,z) = p(t,y|s,x) + \Delta\frac{\partial p}{\partial s} + (z-x)\frac{\partial p}{\partial x} + \frac{1}{2}(z-x)^2\frac{\partial^2 p}{\partial x^2} +$ remainder (the partial derivatives are computed at the point (s,x)), and replace on the previous expression. The result is $u^*(s,x) = u^*(s,x) + \Delta\int h(y)\frac{\partial p}{\partial s}dy + \mathbb{E}_{s,x}[X_{s+\Delta} - x]\int h(y)\frac{\partial p}{\partial x}dy + \frac{1}{2}\mathbb{E}_{s,x}[(X_{s+\Delta} - x)^2]\int h(y)\frac{\partial^2 p}{\partial x^2}dy+$ remainder. Dividing both sides by Δ and letting $\Delta \to 0^+$, one gets, as long as one proves that the remainder term in the resulting expression converges to zero (that is the delicate part of the proof), $0 = \frac{\partial u^*}{\partial s} + a(s,x)\frac{\partial u^*}{\partial x} + \frac{1}{2}b(s,x)\frac{\partial^2 u^*}{\partial x^2} = (\frac{\partial}{\partial s} + D)u^*$. The case $\Delta < 0$ can also be treated with some adjustments. The terminal condition is a consequence of the properties of h and of $\mathbb{E}_{s,x}[h(X_s)] = h(x)$.

from the knowledge of the solution $u^*(s, x) = \mathbb{E}_{s,x}[h(X_t)]$ for all function h.[2] That is, to characterize probabilistically the diffusion process, one only needs its first two infinitesimal moments since these are the only moments that intervene in the BKE. There is, however, an easier way of determining the transition probabilities in case the transition density $p(t, y|s, x)$ exists, is continuous in s, and has first and second partial derivatives in x, also continuous in s. In fact, in this case, the transition density can be obtained directly (for fixed t with $t > s$ and fixed y) as the fundamental solution of the BKE

$$\frac{\partial p}{\partial s} + Dp = 0, \tag{5.8}$$

where the *fundamental solution* means the solution that satisfies the terminal condition

$$\lim_{s \uparrow t} p(t, y|s, x) = \delta(x - y), \tag{5.9}$$

δ being the Dirac delta function (defined in Section 2.7).[3]

If we have a homogeneous diffusion process and use, with $\tau > 0$, the notation $p(\tau, y|x) = p(t, y|t - \tau, x)$ (note that it does not depend on t), since the drift and diffusion coefficients do not depend on time, $D = a(x)\frac{\partial}{\partial x} + \frac{1}{2}b(x)\frac{\partial^2}{\partial x^2}$, and, since $s = t - \tau$ implies $\frac{\partial}{\partial s} = -\frac{\partial}{\partial \tau}$, we obtain for the BKE

$$\left(-\frac{\partial}{\partial \tau} + D\right)p(\tau, y|x) = 0 \tag{5.10}$$

and for the terminal condition

$$\lim_{\tau \downarrow 0} p(\tau, y|x) = \delta(x - y). \tag{5.11}$$

In this homogeneous case, we also have (see (5.5)) $u(\tau, x) := u^*(t - \tau, x) = \mathbb{E}_{t-\tau,x}[h(X_t)] = \mathbb{E}_{0,x}[h(X_{t-(t-\tau)})] = \mathbb{E}_{0,x}[h(X_\tau)]$. So, from (5.6) and (5.7), one sees that

$$u(\tau, x) := \mathbb{E}_{0,x}[h(X_\tau)] \tag{5.12}$$

satisfies for $\tau \geq 0$ the BKE

$$-\frac{\partial u}{\partial \tau} + Du = 0 \tag{5.13}$$

---

2 In fact, we only need to know the solution for a set of functions h that is dense in the space of continuous bounded functions with the uniform morm topology.
3 The proof is similar to the proof of the BKE (5.6), but now we depart from $p(t, y|s, x) = \int p(t, y|s + \Delta, z)p(s + \Delta, z|s, x)dz$. The terminal condition comes from the fact that the transition distribution between time s and itself has the whole probability mass concentrated in x, so the transition d.f. is a Heaviside function and its derivative, the transition density between s and s, is therefore the Dirac delta function (it is not a proper pdf in the ordinary sense).

with terminal condition

$$\lim_{\tau \downarrow 0} u(\tau, x) = h(x). \tag{5.14}$$

There is also a *forward Kolmogorov equation* (FKE) for the transition density $p(t, y|s, x)$ if it exists and has continuous partial derivatives $\dfrac{\partial p}{\partial t}, \dfrac{\partial (a(t, y)p)}{\partial y}$ and $\dfrac{\partial^2 (b(t, y)p)}{\partial y^2}$. For fixed s such that $s < t$ and fixed x, the transition density is the fundamental solution of the FKE, also known as *Fokker–Planck equation* or the *diffusion equation*:[4]

$$\frac{\partial p}{\partial t} + \frac{\partial}{\partial y}(a(t, y)p) - \frac{1}{2}\frac{\partial^2}{\partial y^2}(b(t, y)p) = 0. \tag{5.15}$$

The fundamental solution means the solution that satisfies the initial condition

$$\lim_{t \downarrow s} p(t, y|s, x) = \delta(x - y). \tag{5.16}$$

If the unconditional density $p(t, y) = f_{X_t}(y)$ (pdf of the X_t distribution) exists and has continuous partial derivatives $\dfrac{\partial p}{\partial t}, \dfrac{\partial (a(t, y)p)}{\partial y}$ and $\dfrac{\partial^2 (b(t, y)p)}{\partial y^2}$, that density also satisfies the FKE (5.15) but with the different initial condition

$$\lim_{t \downarrow s} p(t, y) = p(s, y). \tag{5.17}$$

In the case of a homogeneous diffusion process, with $\tau > 0$, using the notation $p(\tau, y|x) = p(s + \tau, y|s, x)$ (note that it does not depend on s), we obtain as FKE

$$\frac{\partial p(\tau, y|x)}{\partial \tau} + \frac{\partial}{\partial y}(a(y)p(\tau, y|x)) - \frac{1}{2}\frac{\partial^2}{\partial y^2}(b(y)p(\tau, y|x)) = 0 \tag{5.18}$$

with initial condition

$$\lim_{\tau \downarrow 0} p(\tau, y|x) = \delta(x - y). \tag{5.19}$$

If the unconditional density $p(t, y) = f_{X_t}(y)$ exists, it satisfies the FKE

$$\frac{\partial p(t, y)}{\partial t} + \frac{\partial}{\partial y}(a(y)p(t, y)) - \frac{1}{2}\frac{\partial^2}{\partial y^2}(b(y)p(t, y)) = 0.$$

It is interesting to see if a homogeneous diffusion process has an *invariant density* or *equilibrium density* $p(y)$ which, as the name indicates, does not vary with

---

4 The technique of the proof is analogous to the BKE (5.8) but starting from $p(t, y|s, x) = \int p(t, y|t - \Delta, z)p(t - \Delta, z|s, x)dz$, i.e. splitting the transition from time s to time t into the transition from s to $t - \Delta$ (with $\Delta > 0$ such that $s < t - \Delta$) and the transition from $t - \Delta$ to t. So, while in the BKE one would look at what was happening in a Δ-neighbourhood of the past time s, in the FKE one looks at what is happening in a Δ-neighbourhood of the present time t. We then use a Taylor expansion of $p(t - \Delta, z|s, x)dz$ about the point (t, y).

time, i.e. when the pdf of X_0 is $p(y)$, then $p(y)$ is also the pdf of X_t for all $t > 0$. If it exists, it will necessarily satisfy (5.15) and, since $\dfrac{\partial p(y)}{\partial t} = 0$, it will be a solution of the ordinary differential equation

$$\frac{d}{dy}(a(y)p(y)) - \frac{1}{2}\frac{d^2}{dy^2}(b(y)p(y)) = 0. \tag{5.20}$$

If an invariant density $p(y)$ exists, it is even more interesting to know what happens to the (assumed existing) density $p(t, y) = f_{X_t}(y)$ of X_t when the initial distribution of X_0 is different from the invariant distribution (the distribution with pdf $p(y)$). We would like to know if, when $t \to +\infty$, X_t converges in distribution to the invariant distribution. If that happens, we will say that $p(y) = f_{X_\infty}(y)$ is the *stationary density* or *asymptotic density* and we may denote by X_∞ a r.v. with distribution having pdf $p(y)$. In this case, X_t *converges in distribution* to X_∞, i.e. $F_{X_t}(y) \to F_{X_\infty}(y)$ (where $F_{X_\infty}(y) = \int_{-\infty}^{y} p(z)dz$ is the d.f. of X_∞). When there is a stationary density and appropriate regularity conditions are satisfied, X_t is even an *ergodic process*, which basically means that the *sample time-moments* (sample means during the time interval $[0, T]$ along one trajectory of a function h of X_t) converge (when $T \to +\infty$) to the corresponding *ensemble moments* of the stationary distribution (which is the mathematical expectation of $h(X_\infty)$). This is very useful. In fact, ordinarily, the estimation of the ensemble moment $\mathbb{E}[h(X_t)]$ for large t would require observing $X_t(\omega)$ for many sample paths (many different ω values) and taking the sample average of the observed $h(X_t)$ as the estimate. However, in many real applications we cannot repeat the experiment, for example short-term interest rates or the time evolution of the size of a natural population cannot be repeated for different randomly 'chosen' market scenarios or states of nature. In such cases, there is only one sample path available (the one corresponding to the market scenario or state of nature ω that actually has occurred). When the process is ergodic, the time-average taken along that single sample path using many large values of t will also be a good estimate of the desired ensemble average. This is quite important in parameter estimation and in prediction.

The standard Wiener process $W(t)$ is a homogeneous diffusion process with zero drift coefficient and unit diffusion coefficient. The BKE and the FKE equations for the transition density $p(\tau, y|x)$ have a similar appearance:

$$\frac{\partial p(\tau, y|x)}{\partial \tau} = \frac{1}{2}\frac{\partial^2 p(\tau, y|x)}{\partial x^2} \quad \text{and} \quad \frac{\partial p(\tau, y|x)}{\partial \tau} = \frac{1}{2}\frac{\partial^2 p(\tau, y|x)}{\partial y^2}. \tag{5.21}$$

These partial differential equations are also known as the *heat equation* (since it gives the time evolution of the heat intensity when heat propagates along an axis). The fundamental solution gives the transition density of $W(t)$

$$p(\tau, y|x) = (2\pi\tau)^{-1/2} \exp\left(-\frac{(y-x)^2}{2\tau}\right); \tag{5.22}$$

note that the solution is determined up to a multiplicative constant, which we can obtain using the fact that this is a pdf and so we must have $\int_{-\infty}^{+\infty} p(t, y|x)\, dy = 1$. Therefore, we retrieve by a different method the result

$$(W(s + \tau) \mid W(s) = x) \frown \mathcal{N}(x, \tau) \tag{5.23}$$

that we had previously obtained in (4.3).

The Kolmogorov equations reduce the probabilistic problem of determining the transition densities of a diffusion process to the deterministic problem of solving a partial differential equation. Since the solutions of SDEs are, under certain regularity conditions, diffusion processes, the probabilistic characterization of their solutions (their transition probabilities) is therefore reduced to a deterministic problem. But also, inversely, some relevant questions of the theory of partial differential equations of the diffusion type have advanced considerably thanks to the study by probabilistic methods of properties of the associated diffusion processes (or solutions of SDEs).

That said, we should mention that it is usually hard to obtain solutions of the Kolmogorov equations, even by numerical methods, and probabilistic methods, recurring to Monte Carlo simulations if necessary, are often preferable. The reader may experiment with obtaining the transition density as the fundamental solution of the corresponding FKE for the process $Z(t) = x_0 + \mu t + \sigma W(t)$ (x_0, μ and σ are constants), called Brownian motion with drift, which is a homogeneous diffusion process with drift coefficient μ and diffusion coefficient σ^2. In this case it is possible to obtain that solution explicitly but it is not an easy task. If the reader just knew the drift and diffusion coefficients of $Z(t)$ and had no idea of the relation of $Z(t)$ to the Wiener process, (s)he might be tempted to use that methodology. After finding the solution $p(\tau, y|x)$, the reader would immediately notice that it is just the pdf of a Gaussian distribution with mean $x_0 + \mu\tau$ and variance $\sigma^2\tau$. However, the reader could easily reach this same conclusion by using the properties of the Wiener process (we leave it as an exercise). Now, of course, already knowing the solution, it is trivial to check that it is indeed the fundamental solution of the FKE. Anyway, the Kolmogorov equations, even if not the most amenable for computational purposes, are important tools for the theoretical study of diffusion processes and, therefore, of SDEs.

5.3 Multidimensional case

The study can be easily extended to a *n-dimensional diffusion process* $\mathbf{X}(t) = [X_1(t), ..., X_n(t)]^T$. The drift coefficient is now a *n*-dimensional vector $\mathbf{a}(s, \mathbf{x})$, whose coordinates are the drift coefficients of the corresponding unidimensional processes. The diffusion coefficient is a matrix $\mathbf{b}(s, \mathbf{x})$ with entries

$$b_{i,j}(s, \mathbf{x}) = \lim_{\Delta \to 0^+} \mathbb{E}_{s,\mathbf{x}} \left[\frac{(X_i(s + \Delta) - x_i)(X_j(s + \Delta) - x_j)}{\Delta} \right].$$

A diagonal element $b_{i,i}(s, \mathbf{x})$ is the diffusion coefficient of $X_i(t)$. The diffusion operator is

$$D = \sum_{i=1}^{n} a_i(s, \mathbf{x}) \frac{\partial}{\partial x_i} + \frac{1}{2} \sum_{i=1}^{n} \sum_{j=1}^{n} b_{ij}(s, \mathbf{x}) \frac{\partial^2}{\partial x_i \partial x_j}.$$

The FKE takes the form

$$\frac{\partial p}{\partial t} + \sum_{i=1}^{n} \frac{\partial}{\partial y_i} (a_i(t, \mathbf{y})p) - \frac{1}{2} \sum_{i=1}^{n} \sum_{j=1}^{n} \frac{\partial^2}{\partial y_i \partial y_j} (b_{ij}(t, \mathbf{y})p) = 0.$$

Multidimensional SDEs appear in many applications of natural or financial phenomena involving several dynamical values. To give just one example, suppose that particles of certain pollutant are spread in the atmosphere. Due to collisions with air molecules, the particles diffuse (Brownian motion). Let $b_{ij}(s, \mathbf{x})$ be the speed of change of the covariance between the diffusion movements in the directions of the x_i and x_j axes at time s and location $\mathbf{x}$ (if $i = j$, that would just be the speed of change of the variance of the diffusion movement in the x_i axis). Frequently, it is reasonable to assume that matrix $\mathbf{b}(s, \mathbf{x})$ is the product of a scalar function $\sigma(s, \mathbf{x})$ by the identity matrix (isotropy of the diffusion of the particle with respect to the different space directions); $\sigma(s, \mathbf{x})$ may depend on time s and location $\mathbf{x}$ if temperature or other variables that affect the diffusion are different at different times and locations. If there is wind, besides diffusing, the particles have a tendency to move in the wind direction at the wind speed. Let $\mathbf{a}(s, \mathbf{x})$ be the wind speed vector at time s and location $\mathbf{x}$. Then, the position $\mathbf{X}(t)$ of the particle at time $t \geq 0$ will be a three-dimensional diffusion process with drift coefficient $\mathbf{a}(s, \mathbf{x})$ and diffusion coefficient $\mathbf{b}(s, \mathbf{x})$. Solving one of the Kolmogorov equations, one can obtain the pdf $p(t, \mathbf{y})$ of $\mathbf{X}(t)$, i.e. the probability density of finding a given particle in location $\mathbf{y}$ at time t. If a large number N of particles is dropped, then, if we consider a small space volume ΔV around location $\mathbf{y}$, the number of particles to be found in that volume at time t will be approximately equal to $p(t, \mathbf{y})N\Delta V$. A similar model can be applied when oil is spilled in the ocean.

6

Stochastic integrals

6.1   Informal definition of the Itô and Stratonovich integrals

The Black–Scholes model (3.4) is a particular case of a stochastic differential equation, but we are going to study in this book the more general case of *stochastic differential equations* (SDEs) in a time interval $[0, T]$ ($T > 0$) of the form

$$dX(t) = f(t, X(t))dt + g(t, X(t))dW(t), \qquad X(0) = X_0, \qquad (6.1)$$

where $f(s, x)$ and $g(s, x)$ are real functions with domain $[0, T] \times \mathbb{R}$ and X_0 (initial condition) is a r.v. independent of the Wiener process $W(t)$. The initial condition can, in particular, be deterministic, i.e. assume a constant real value x_0 (in which case we may say that the r.v. X_0 is degenerate and assumes the value x_0 with probability one). For convenience, we have labelled our starting time as time 0, but nothing prevents one giving it a different label, say t_0, and working in a time interval of the form $[t_0, T]$.

Equations of this type appear in a wide range of scientific and technological domains whenever we want to model a dynamical phenomenon described by a differential equation perturbed by random fluctuations, as long as such perturbations can, to a good approximation, be considered continuous in time and having independent increments.

In the same way as for ordinary differential equations (ODE), the *Cauchy problem* (6.1) (i.e. the problem involving the equation and the initial condition) is nothing else than an integral equation in 'disguise', namely the *stochastic integral equation*:

$$X(t) = X_0 + \int_0^t f(s, X(s))ds + \int_0^t g(s, X(s))dW(s), \qquad (6.2)$$

which is obtained by integrating both sides of the SDE between times 0 and $t \in [0, T]$ and is known as the integral form of the SDE (6.1). By *solution*

Introduction to Stochastic Differential Equations with Applications to Modelling in Biology and Finance,
First Edition. Carlos A. Braumann.
© 2019 John Wiley & Sons Ltd. Published 2019 by John Wiley & Sons Ltd.
Companion Website: www.wiley.com/go/braumann/stochastic-differential-equations

$X(t) = X(t, \omega)$ of (6.1) we mean a stochastic process satisfying (6.2). It is more natural to define the solution of an SDE as the solution of its integral form since the derivatives of $W(t)$ do not exist in the ordinary sense (only as generalized stochastic processes) and the same happens to the derivatives of $X(t)$. This definition of a solution requires that the integrals in (6.2) are well defined.

As for the first integral $\int_0^t f(s, X(s, \omega))ds$, we can fix a particular 'chance' state ω (i.e. choose a particular market scenario or state of nature), so that (assuming f to be a sufficiently well-behaved function) the integral becomes a common Riemann integral $\int_0^t F(s, \omega)ds$ of a function $F(s, \omega) = f(s, X(s, \omega))$ of time alone (since ω is fixed) with respect to time. However, the value of this integral will depend on the value of ω that has been chosen. Considering all possible $\omega \in \Omega$ values, the corresponding values of the integral will vary with ω and this variable is indeed a r.v. (a measurable function of ω). So, $\int_0^t f(s, X(s, \omega))ds$ is well defined in a very simple manner as long as f is well behaved.

Will this idea of fixing ω work for the second integral $\int_0^t g(s, X(s, \omega))dW(s, \omega)$, using now 'Riemann–Stieltjes integrals' $\int_0^t G(s, \omega)dW(s, \omega)$, with $G(s, \omega) = g(s, X(s, \omega))$? The answer is negative, since these alleged *Riemann–Stieltjes integrals* (RS integrals) do not exist as such. In fact, the usual definition of an RS integral as a joint limit of all *Riemann–Stieltjes sums* (RS sums) as the mesh of the $[0, t]$ tagged partitions converges to zero does not work. Indeed, different choices of the tags (the intermediate time points chosen in each subinterval of the partition where the integrand function G is computed) usually result in different limits, not in a joint one. The reason for that behaviour resides in the fact that the integrator function, which is the Wiener process $W(t)$, has a.s. unbounded variation. So, we need a different way of defining the second integral in (6.2).

In this chapter we are going to define integrals of the form $\int_0^t G(s, \omega)dW(s, \omega)$, abbreviated

$$\int_0^t G(s)dW(s),$$

for quite general integrand functions G.

But, before proceeding to such a definition, let us first show the reader, with an example, namely the particular case $G(s) = W(s)$ where the integrand function coincides with the integrator (the Wiener process), that indeed the naive definition as RS integral does not work. So, let us try to determine the integral $\int_0^t W(s)dW(s)$. If ordinary calculus would apply, the reader, noticing that $W(0) = 0$, would immediately say that the integral would be equal to

$$\frac{1}{2}W^2(t).$$

Is it? Let us consider a sequence of tagged partitions

$$0 = t_{n,0} < t_{n,1} < \cdots < t_{n,n-1} < t_{n,n} = t \quad (n = 1, 2, \ldots) \tag{6.3}$$

of the integration interval $[0, t]$ with a mesh $\delta_n = \max_{k=1,\ldots,n}(t_{n,k} - t_{n,k-1}) \to 0$ as $n \to +\infty$, in which, for each subinterval $[t_{n,k-1}, t_{n,k}]$ of the partition, we choose a tag (we can also call it the 'intermediate point') $\tau_{n,k}$ located in that subinterval. The RS sums take the form

$$\sum_{k=1}^{n} W(\tau_{n,k})(W(t_{n,k}) - W(t_{n,k-1})). \tag{6.4}$$

Let us determine the limits of these RS sums when $n \to +\infty$. Since the sums are r.v., we will work for convenience with mean square limits (abbreviated to m.s. limits), which are limits in the L^2 norm, and will represent such limits by 'l.i.m.'.

If we choose as tags the initial points of the partition subintervals, $\tau_{n,k} = t_{n,k-1}$, the resulting RS sums are

$$S_n = \sum_{k=1}^{n} W(t_{n,k-1})(W(t_{n,k}) - W(t_{n,k-1})) \tag{6.5}$$

and the m.s. limit is the *non-anticipative integral*, called the *Itô integral*:

$$I = \int_0^t W(s)dW(s) = \text{l.i.m. } S_n. \tag{6.6}$$

Note that, in (6.5), the values of the integrand function (which measure, when we work with SDE, the strength of the impact of random fluctuations on the phenomenon being studied) are independent of the future increments $W(t_{n,k}) - W(t_{n,k-1})$ of the integrator (which describe the random perturbations, say in the market or in the environment, on the subintervals $]t_{n,k-1}, t_{n,k}]$). Being independent means that the values of the integrand G do not anticipate (do not 'guess') such random perturbations. Using the Itô integral is therefore a way of saying that the current behaviour of the phenomenon has the nice property of not depending on the future random perturbations (absence of 'clairvoyance'). However, the Itô integral does not follow the ordinary rules of calculus.

In fact, we have

$$I = \int_0^t W(s)dW(s) = \frac{1}{2}(W^2(t) - t). \tag{6.7}$$

Let us prove this result, i.e. let us prove that

$$\mathbb{E}\left[\left(S_n - \frac{1}{2}(W^2(t) - t)\right)^2\right] \to 0$$

when $n \to +\infty$. Note that $W^2(t) = \sum_{k=1}^n (W^2(t_{n,k}) - W^2(t_{n,k-1}))$ and therefore $S_n - \frac{1}{2}(W^2(t) - t) = -\frac{1}{2}\left(\sum_{k=1}^n (W(t_{n,k}) - W(t_{n,k-1}))^2 - t\right) = -\frac{1}{2}\sum_{k=1}^n h_{n,k}$, with $h_{n,k} = (W(t_{n,k}) - W(t_{n,k-1}))^2 - (t_{n,k} - t_{n,k-1})$. Since the $h_{n,k}$ $(k = 1, 2, ..., n)$ are independent and $\mathbb{E}[h_{n,k}] = 0$, we get $4\mathbb{E}\left[\left(S_n - \frac{1}{2}(W^2(t) - t)\right)^2\right] = \mathbb{E}\left[\left(\sum_k h_{n,k}\right)^2\right] = VAR\left[\sum_k h_{n,k}\right] = \sum_k VAR[h_{n,k}] = 2\sum_k (t_{n,k} - t_{n,k-1})^2 \le 2\delta_n \sum_k (t_{n,k} - t_{n,k-1}) \le 2\delta_n t \to 0$, as intended.

If, however, we choose as tags of the subintervals $[t_{n,k-1}, t_{n,k}]$ the terminal points $\tau_{n,k} = t_{n,k}$, we obtain the RS sums

$$S_n^+ = \sum_{k=1}^n W(t_{n,k})(W(t_{n,k}) - W(t_{n,k-1})) \tag{6.8}$$

and, in the m.s. limit, the (anticipative) integral

$$I^+ = \text{l.i.m. } S_n^+ = \frac{1}{2}(W^2(t) + t). \tag{6.9}$$

This integral does not follow the ordinary rules of calculus either.

Exercise 6.1 (*) *Prove (6.9). Obtain also the integrals $I^{(\alpha)}$ corresponding to the m.s. limits of RS sums when we choose as tags $\tau_{n,k} = (1 - \alpha)t_{n,k-1} + \alpha t_{n,k}$, with $0 \le \alpha \le 1$. Note that $I = I^{(0)}$ and $I^+ = I^{(1)}$.*

The important message is that different RS sums (corresponding to different choices of the tag points) produce different m.s. limits and so there is no joint limit and, consequently, the RS integral does not exist. Would that be the case had we worked with a different type of limit (such as limit in probability or the a.s. limit) instead of the m.s. limit? The answer is negative since, if such limits did exist, they would coincide with the m.s. limit and so they also would vary with the choice of the tag points. There are, therefore, an infinite variety of stochastic integrals, of which the Itô integral, despite the fact that it does not follow ordinary calculus rules, is usually the preferred one in the literature. The reason is its absence of 'clairvoyance', which seems to be more appropriate to model natural phenomena, and also its nice probabilistic properties, which we will study in this chapter.

Another variety of the stochastic integral that we will also take a look at is the *Stratonovich integral*

$$(S) \int_0^t W(s)dW(s) = \int_0^t W(s) \circ dW(s) = \text{l.i.m. } \frac{S_n + S_n^+}{2} = \frac{1}{2}W^2(t). \tag{6.10}$$

It is anticipative and does not have the nice probabilistic properties of the Itô integral, but it does follow the ordinary rules of calculus and seems to be appropriate to model certain circumstances. It has a more restricted range of integrand functions G for which it is defined but, under appropriate regularity conditions, that range is enough for SDEs. We will look at the Stratonovich integral later and will concentrate now on the study of the Itô integral.

We have obtained the Itô integral for a very special integrand function $W(t)$, but the same ideas can be used to define it for arbitrary integrand functionss $G(s, \omega)$ as long as they are *non-anticipative* (i.e. at each time t, the function is independent of future increments of the Wiener process) and *mean square continuous* (m.s. continuous). For such functions, we can define the *Itô integral* as

$$I(G) = \int_0^t G(s)dW(s) = \text{l.i.m.} \sum_{k=1}^n G(t_{n,k-1})(W(t_{n,k}) - W(t_{n,k-1})). \quad (6.11)$$

This class of integrand functions is, however, a bit restrictive for some purposes and we need to generalize this definition so that the Itô integral can be defined for a wider class of non-anticipative integrand functions that may not be m.s. continuous. That is what will be done in Section 6.2. The reader not so concerned with these technical issues may very well be happy with the definition of the Itô integral for the m.s. continuous function just given, which has the advantage of being quite intuitive; if that is the case, when reading Section 6.2 you can limit your attention to the main conclusions concerning the properties of the integral.

6.2 Construction of the Itô integral

Consider a standard Wiener process $W(t) = W(t, \omega)$ $(t \geq 0)$ on a complete probability space $(\Omega, \mathcal{F}, P)$ and let $\mathcal{M}_s = \sigma(W(u), 0 \leq u \leq s)$ $(s \geq 0)$ be its natural filtration. Our purpose is to define the Itô integral on a time interval $[0, t]$ $(t \geq 0)$. Of course, the starting time was labelled 0 just for convenience and nothing prevents one using integrals on other intervals, say $[t_0, t]$ with $t \geq t_0$.

We say that $\{\mathcal{A}_s\}_{s \in [0,t]}$ is a *non-anticipating* or *non-anticipative filtration* if, for $0 \leq s \leq t$, $\mathcal{A}_s \supset \mathcal{M}_s$ and $\mathcal{A}_s$ is independent of the future increments $W(u) - W(s)$ $(u \geq s)$ of the Wiener process (which means that it is independent of the σ-algebra $\sigma(W(u) - W(s), u \geq s)$). Usually, one chooses for $\{\mathcal{A}_s\}_{s \in [0,t]}$ the natural filtration $\{\mathcal{M}_s\}_{s \in [0,t]}$ itself, but sometimes there is a need to include additional information besides the Wiener process information (e.g. when we consider the information on the initial condition of an SDE). If that is the case, we can choose a larger filtration, as long as it contains the natural filtration and does not anticipate the future increments of the Wiener process.

Let us now define the class of integrand functions G we will be working with.

Let λ be the Lebesgue measure on the interval $[0, t]$;[1] an interval $]a, b] \subset [0, t]$ has as Lebesgue measure $\lambda(]a, b]) = b - a$ its length and the same works for all types of intervals, but λ allows us to determine the length of many other sets contained in $[0, t]$. Let $G(s, \omega)$ (one usually abbreviates the notation to $G(s)$) be a function with domain $[0, t] \times \Omega$ having values in $\mathbb{R}$ that is jointly measurable with respect to both variables (s, ω).[2] Obviously, $\{G(s)\}_{s \in [0,t]}$ is a stochastic process. It is customary to abuse language and identify two jointly measurable functions that are almost equal with respect to the product measure $\lambda \times P$ (this means that the set of points (s, ω) for which the two functions might differ has zero $\lambda \times P$ measure).[3]

In so doing, we can define the $L^{2*}[0, t]$ space of functions[4] G jointly measurable such that $\int_{[0,t] \times \Omega} |G(s, \omega)|^2 d(\lambda \times P) < +\infty$; note that by *Fubini's theorem*, $\int_{[0,t] \times \Omega} |G(s, \omega)|^2 d(\lambda \times P) = \int_0^t \left(\int_\Omega |G(s, \omega)|^2 dP(\omega) \right) ds = \int_0^t \mathbb{E}[|G(s)|^2] ds = \mathbb{E}\left[\int_0^t |G(s)|^2 ds \right]$. This is a L^2 type space, but now with respect to the product measure $\lambda \times P$, and so is a Hilbert space with $L^{2*}[0, t]$ *norm*

$$\|G\|_{2*} := \left(\int_0^t \mathbb{E}[|G(s)|^2] ds \right)^{1/2} = \left(\mathbb{E}\left[\int_0^t |G(s)|^2 ds \right] \right)^{1/2} \tag{6.12}$$

and *inner product* $\langle G_1, G_2 \rangle_* = \left(\int_0^t \mathbb{E}[G_1(s)G_2(s)] ds \right)$.

We say that a jointly measurable function G is $\mathcal{A}_s$-*non-anticipative* or $\mathcal{A}_s$-*non-anticipating* or that it is *adapted to the filtration* $\mathcal{A}_s$, if, for any fixed $s \in [0, t]$, $G(s, .)$ is, as a function of ω, $\mathcal{A}_s$-measurable. If it is clear from context which is the filtration $\mathcal{A}_s$ or in the case where it is the natural filtration of the Wiener process, we may not mention it and say simply that G is non-anticipating.

---

1 A *measure* μ defined on a measurable space $(\Omega, \mathcal{F})$ is a function of $\mathcal{F}$ into $[0, +\infty]$ which is σ-additive and such that $\mu(\emptyset) = 0$. $(\Omega, \mathcal{A}, \mu)$ is called a *measure space*. A probability is a particular case of measure that satisfies the supplementary property of being normed, i.e. $P(\Omega) = 1$. The Lebesgue measure λ is an extension of the concept of length. The Lebesgue measure λ on $[0, t]$ can be defined for Borel sets $B \in \mathcal{B}_{[0,t]}$ of $[0, t]$ (one can prove that there a unique measure on these Borel sets such that the measure of any interval contained in $[0, t]$ is the interval's length). We have the measure space $([0, t], \mathcal{B}_{[0,t]}, \lambda)$ and we can extend it by completion to a measure space $([0, t], \mathcal{M}_{[0,t]}, \lambda)$. The completion is the same procedure used for probability spaces in Section 2.1 and consists of taking for $\mathcal{M}_{[0,t]}$ the class of sets of the form $B \cup N$ (with $B \in \mathcal{B}_{[0,t]}$ and N any subset of a Borel set with zero λ measure) and extending the measure to such sets putting $\lambda(B \cup N) = \lambda(B)$.

2 This means that the inverse image by G of any Borel set belongs to the product σ-algebra $\mathcal{B}_{[0,t]} \times \mathcal{F}$.

3 This identification means that, in reality, we are working with the equivalence classes of these functions (with respect to the almost equality relation) but, in order to simplify the language, instead of speaking of an equivalence class we speak of a representative function (any function in that class).

4 Rigorously, we should speak of equivalence classes of functions.

From the definition of non-anticipative function, we achieve our main purpose of the function being, at each time s, independent of future increments $W(u) - W(s)$ $(u \geq s)$ of the Wiener process. Remember that, when in Section 6.1 we defined the Itô integral $\int_0^t W(t)dW(t)$ as m.s limit of the special RS sums (6.5), the essential feature of the integral being non-anticipative resulted from the value of the integrand at the tag point (in that example $W(t_{n,k-1})$) being independent of future increments $W(t_{n,k}) - W(t_{n,k-1})$ of the integrator Wiener process. If G is non-anticipative, this property remains valid when we replace, in (6.5), $W(t_{n,k-1})$ by $G(t_{n,k-1})$. This opens the way to the definition of the Itô integral $\int_0^t G(s)dW(s)$.

The family of integrand functions that we will adopt is precisely the family G of jointly measurable functions[5] that belong to $L^{2*}[0, t]$ and are $\mathcal{A}_s$-non-anticipative. This family is also a Hilbert space and we will denote it by $H^2[0, t]$.

We will follow the traditional route in constructing the Itô integral, which is to define the integral for simpler functions and then extend the definition by continuity.

We start by defining the Itô integral for *step functions* (sometimes also called simple functions), which are functions $G \in H^2[0, t]$ that are constant in time s on each subinterval of some partition $0 = t_0 < t_1 < \ldots < t_n = t$ of $[0, t]$, i.e. $G(s, \omega) = G(t_{k-1}, \omega)$ for all $s \in [t_{k-1}, t_k[$ $(k = 1, \ldots, n)$. Note, however, that the constancy on the subintervals of that partition is only with respect to the time variable s, not with respect to the 'chance' variable ω; so, for $s \in [t_{k-1}, t_k[$, $G(s, \omega)$

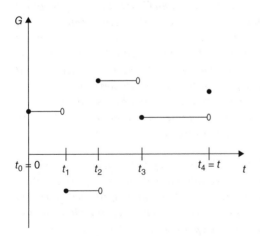

Figure 6.1 Example of a step function $G(s, \omega)$ with $n = 4$ for a fixed ω (one trajectory). Choosing a different trajectory (i.e. a different ω), the height of the steps would change but the separation points t_0, t_1, t_2, t_3, t_4 would remain.

---

5 Idem.

is a r.v. that does not change with s but may take different values for different $\omega \in \Omega$. Figure 6.1 shows an example of one trajectory (for a particular fixed ω) of a step function; if we had chosen a different ω, the height of the steps would change, but the projection of the steps on the time axis would be the same partition of $[0, t]$.

Let $H_S^2[0, t]$ be the space of step functions, a vectorial subspace of $H^2[0, t]$. The definition of the Itô integral for step functions is obvious:

Definition 6.1 (Itô integral of a step function)   *Given a step function* $G \in H_S^2[0, t]$, *its Itô integral* $I(G) = \int_0^t G(s, \omega)dW(s, \omega)$ *(abbreviated* $\int_0^t G(s)dW(s)$*) is defined by*

$$I(G) = \int_0^t G(s)dW(s) := \sum_{k=1}^n G(t_{k-1})(W(t_k) - W(t_{k-1})). \qquad (6.13)$$

Note that, since step functions are constant in each subinterval, we could have used in (6.13) as tag point for the computation of the integrand G any other point inside the subinterval, obtaining exactly the same result. So, for step functions the choice of the tag point is not relevant.

Observation 6.1   *Note that the integral depends on ω, so it is a r.v. Since the integral is defined not just for the function G (this is an abuse of language) but for all functions that are identified with G (the ones that are almost equal to G), it is important that the integral does not change when G is replaced by an almost equal function G^*. In fact, if we replace G by G^*, the resulting Itô integral may not be exactly equal but is almost equal (may differ only on a set of ω values having zero probability) and, since we also made the convention (abuse of language) of identifying random variables that are almost surely equal, we may say that we get the same integral. So, the definition is univocal as long as we identify almost equal integrands and almost equal random variables.[6] The Itô integral of step functions is a mapping that, applied to an integrand function $G \in H_S^2[0, t]$, produces its Itô integral $I(G)$, a r.v. defined by (6.13).*

Exercise 6.2   *Of course, one can define the same step function G using different partitions. For example, if in Figure 6.1 we split one subinterval of the partition into two smaller subintervals without changing the function, the new partition also works. Show that the Definition 6.1 is consistent, i.e. we get the same integral if we use different partitions for the same step function.*

Suggestion: You can merge the two initial partitions into a new more refined partition consisting of all the partition points of the initial partitions.

---

6 This is to say that in fact we are working with equivalence classes for the integrands G and with equivalence classes of r.v. for the integrals.

Exercise 6.3 *Show that the Itô integral of step functions is a* linear map, *i.e. given real constants* α_1 *and* α_2 *and* $G_1, G_2 \in H_S^2[0, t]$, *we have*

$$I(\alpha_1 G_1 + \alpha_2 G_2) = \alpha_1 I(G_1) + \alpha_2 I(G_2). \tag{6.14}$$

Suggestion: Merge the two partitions associated with G_1 and G_2. This new partition serves both G_1 and G_2 and also serves the linear combination $\alpha_1 G_1 + \alpha_2 G_2$.

For a step function $G \in H_S^2[0, t]$, we have

$$\mathbb{E}[I(G)] = 0. \tag{6.15}$$

In fact, each summand in the sum of the definition (6.13) has mathematical expectation $\mathbb{E}[G(t_{k-1})(W(t_k) - W(t_{k-1}))] = \mathbb{E}[G(t_{k-1})]\mathbb{E}[(W(t_k) - W(t_{k-1}))] = \mathbb{E}[G(t_{k-1})] \times 0 = 0$ due to independence between $G(t_{k-1})$ and the increment $W(t_k) - W(t_{k-1})$. So, the sum also has zero expectation. We just need to check that $\mathbb{E}[G(t_{k-1})]$ exists and is finite, i.e. $\mathbb{E}[|G(t_{k-1})|] < +\infty$. From the definition, if $G \in H^2[0, T]$ (and so this is also true for non-step functions in that space), we have $\int_0^t \mathbb{E}[|G|^2(s)]ds < +\infty$, which implies that $\mathbb{E}[|G(s)|^2]$ is finite for almost all $s \in [0, t]$ (it may only fail for a set of times with zero Lebesgue measure), and so the same happens to $\mathbb{E}[|G(s)|]$).

Let me remind the reader that, given the probability space $(\Omega, \mathcal{F}, P)$, we have denoted by L^2 the Hilbert space of r.v. X (identifying r.v. that are almost equal) with finite L^2-norm $\|X\|_2 = (\mathbb{E}[|X|^2])^{1/2}$. For a step function $G \in H_S^2[0, t]$, the integral $I(G)$ belongs to L^2. In fact, we will show that

$$(\|I(G)\|_2)^2 = \mathbb{E}[(I(G))^2] = VAR\ [I(G)] = \int_0^t \mathbb{E}[|G(s)|^2]ds = (\|G\|_{2*})^2, \tag{6.16}$$

which is finite. This shows that the integral $I(.)$, which is a *linear map* from $H_S^2[0, t]$ into L^2, is a map that *preserves the norm*; more precisely, the L^2-norm of the integral in the L^2-space is equal to the $L^{2*}[0, t]$-norm of the integrand in the $H^2[0, t]$ space. Therefore, $I(.)$ is a continuous map.

(6.16) also tells us that the variance of the Itô stochastic integral is simply a deterministic Riemann integral of the second-order moment of the integrand. To show that (6.16) is true, we just need to use the fact that G is independent of the future increments of the Wiener process and that this process has independent increments; in fact,

$$\mathbb{E}[(I(G))^2] = \sum_k \mathbb{E}[G^2(t_{k-1})]\mathbb{E}[(W(t_k) - W(t_{k-1}))^2]+$$
$$2 \sum_k \sum_{i>k} \mathbb{E}[G(t_{k-1})G(t_{i-1})(W(t_k) - W(t_{k-1}))]\ \mathbb{E}[W(t_i) - W(t_{i-1})]$$
$$= \sum_k \mathbb{E}[G^2(t_{k-1})](t_k - t_{k-1}) + 0 = \int_0^t \mathbb{E}[G^2(s)]ds.$$

To define the Itô integral for a general function $G \in H^2[0, t]$ one uses an *approximating sequence* of step functions $G_n \in H_S^2[0, t]$ ($n = 1, 2, ...$) converging in the $L^{2*}[0, t]$-norm to G and defines the Itô integral $I(G)$ of G as the limit in the L^2-norm (the m.s. limit) of the sequence of integrals $I(G_n)$ of the approximating step functions.

By *approximating sequence* we mean a sequence G_n that converges to G in the $L^{2*}[0, t]$-norm of the $H^2[0, t]$ space, i.e. such that

$$\int_0^t \mathbb{E}[|G_n(s) - G(s)|^2] ds \to 0 \text{ as } n \to +\infty. \tag{6.17}$$

This method of constructing the Itô integral requires proving that there is an approximating sequence $G_n \in H_S^2[0, t]$, that $I(G_n)$ converges in the L^2-norm, and that the definition of the Itô integral $I(G)$ of G is consistent (i.e. does not depend on the choice of the approximating sequence G_n). We will prove these properties and adopt the following definition of the Itô integral:

Definition 6.2 (Itô integral) *The Itô integral $I(G) = \int_0^t G(s, \omega) dW(s, \omega)$, abbreviated $\int_0^t G(s) dW(s)$, of a function $G \in H^2[0, t]$ is the a.s. unique m.s. limit of the Itô integrals $I(G_n) = \int_0^t G_n(s) dW(s)$ of any approximating sequence of step functions $G_n \in H_S^2[0, t]$ ($n = 1, 2, ...$).*

This definition, in the particular case when $G \in H^2[0, t]$ is a m.s. continuous function, is equivalent to the definition given for that case by (6.11) (this can be confirmed by part (a) of the proof below).

For interested readers, we show proofs of the mentioned properties; the reader can also see, for instance, Arnold (1974) or Øksendal (2003). They are quite technical and non-interested readers may skip them.

Proof that, given $G \in H^2[0, t]$, there is an approximating sequence of step functions $G_n \in H_S^2[0, t]$. This means that $H_S^2[0, t]$ is dense in $H^2[0, t]$

(a) *Case of G m.s. continuous*

This is precisely the case considered at the end of Section 6.1 in which the Itô integral could be defined by (6.11). A sequence of partitions $0 = t_{n,0} < t_{n,1} < \cdots < t_{n,n-1} < t_{n,n} = t$ ($n = 1, 2, ...$) of $[0, t]$ with mesh $\delta_n = \max_{k=1,...,n}(t_{n,k} - t_{n,k-1}) \to 0$ was chosen. What was done there is equivalent to choosing as approximating step function G_n the function defined by

$$G_n(s) = G(t_{n,k-1}) \text{ for } s \in [t_{n,k-1}, t_{n,k}[\quad (k = 1, ..., n),$$

i.e. the step function obtained by replacing G in each subinterval of the partition by its value at the initial point of the subinterval.

Now we show that G_n converges to G in the $L^{2*}[0, t]$-norm of the space $H^2[0, t]$. Given an arbitrary $\varepsilon > 0$, since G is m.s. uniformly continuous on the closed interval $[0, t]$, one has $\mathbb{E}[(G(u) - G(s))^2] < \varepsilon/t$ for $|u - s| < \delta$, where $\delta > 0$ is sufficiently small and does not depend on s. One can choose a sufficiently large n not dependent on s such that $\delta_n < \delta$. Then, one determines k such that $t_{n,k-1} \leq s < t_{n,k}$ and chooses $u = t_{n,k-1}$. We have $|u - s| < \delta$ and $\mathbb{E}[(G_n(s) - G(s))^2] = \mathbb{E}[(G(u) - G(s))^2] < \varepsilon/t$. So, $\int_0^t \mathbb{E}[(G_n(s) - G(s))^2]ds < \varepsilon$, which shows that our G_n converges to G in the $L^{2*}[0, t]$ norm.

(b) *Case of G bounded*

Let $|G| \leq M$ (M finite constant). For ease of notation, we will extend $G(s)$ to the time interval $(-\infty, +\infty)$ by putting $G(s) = 0$ for $s \notin [0, t]$. Define $H_n(s) = \int_0^{+\infty} e^{-\tau} G(s - \tau/n)d\tau$, which is not usually a step function, but is jointly measurable and non-anticipating (its value at s only depends on the values of G up to time s). Since $|H_n(s)| \leq M \int_0^{+\infty} e^{-\tau}d\tau = M$, we have $\mathbb{E}\left[\int_0^t H_n^2(s)ds\right] \leq M^2 t < +\infty$ and therefore $H_n \in H^2[0, t]$.

For $s \in [0, t]$, let $h \to 0$. We have $H_n(s + h) = \int_0^{+\infty} e^{-\tau} G(s + h - \tau/n)d\tau = e^{-nh} \int_{-nh}^{+\infty} e^{-\theta} G(s - \theta/n)d\theta$, so that, using the inequality $(a + b)^2 \leq 2(a^2 + b^2)$, we get as $h \to 0$,

$$(H_n(s + h) - H_n(s))^2$$

$$= \left((e^{-nh} - 1) \int_{-nh}^{+\infty} e^{-\tau} G(s - \tau/n)d\tau + \int_{-nh}^0 e^{-\tau} G(s - \tau/n)d\tau \right)^2$$

$$\leq 2\left((e^{-nh} - 1) \int_{-nh}^{+\infty} e^{-\tau} G(s - \tau/n)d\tau \right)^2 + 2\left(\int_{-nh}^0 e^{-\tau} G(s - \tau/n)d\tau \right)^2$$

$$\leq 2M^2 \left(\left((e^{-nh} - 1) \int_{-nh}^{+\infty} e^{-\tau}d\tau \right)^2 + \left(\int_{-nh}^0 e^{-\tau}d\tau \right)^2 \right)$$

$$= 2M^2(e^{2nh}(e^{-nh} - 1)^2 + (e^{nh} - 1)^2) \to 0.$$

This shows that H_n are continuous functions on $[0, t]$ and, using the dominated convergence theorem, we get $\mathbb{E}[(H_n(s + h) - H_n(s))^2] \to 0$, so that H_n are m.s. continuous on $[0, t]$.

We also have $H_n(s) \to G(s)$ for almost all $s \in [0, t]$ (exceptions may occur for a set of s values with zero Lebesgue measure).[7] So, applying the dominated convergence theorem twice (first for the Lebesgue measure on $[0, t]$ and then for the probability P), we get $\int_0^t (H_n(s) - G(s))^2 ds \to 0$ a.s.

---

7 From *Lusin's theorem* (see Rudin (1987), p. 56–57), since we are working with bounded measurable functions that are null outside the interval $[0, t]$ (which has finite Lebesgue measure), given a positive integer k, there is a continuous function $J_k(s)$ coinciding with $G(s)$ with the possible exception of s values in a set N_k with Lebesgue measure smaller than $1/2^k$. Let us remind ourselves of the *Borel–Cantelli lemma* for probabilities P, which, given a sequence A_n ($n = 1, 2, ...$) of events and the event $A = \{\omega : \omega \in A_n$ for an infinite number of n values$\}$, states that (a) if $\sum_n P(A_n) < +\infty$, then $P(A) = 0$ and (b) if $\sum_n P(A_n) = +\infty$ and the events A_n are independent, then $P(A) = 1$. Part (a) of the lemma also applies to finite measures (such as the Lebesgue measure λ on the interval $[0, t]$) and so, since $\sum_k 1/2^k < +\infty$, we conclude that it is null the Lebesgue measure of the set of s values that are in an infinite number of sets N_k. Therefore, for almost all s, we have $J_k(s) = G(s)$ for sufficiently large k. If G was continuous, it would be obvious that $H_n(s)$ would converge to $G(s)$ for all s. Since G may not be continuous, we can only guarantee convergence for almost all s.

and $\mathbb{E}\left[\int_0^t (H_n(s) - G(s))^2 ds\right] \to 0$. So, there is a sequence $H_n \in H^2[0, t]$ of m.s. continuous functions converging to G in the $L^{2*}[0, t]$-norm.

Since each H_n is m.s. continuous, it can be approximated, according to (a), by a sequence of step functions $G_{n,m}$ $(m = 1, 2, ...)$. Given n, let $H_{k(n)}$ be the first term of the sequence H_k $(k = 1, 2, ...)$ located in the $1/2^{n+1}$-neighbourhood of G (for the $L^{2*}[0, t]$-norm), so that $\|H_{k(n)} - G\|_{2*} < 1/2^{n+1}$. Let $G_n := G_{k(n),m(k(n))}$ be the first term of the sequence $G_{k(n),m}$ $(m = 1, 2, ...)$ located in the $1/2^{n+1}$-neighbourhood of $H_{k(n)}$, so that $\|G_{k(n),m(k(n))} - H_{k(n)}\|_{2*} < 1/2^{n+1}$. Then $\|G_n - G\|_{2*} < 1/2^n$, and so *the sequence of step functions G_n $(n = 1, 2, ...)$ converges to G in the $L^{2*}[0, t]$-norm.*

Since $\mathbb{E}\left[\int_0^t (G_n(s) - G(s))^2 ds\right] < 1/2^n$ and $\sum_n 1/2^n < +\infty$, we can even conclude that $\int_0^t (G_n(s) - G(s))^2 ds \to 0$ a.s.

(c) *Case of general $G \in H^2[0, t]$*

For each $n = 1, 2, ...$, use the truncated functions

$$F_n(s, \omega) = \begin{cases} -n & \text{if } G(s, \omega) < -n \\ G(s, \omega) & \text{if } |G(s, \omega)| \le n \\ n & \text{if } G(s, \omega) > n. \end{cases}$$

They are in $H^2[0, t]$, they are bounded, and obviously $|F_n(s) - G(s)|^2 \to 0$ and $|F_n(s) - G(s)|^2 \le 2G^2(s)$. Since $G^2 \in H^2[0, t]$, it is integrable with respect to $\lambda \times P$. Applying the dominated convergence theorem twice (with respect to the Lebesgue measure λ and with respect to P), we obtain $\mathbb{E}\left[\int_0^t (F_n(s) - G(s))^2 ds\right] \to 0$, which shows that F_n converges in the $L^{2*}[0, t]$-norm to G. By (b), each F_n can be approximated by a sequence of step functions. We can follow a reasoning similar to that used at the end of (b) to show that *there is a sequence of step functions G_n converging in the $L^{2*}[0, t]$-norm to G.*

We can, like in (b), conclude that $\int_0^t (G_n(s) - G(s))^2 ds \to 0$ a.s. ∎

Observation 6.2 *At the end of the proof, we concluded that we can even construct the approximating sequence G_n in such a way that*

$$\int_0^t (G_n(s) - G(s))^2 ds \to 0 \quad a.s. \tag{6.18}$$

Proof that, given $G \in H^2[0, t]$ and an approximating sequence $G_n \in H_S^2[0, t]$, the sequence of Itô integrals $I(G_n)$ converges in the L^2-norm (i.e. in m.s.)

Since G_n converges to G in the space $H^2[0, t]$ and this is a complete space, G_n is a Cauchy sequence with respect to the norm $L^{2*}[0, t]$ of that space. So, the sequence $I(G_n)$ is also a Cauchy sequence with respect to the L^2-norm, due to the norm preservation property (6.16) of the Itô integral for step functions. Since L^2 is a complete space, $I(G_n)$ converges in the L^2-norm, i.e. converges in m.s. This m.s. limit is, by definition, the Itô integral $\int_0^t G(s)dW(s)$ of G. ∎

Exercise 6.4 (*) *[Proof of consistency of the Definition 6.2 of Itô integral] Prove that the integral does not depend on the approximating sequence, i.e. given two approximating sequences G_n and G_n^* of step functions, we have l.i.m. $I(G_n) = $ l.i.m. $I(G_n^*)$ a.s. You can also see that, if you replace G by an almost equal function, the integral remains a.s. the same.*

Suggestion: For the first part, combine the two sequences into a single sequence by alternating terms of the two and use the fact that the limit is unique.

Property 6.1 (Main properties of the Itô integral of $G \in H^2[0, t]$) *The Itô integral $I(G) = \int_0^t G(s)dW(s)$ for functions $G \in H^2[0, t]$ satisfies the following properties:*

a) *Is a linear map, i.e. if G_1, $G_2 \in H^2[0, t]$ and α_1, α_2 are real constants, then*

$$I(\alpha_1 G_1 + \alpha_2 G_2) = \alpha_1 I(G_1) + \alpha_2 I(G_2). \tag{6.19}$$

b) *Has null mathematical expectation:*

$$\mathbb{E}[I(G)] = 0. \tag{6.20}$$

c) *Preserves the norm:*

$$(\|I(G)\|_2)^2 = \mathbb{E}[(I(G))^2] = VAR[I(G)] = \int_0^t \mathbb{E}[|G(s)|^2]ds = (\|G\|_{2*})^2. \tag{6.21}$$

d) *The r.v. $I(G)$ is an $\mathcal{A}_t$-measurable function.*

e) *Is a continuous map, i.e. if the sequence $G_n \in H^2[0, t]$ $(n = 1, 2, ...)$ converges to $G \in H^2[0, t]$ (in the $L^{2*}[0, t]$-norm of this space), then the sequence $I(G_n)$ converges in m.s. (i.e. in the L^2-norm) to $I(G)$.*

f) *In the particular case of a deterministic function G (i.e. a function that does not depend on chance ω), $I(G)$ is normally distributed with mean zero and variance $\int_0^t G^2(s)ds$, i.e.*

$$\int_0^t G(s)dW(s) \frown \mathcal{N}\left(0, \int_0^t G^2(s)ds\right) \quad \text{if } G \text{ deterministic.} \tag{6.22}$$

Note: For a random function G, the integral is usually not normally distributed.

Exercise 6.5

(a) (*) *Prove the results stated in Property 6.1.*
(b) *For a real constant c and the time interval [a, b] with $0 \leq a \leq b$, determine the distribution of $\int_a^b c\,dW(s)$ and show that this integral is equal to $c(W(b) - W(a))$.*

Suggestions for (a): Since the Itô integral of functions in $H^2[0, t]$ is the m.s. limit of the integrals of an approximating sequence of step functions (in $H_S^2[0, t]$), some properties already proven for step functions are carried for general functions in $H^2[0, t]$ when you take the limit. This is trivially the case for (a), (b), and (c). The same applies for property (d) since the integral of a step function G_n is by (6.13) also an $\mathcal{A}_t$-*measurable function*; in fact, it is a finite sum of random variables $G(t_{k-1})(W(t_k) - W(t_{k-1}))$, which are (show why) $\mathcal{A}_t$-measurable. Property (c) of norm preservation implies the continuity of the map and therefore (e) holds. To prove (f), notice that, since G is deterministic, we can choose an approximating sequence G_n of deterministic step functions and, from (6.13), we immediately see that the integrals $I(G_n) = \int_0^t G_n(s)dW(s)$ are normally distributed, so the same happens to their m.s. limit $I(G)$; the zero mean comes from (b) and we can simplify the expression given in (c) for the variance since in this case $\mathbb{E}[|G(s)|^2] = G^2(s)$.

Suggestions for (b): Notice that this is a step function with just one step on $[a, b]$.

6.3 Study of the integral as a function of the upper limit of integration

Let $G \in H^2[0, T]$ $(T > 0)$ and let $[a, b] \subset [0, T]$. Since $G(s)I_{[a,b]}(s) \in H^2[0, T]$, the Itô integral can be defined on $[a, b]$ by

$$\int_a^b G(s)dW(s) := \int_0^T G(s)I_{[a,b]}(s)dW(s). \qquad (6.23)$$

One easily recognizes that, for $0 \leq a \leq b \leq c \leq T$,

$$\int_a^c G(s)dW(s) = \int_a^b G(s)dW(s) + \int_b^c G(s)dW(s). \qquad (6.24)$$

To study the *Itô integral as a function of the upper integration limit t*, we will assume t to vary in the interval $[0, T]$. Taking $G \in H^2[0, T]$, we also have $G \in H^2[0, t]$ for all $t \in [0, T]$ (denoting the restriction of $G = G(s, \omega)$ to $[0, t] \times \Omega$ by the same letter G). Let

$$Z(t) = \int_0^t G(s)dW(s) = \int_0^T G(s)I_{[0,t]}(s)dW(s). \qquad (6.25)$$

Property 6.2 *Since $Z(t) = \int_0^t G(s)dW(s)$, with $G \in H^2[0,T]$, is A_t-measurable, the stochastic process $\{Z(t)\}_{[0,T]}$ is adapted to the filtration $\{A_t\}_{t\in[0,T]}$. Furthermore, it is also an A_t-martingale.*

The proof is a bit technical and non-interested readers may wish to skip it.

Proof: Consider an approximating sequence $G_n \in H_S^2[0,t]$ $(n = 1, 2, \dots)$ of step functions. Let $0 = t_0^{(n)} < t_1^{(n)} < \cdots < t_{m_n}^{(n)} = T$ be a tagged partition on which subintervals $G_n(s)$ is constant with respect to s. We now show that each $Z_n(t) = \int_0^t G_n(s)dW(s)$ is an A_t-martingale. In fact, given $s < t$ and putting $N_n = \{k \in \{1, 2, \dots, m_n\} : s \leq t_{k-1}^{(n)}, t_k^{(n)} \leq t\}$, we have

$$\mathbb{E}[Z_n(t)|A_s] = \mathbb{E}\left[\int_0^s G_n(s)dW(s) + \int_s^t G_n(s)dW(s)|A_s\right]$$

$$= \int_0^s G_n(s)dW(s) + \sum_{k\in N_n} \mathbb{E}\left[G_n(t_{k-1}^{(n)})(W(t_k^{(n)}) - W(t_{k-1}^{(n)}))|A_s\right]$$

$$= \int_0^s G_n(s)dW(s)$$

$$+ \sum_{k\in N_n} \mathbb{E}\left[\mathbb{E}\left[G_n(t_{k-1}^{(n)})(W(t_k^{(n)}) - W(t_{k-1}^{(n)}))|A_{t_{k-1}^{(n)}}\right]|A_s\right]$$

$$= \int_0^s G_n(s)dW(s)$$

$$+ \sum_{k\in N_n} \mathbb{E}\left[G_n(t_{k-1}^{(n)})\mathbb{E}[W(t_k^{(n)}) - W(t_{k-1}^{(n)})|A_{t_{k-1}^{(n)}}]|A_s\right]$$

$$= \int_0^s G_n(s)dW(s) + \sum_{k\in N_n} \mathbb{E}[0|A_s] = Z_n(s).$$

Using *Schwarz's inequality*[8] and since $Z_n(s)$ converges in m.s. to $Z(s)$, we get

$$\mathbb{E}[(\mathbb{E}[Z_n(t)|A_s] - \mathbb{E}[Z(t)|A_s])^2] = \mathbb{E}[(\mathbb{E}[1 \times (Z_n(t) - Z(t))|A_s])^2]$$

$$\leq \mathbb{E}[\mathbb{E}[1^2|A_s]\mathbb{E}[(Z_n(t) - Z(t))^2|A_s]] = \mathbb{E}[(Z_n(t) - Z(t))^2] \to 0.$$

Therefore, $\mathbb{E}[Z_n(t)|A_s] = Z_n(s)$ converges in m.s. to $\mathbb{E}[Z(t)|A_s]$. Since the m.s. limit is a.s. unique, we get $\mathbb{E}[Z(t)|A_s] = Z(s)$ a.s., thus proving that $Z(t)$ is an A_t-martingale. ∎

Let $a, b \in [0, T]$ with $a \leq b$. From Property 6.2, $Z(t) - Z(a)$ is an A_t-martingale for $t > a$. By the martingale maximum inequalities (2.2)-(2.3),

---

8 *Schwarz's inequality* states that $(\int fgd\mu)^2 \leq (\int fd\mu)^2(\int gd\mu)^2$, where μ is a measure and f, g are square integrable functions. This is true, in particular, when the measure is a probability P, in which case we may write $(\mathbb{E}[XY])^2 \leq \mathbb{E}[X^2]\mathbb{E}[Y^2]$.

we get

$$P\left[\sup_{a \leq t \leq b} |Z(t) - Z(a)| \geq c\right] \leq \frac{1}{c^2}\mathbb{E}[|Z(b) - Z(a)|^2] = \frac{1}{c^2}\int_a^b \mathbb{E}[G^2(s)]ds$$

(6.26)

and

$$\mathbb{E}\left[\sup_{a \leq t \leq b} |Z(t) - Z(a)|^2\right] \leq 4\mathbb{E}[|Z(b) - Z(a)|^2] = 4\int_a^b \mathbb{E}[G^2(s)]ds. \quad (6.27)$$

Property 6.3 *The process $Z(t) = \int_0^t G(s)dW(s)$, $t \in [0, T]$, with $G \in H^2[0, t]$, has a continuous version, i.e. a version with continuous trajectories, and we shall assume to be working with that version.*

The proof is a bit technical and non-interested readers may wish to skip it.

Proof: From (6.13), we see that $Z_n(t) = \int_0^t G_n(s)dW(s)$ (G_n, $n = 1, 2, ...$, being a sequence of step functions approximating G) is a.s. a continuous function of t on $[0, T]$. Since $Z_n(t) - Z_m(t)$ is a martingale, we obtain from (2.2)

$$P[\sup_{0 \leq t \leq T} |Z_n(t) - Z_m(t)| > \varepsilon] \leq \frac{1}{\varepsilon^2}\mathbb{E}[|Z_n(T) - Z_m(T)|^2].$$

When $n, m \to +\infty$, the right-hand side converges to zero. Choose $\varepsilon = 1/2^k$ and a subsequence $n_k \to +\infty$ such that

$$P[\sup_{0 \leq t \leq T} |Z_{n_{k+1}}(t) - Z_{n_k}(t)| > 1/2^k] < 1/2^k.$$

Since $\sum_k 1/2^k < +\infty$, the Borel–Cantelli lemma implies

$$P[\sup_{0 \leq t \leq T} |Z_{n_{k+1}}(t) - Z_{n_k}(t)| > 1/2^k \text{ for infinite values of } k] = 0.$$

Consequently, with probability one, $\sup_{0 \leq t \leq T} |Z_{n_{k+1}}(t) - Z_{n_k}(t)| \leq 1/2^k$ for every sufficiently large k, which implies that $Z_{n_k}(t)$ is a.s. uniformly convergent on $[0, T]$. Denote by $J(t)$ the corresponding a.s. limit when $k \to +\infty$. Being the uniform limit of a.s. continuous functions, it is a.s. continuous. Since $Z_{n_k}(t)$ converges in m.s. to $Z(t)$, then $Z(t) = J(t)$ a.s., which concludes the proof. ∎

Exercise 6.6 (*) *The process $Z(t) = \int_0^t G(s)dW(s)$, $t \in [0, T]$, with $G \in H^2[0, t]$, has uncorrelated increments (this does not mean that they are independent), i.e. given non-overlapping intervals $[s, t]$ and $[u, v]$ with $0 \leq s \leq t \leq u \leq v \leq T$,*

$$
\begin{aligned}
&COV\,[Z(t) - Z(s), Z(v) - Z(u)] \\
&= E[(Z(t) - Z(s))\,(Z(v) - Z(u))] \\
&= E\left[\left(\int_s^t G(z)dW(z)\right)\left(\int_u^v G(z)dW(z)\right)\right] = 0.
\end{aligned}
$$

(6.28)

Suggestion: Start by showing it for step functions and then consider an approximating sequence of step functions and go to the m.s. limit.

6.4 Extension of the Itô integral

This section can be skipped on a first reading.

The Itô integral can be extended to the larger space $M^2[0, t]$ of integrand functions G jointly measurable and $\mathcal{A}_s$-non-anticipative such that

$$\int_0^t |G(s)|^2 ds < +\infty \quad \text{a.s.} \tag{6.29}$$

Since this condition is weaker than the condition $\int_0^t E[|G(s)|^2]ds < +\infty$, we have $M^2[0, t] \supset H^2[0, t]$.

For step functions G in $M^2[0, t]$, the definition of the Itô integral coincides with Definitions 6.1 and (6.13). However, now there is no guarantee that the mathematical expectations of G and of G^2 exist, and so we have no guarantee of existence of the expected value and the variance of the integral. Of course, the same applies to arbitrary functions G in $M^2[0, t]$. The definition of the Itô integral for an arbitrary function $G \in M^2[0, t]$ uses the same technique of an *approximating sequence* of step functions G_n ($n = 1, 2, \ldots$) in $M^2[0, t]$. The approximation is not, however, with respect to the $L^{2*}[0, t]$ norm (which requires the existence of second order moments) since such norm may not be defined for G. We use instead a weaker form of convergence defined by

$$\int_0^t (G_n(s) - G(s))^2 ds \to 0 \quad \text{a.s. as } n \to +\infty. \tag{6.30}$$

Proof that G can be approximated by step functions in the sense of (6.30)

The technique to show the existence of approximating sequences is similar to the one used in Section 6.2.

One first shows that it is true for continuous functions $G \in M^2[0, t]$, using the same method of defining the approximating sequence by using a sequence of partitions $0 = t_{n,0} < t_{n,1} < \cdots < t_{n,n-1} < t_{n,n} = t$ ($n = 1, 2, \ldots$) of $[0, t]$ with mesh $\delta_n = \max_{k=1,\ldots,n}(t_{n,k} - t_{n,k-1}) \to 0$ and choosing the G_n defined by $G_n(s) = G(t_{n,k-1})$ for $s \in [t_{n,k-1}, t_{n,k}[$ ($k = 1, \ldots, n$) (the step function is obtained by replacing G in each subinterval of the partition by its value at the initial point of the subinterval). The proof is similar to the one used in Section 6.2 for m.s. continuous functions $G \in H^2[0, t]$ with some adaptations that we leave to the reader; of course, now one uses continuity instead of m.s. continuity and mathematical expectations have to de dropped.

After that, one extends to bounded functions $G \in M^2[0, t]$, by showing that such a function can be approximated by a sequence of continuous functions

$H_n \in M^2[0, t]$ using a similar, although slightly simpler, proof to the one used in Section 6.2. The reader can make the required adaptations.

Finally, to extend to arbitrary $G \in M^2[0, t]$, again one mimics the proof in Section 6.2 using a truncation to obtain a sequence of bounded functions approaching G; the proof is even simpler since one only needs to apply the dominated convergence theorem once instead of twice. ∎

The approximation to $G \in M^2[0, t]$ by a sequence of step function G_n uses now a weaker convergence than before and so it is no longer possible to show that the sequence of integrals $I(G_n) = \int_0^t G_n(s)dW(s)$ converges in mean square. However, this sequence *converges in probability.*[9]

Proof that $I(G_n)$ converges in probability as $n \to +\infty$

The proof follows Arnold (1974) and starts by showing that, for any step function J in $M^2[0, t]$, we have, for all $N > 0$ and all $\delta > 0$,

$$P\left[\left|\int_0^t J(s)dW(s)\right| > \delta\right] \leq \frac{N}{\delta^2} + P\left[\int_0^t J^2(s)dW(s) > N\right]. \tag{6.31}$$

Let $0 = t_0 < t_1 < \cdots < t_n = t$ be a partition of $[0, t]$ on which subintervals J is constant, i.e. $J(s, \omega) = J(t_{k-1}, \omega)$ for all $s \in [t_{k-1}, t_k[$ $(k = 1, \ldots, n)$. For $s \in [t_{i-1}, t_i[$, define the truncation

$$J_N(s) := \begin{cases} J(s) & \text{if } \int_0^{t_i} J^2(s)ds \leq N \\ 0 & \text{if } \int_0^{t_i} J^2(s)ds > N, \end{cases}$$

which is a step function in $M^2[0, t]$ such that $\int_0^{t_i} J_N^2(s)ds \leq N$. Therefore,

$$\|J_N\|_{2*}^2 = \mathbb{E}\left[\int_0^{t_i} J_N^2(s)ds\right] \leq N$$

and so $J_N \in H_S^2[0, t]$. Therefore, by the norm preservation property, $\mathbb{E}[(I(J_N))^2] = (\|I(J_N)\|_2)^2 \leq N$. Since $J \equiv J_N$ if and only if $\int_0^t J^2(s)ds \leq N$, we get

$$P\left[\int_0^t J^2(s)ds > N\right] = P\left[\sup_{0 \leq s \leq t} |J(s) - J_N(s)| > 0\right]$$

$$\geq P\left[\left|\int_0^t (J(s) - J_N(s))dW(s)\right| > 0\right].$$

---

9 See the concept of convergence in probability in Section 2.1. As we have mentioned there, convergence in probability is weaker than m.s. convergence and than a.s. convergence (also known as convergence with probability one).

So, using *Tchebyshev's inequality*,

$$P[|I(J)| > \delta] \leq P[|I(J_N)| > \delta] + P[|I(J) - I(J_N)| > 0]$$

$$\leq \mathbb{E}[(I(J_N))^2]/\delta^2 + P\left[\int_0^t J^2(s)ds > N\right] \leq N/\delta^2 + P\left[\int_0^t J^2(s)ds > N\right],$$

which shows that (6.31) holds.

Since $\int_0^t (G_n(s) - G(s))^2 ds \to 0$ a.s., then $\int_0^t (G_n(s) - G(s))^2 ds \to 0$ in probability, and so $\int_0^t (G_n(s) - G_m(s))^2 ds \leq 2\int_0^t (G_n(s) - G(s))^2 ds + 2\int_0^t (G_m(s) - G(s))^2 ds \to 0$ in probability as $n, m \to +\infty$. Let $\delta > 0$ be arbitrary. Applying (6.31) to $G_n - G_m$, we get

$$\limsup_{n,m \to +\infty} P[|I(G_n) - I(G_m)| > \varepsilon]$$

$$\leq \delta/\varepsilon^2 + \limsup_{n,m \to +\infty} P\left[\int_0^t |G_n(s) - G_m(s)|^2 > \delta\right] \leq \delta/\varepsilon^2 + 0.$$

Since δ is arbitrary, we conclude that $\lim_{n,m \to +\infty} P[|I(G_n) - I(G_m)| > \varepsilon] = 0$ and, therefore, $I(G_n)$ is a Cauchy sequence in probability, which implies that it converges in probability. ∎

We can therefore state the definition of Itô integrals of $G \in M^2[0, t]$ as:

Definition 6.3 (Itô integral extension) *For a function $G \in M^2[0, t]$, the Itô integral $I(G) = \int_0^t G(s, \omega)dW(s, \omega)$, abbreviated $\int_0^t G(s)dW(s)$, is the a.s. unique limit in probability of the Itô integrals $I(G_n) = \int_0^t G_n(s)dW(s)$ of any approximating sequence (in the sense of (6.30)) of step functions G_n in $M^2[0, t]$ ($n = 1, 2, \ldots$).*

Exercise 6.7 (*) *The only thing missing, which we leave to the reader, is to show the consistency of this definition, in the sense that different approximating sequences of step functions lead a.s. to the same value of the integral.*

Exercise 6.8 *Show that the Itô integral for integrands in $M^2[0, t]$ is a linear map.*

Exercise 6.9 (*) *Show that (6.31) remains valid for any $J \in M^2[0, t]$, even if J is not a step function. Using this, prove that, given any sequence $G_n \in M^2[0, t]$ ($n = 1, 2, \ldots$) (no need for the G_n to be step functions) converging to $G \in M^2[0, t]$ in the sense of $\int_0^t (G_n(s) - G(s))^2 ds \to 0$ a.s., then $\int_0^t G_n(s)dW(s)$ converges in probability to $\int_0^t G(s)dW(s)$. You can mimic the proof for step functions G_n given above.*

The properties (6.20) and (6.21) cannot be guaranteed for $G \in M^2[0, t]$ since the moments involved may not exist. For the same reason, the integral, as a function of the upper limit of integration, may not be a martingale.

However, we have the following two properties, the first of which is obvious:

Property 6.4 $Z(t) = \int_0^t G(s)dW(s)\ (t \in [0, T])$, *with* $G \in M^2[0, T]$, *is adapted to the non-anticipative filtration* $\mathcal{A}_t$.

Property 6.5 $Z(t) = \int_0^t G(s)dW(s)\ (t \in [0, T])$, *with* $G \in M^2[0, T]$, *has continuous trajectories a.s.*

Proof: Consider the truncated functions

$$G_N(t) := \begin{cases} G(t) & \text{if } \int_0^t G^2(u)du \le N \\ 0 & \text{if } \int_0^t G^2(u)du > N \end{cases},$$

which are in $H^2[0, T]$, and let $Z_N(t) = \int_0^t G_N(t)dW(t)$. From Property 6.3, $Z_N(t)$ has a.s. continuous trajectories. We have $Z(t, \omega) = Z_N(t, \omega)$ for all $t \in [0, T]$ and for $\omega \in A_N := \{\omega : \int_0^t G^2(s, \omega)ds) \le N\}$. Choosing a sufficiently large N, $P(A_N)$ can become as close to one as desired, so that the set of ω values for which $Z(t, \omega)$ is continuous has probability one. ∎

6.5 Itô theorem and Itô formula

In Section 6.2 the Itô integral was defined for integrand functions in the class $H^2[0, t]$ of jointly measurable non-anticipative functions (i.e. adapted to the chosen non-anticipative filtration $\mathcal{A}_s$) with finite $L^{2*}[0, t]$-norm. In Section 6.4, the definition was extended to the class $M^2[0, t]$ of jointly measurable non-anticipative functions G satisfying (6.29). We now proceed to a further extension by defining *Itô processes*.

Definition 6.4 *Given a complete probability space* $(\Omega, \mathcal{F}, P)$, *a time interval* $[0, T]$, *and a non-anticipative filtration* $\mathcal{A}_s$ $(s \in [0, T])$, *a stochastic process* $\{X(t)\}_{t \in [0, T]}$ *on that probability space is an* Itô *process if it is defined by*

$$X(t, \omega) = X_0(\omega) + \int_0^t F(s, \omega)ds + \int_0^t G(s, \omega)dW(s, \omega), \qquad (6.32)$$

where:

- X_0 *is* $\mathcal{A}_0$-*measurable (and so is independent of the Wiener process); it can, in particular, be a deterministic constant.*
- F *is a joint measurable function adapted to the filtration and such that* $\int_0^T |F(s)|ds < +\infty$ *a.s.; we will say that* $F \in M^1[0, T]$.
- $G \in M^2[0, T]$, *i.e.* G *is a joint measurable function adapted to the filtration and such that* $\int_0^T |G(s)|^2ds < +\infty$ *a.s.*

The stochastic integral can also be written in the differential form, i.e.

$$dX(t, \omega) = F(t, \omega)dt + G(t, \omega)dW(t, \omega)$$

abbreviated $\qquad\qquad\qquad\qquad\qquad\qquad\qquad$ (6.33)

$$dX(t) = F(t)dt + G(t)dW(t),$$

together with the initial condition $X(0, \omega) = X_0(\omega)$ (abbreviated $X(0) = X_0$). This looks almost like an SDE

$$dX(t, \omega) = f(t, X(t, \omega))dt + g(t, X(\omega))dW(t, \omega). \qquad (6.34)$$

Indeed, if we put

$$F(t, \omega) = f(t, X(t, \omega)) \quad \text{and} \quad G(t, \omega) = g(t, X(t, \omega)), \qquad (6.35)$$

we see that, when F and G satisfy the required properties (and that will be the tricky part to be dealt with in Chapter 7), a solution of the SDE is a particular case of an Itô process, in which the dependence of F and G on ω is not a direct but rather an indirect (more restricted) dependence since f and g do depend on $X(t)$, which depends on ω.

We have seen in Section 6.1, through an example, that the Itô integrals do not follow the ordinary rules of calculus. The same therefore happens also with Itô processes, no matter if they are written in the integral or the differential form. Consequently, we need to study the rules of differential and integral calculus that do apply, the so-called *Itô calculus*. To characterize Itô calculus, we just need the Itô calculus *chain rule*, i.e. the rule that tells us how to differentiate a composite function. This is known as the *Itô formula*, and it is the object of the so-called *Itô theorem*.

Theorem 6.1 (*Itô theorem*) *Let $X(t) = X(t, \omega)$ ($t \in [0, T]$) be an Itô process given by (6.32) or (6.33), abbreviated $dX(t) = F(t)dt + G(t)dW(t)$ with initial condition $X(0) = X_0$. Consider the composite function $Y(t) = h(t, X(t))$, where $h(t, x)$ is a continuous function with continuous first-order partial derivatives in t and continuous first- and second-order partial derivatives in x (we will abbreviate by saying that h is a C^{12} function). Then $Y(t) = Y(t, \omega)$ ($t \in [0, T]$) is also an Itô process with initial condition $Y_0 = h(0, X_0)$ given, in the differential form, by the* Itô *formula:*

$$dY(t) = \left(\frac{\partial h(t, X(t))}{\partial t} + \frac{\partial h(t, X(t))}{\partial x} F(t) + \frac{1}{2} \frac{\partial^2 h(t, X(t))}{\partial x^2} G^2(t) \right) dt$$

$$+ \frac{\partial h(t, X(t))}{\partial x} G(t)dW(t). \qquad (6.36)$$

We can also write the Itô formula in integral form:

$$Y(t) = Y_0 + \int_0^t \left(\frac{\partial h(s, X(s))}{\partial s} + \frac{\partial h(s, X(s))}{\partial x} F(s) + \frac{1}{2} \frac{\partial^2 h(s, X(s))}{\partial x^2} G^2(s) \right) ds$$
$$+ \int_0^t \frac{\partial h(s, X(s))}{\partial x} G(s) dW(s). \tag{6.37}$$

Observation 6.3 (Mnemonic for the Itô rule)   *A good mnemonic to obtain (6.36) is to use a Taylor expansion up to first order in t and to second order in x*

$$dY(t) = \frac{\partial h(t, X(t))}{\partial t} dt + \frac{\partial h(t, X(t))}{\partial x} dX(t) + \frac{1}{2} \frac{\partial^2 h(t, X(t))}{\partial x^2} (dX(t))^2$$

followed by the substitution of $dX(t)$ by $F(t)dt + G(t)dW(t)$ and the substitution of $(dX(t))^2$ by $G^2(t)dt$; this latter substitution results from $(dX(t))^2 = (F(t)dt + G(t)dW(t))^2$ by applying the following rules for multiplication of differentials

×	dt	dW
dt	0	0
dW	0	dt

This second-order term $(dX(t))^2$ would be equal to zero if ordinary calculus rules applied; in fact, in ordinary calculus we would have the multiplication rule $dW \times dW = 0$ instead of the Itô multiplication rule $dW \times dW = dt$. This different behaviour is not surprising because $W(t)$ is very irregular and we cannot apply ordinary rules since the derivative $\frac{dW(t)}{dt}$ does not exist as a proper function.

Let us give an example of the application of the Itô theorem.

Let $X(t) = W(t)$ and $h(t, x) = x^2$. Since $X(0) = 0$ and $X(t) = W(t) = \int_0^t 0 ds + \int_0^t 1 dW(s)$, we see that $X(t)$ is an Itô process with $F(t, \omega) \equiv 0$ and $G(t, \omega) \equiv 1$. We have used the fact that the integrand of the second integral is a step function with a single step (where it takes the value 1 for all ω), so we can use the trivial partition $0 = t_0 < t_1 = t$ of $[0, t]$; this gives $\int_0^t 1 dW(s) = 1(W(t_1) - W(t_0)) = W(t)$. In the differential form, we can write $dX(t) = 0 dt + 1 dW(t) = dW(t)$. Therefore $Y(t) = h(t, X(t)) = W^2(t)$. By Itô formula, one obtains (you can see the mnemonic of Observation 6.3) $dY(t) = 0 dt + 2X(t)dX(t) + \frac{1}{2} 2(dX(t))^2 = 2W(t)dW(t) + dt$. In the integral form, considering that $Y(0) = 0$, that is written as $Y(t) = 2 \int_0^t W(s)dW(s) + \int_0^t dt = 2 \int_0^t W(s)dW(s) + t$. This leads to the already familiar result (see (6.7))

$$\int_0^t W(s)dW(s) = \frac{1}{2}(W^2(t) - t).$$

Exercise 6.10   *Determine $d(tW(t))$ and use the result to show that $\int_0^t s dW(s) = tW(t) - \int_0^t W(s)ds$.*

Exercise 6.11 *Show that the SDE $dY(t) = Y(t)dW(t)$ with $Y(0) = 1$ has solution $Y(t) = \exp\left(W(t) - \frac{t}{2}\right)$ for any $t \geq 0$.*

Suggestion: Apply the Itô rule using the function $h(t, x) = \exp\left(x - \frac{t}{2}\right)$ and $X(t) = W(t)$.

We conclude this section by presenting a sketch of the proof of the Itô theorem 6.1, which can be skipped by non-interested readers. More interested readers can find proofs in, for instance, Arnold (1974), Øksendal (2003) or Gihman and Skorohod (1972).

Sketch of the proof of Itô theorem

From Section 6.4, G has an approximating sequence of step functions $G_n \in M^2[0, t]$ ($n = 1, 2, \ldots$) such that $\int_0^t (G(s) - G_n(s))^2 ds \to 0$ a.s. and such that $\int_0^t G_n(s)dW(s)$ converges in probability to $\int_0^t G(s)dW(s)$. Similar reasoning could be used to prove that F has an approximating sequence of step functions $F_n \in M^1[0, t]$ ($n = 1, 2, \ldots$) such that $\int_0^t |F(s) - F_n(s)|ds \to 0$ a.s., and, since we are dealing with ordinary integrals, this shows that $\int_0^t F_n(s)ds$ converges to $\int_0^t F(s)ds$ a.s. and therefore also converges in probability. So, we only need to prove Itô theorem for step functions, after which applying limits in probability to the approximating sequences we would get the same result a.s. for F and G since limits in probability are a.s. unique. So, without loss of generality, we may assume for the purpose of this proof that F and G are step functions.

These functions are constant with respect to time on the subintervals of some partition and, if we merge the partitions associated to F and G, they will both be simultaneously constant (with respect to time) on the subintervals of that partition.

So, it is enough to prove Itô formula for each subinterval. Let $[a, b]$ be a subinterval of that partition. For $s \in [a, b]$, we have $F(s, \omega) \equiv F(\omega)$ (a r.v. which, according to the usual convention, we abbreviate to F) and $G(s, \omega) \equiv G(\omega)$ (abbreviated to G). For $t \in [a, b]$, we have

$$X(t) = X(a) + (t - a)F + (W(t) - W(a))G,$$

with $X(a)$, F, and G $\mathcal{A}_a$-measurable, therefore independent of future increments $W(t) - W(a)$ ($t > a$) of the Wiener process. Therefore,

$$Y(t) = h(t, X(t)) = h(a, X(a) + (t - a)F + (W(t) - W(a))G)$$

with initial condition $Y_a = h(a, X(a))$. For each $n = 1, 2, \ldots$, consider a partition $a = t_{n,0} < t_{n,1} < \ldots < t_{n,n} = b$ of $[a, b]$ with mesh $\delta_n = \max_{1 \leq k \leq n}(t_{n,k} - t_{n,k-1}) \to 0$ as $n \to +\infty$. We will abbreviate notation and write $t_k = t_{n,k}$, $X_k = X(t_{n,k})$, $\Delta t_k = t_{n,k} - t_{n,k-1}$, $\Delta X_k = X(t_{n,k}) - X(t_{n,k-1})$, and $\Delta W_k = W(t_{n,k}) - W(t_{n,k-1})$.

We get

$$Y(t) - Y(a) = \sum_{k=1}^{n} (h(t_k, X_k) - h(t_{k-1}, X_{k-1})).$$

So, by Taylor formula,

$$h(t_k, X_k) - h(t_{k-1}, X_{k-1}) = h_t(t_{k-1} + \theta_{n,k}\Delta t_k, X_{k-1})\Delta t_k \\ + h_x(t_{k-1}, X_{k-1})\Delta X_k + \frac{1}{2}h_{xx}(t_{k-1}, X_{k-1} + \overline{\theta}_{n,k}\Delta X_k)(\Delta X_k)^2,$$

with $h_t = \partial h/\partial t$, $h_x = \partial h/\partial x$, $h_{xx} = \partial h^2/\partial x^2$, and $0 < \theta_{n,k}, \overline{\theta}_{n,k} < 1$, due to the continuity assumptions on the partial derivatives, continuity which is also uniform since the time interval is closed. Therefore, there are upper bounds

$$\max_{1\leq k \leq n} |h_t(t_{k-1} + \theta_{n,k}\Delta t_k, X_{k-1}) - h_t(t_{k-1}, X_{k-1})| \leq \alpha_n$$

$$\max_{1\leq k \leq n} |h_{xx}(t_{k-1}, X_{k-1} + \overline{\theta}_{n,k}\Delta X_k) - h_{xx}(t_{k-1}, X_{k-1})| \leq \beta_n,$$

with $\alpha_n, \beta_n \to 0$ a.s. The errors involved in replacing $\theta_{n,k}$ and $\overline{\theta}_{n,k}$ by zero are controlled by such upper bounds and we get

$$Y(t) - Y(a) = \sum_{k=1}^{n} (h(t_k, X_k) - h(t_{k-1}, X_{k-1}))$$

$$= \sum_{k=1}^{n} h_t(t_{k-1}, X_{k-1})\Delta t_k + \sum_{k=1}^{n} h_x(t_{k-1}, X_{k-1})\Delta X_k$$

$$+ \frac{1}{2}\sum_{k=1}^{n} h_{xx}(t_{k-1}, X_{k-1})(\Delta X_k)^2 + \text{error}.$$

We have $|\text{error}| \leq \alpha_n(b - a) + \frac{1}{2}\beta_n \sum_{k=1}^{n}(\Delta X_k)^2$ converging in probability to zero since $\sum_{k=1}^{n}(\Delta X_k)^2$ converges in probability to $(b - a)G^2$ (which is a.s. finite).

We just need to prove that

$$\sum_{k=1}^{n}(\Delta X_k)^2 \to (b - a)G^2 \text{ in probability} \tag{6.38}$$

and that

$$\sum_{k=1}^{n} h_t(t_{k-1}, X_{k-1})\Delta t_k$$

$$+ \sum_{k=1}^{n} h_x(t_{k-1}, X_{k-1})(\Delta X_k) \tag{6.39}$$

$$+ \frac{1}{2}\sum_{k=1}^{n} h_{xx}(t_{k-1}, X_{k-1})(\Delta X_k)^2$$

converges in probability to

$$
\begin{aligned}
&\int_a^b h_t(s, X(s))ds \\
&+ \int_a^b h_x(s, X(s))Fds + \int_a^b h_x(s, X(s))GdW(s) \\
&+ \frac{1}{2}\int_a^b h_{xx}(s, X(s))G^2 ds,
\end{aligned}
\tag{6.40}
$$

with each line in (6.39) converging to the corresponding line in (6.40). The result is obvious for the first line. For the second line, the result is also obvious by noting that $\Delta X_k = F\Delta t_k + G\Delta W_k$. We will now deal with the third line.

Note that

$$
\begin{aligned}
&\sum_{k=1}^n h_{xx}(t_{k-1}, X_{k-1})(\Delta X_k)^2 \\
&= F^2 \sum_{k=1}^n h_{xx}(t_{k-1}, X_{k-1})(\Delta t_k)^2 \\
&\quad +2FG \sum_{k=1}^n h_{xx}(t_{k-1}, X_{k-1})\Delta t_k \Delta W_k \\
&\quad +G^2 \sum_{k=1}^n h_{xx}(t_{k-1}, X_{k-1})(\Delta W_k)^2;
\end{aligned}
$$

the first two terms of the right-hand side of the equality converge in probability to zero due to the continuity of h_{xx} and W. We only need to prove that (notice that here G is constant)

$$
\sum_{k=1}^n h_{xx}(t_{k-1}, X_{k-1})(\Delta W_k)^2 \to \int_a^b h_{xx}(s, X(s))ds \quad \text{in probability.} \tag{6.41}
$$

In fact, the other still missing proof of the validity of (6.38) is a particular case of this one (the particular case of $h(s, x) \equiv 1$).

Let

$$
H_{nk} := (W_k - W_{k-1})^2 - (t_k - t_{k-1}) = (\Delta W_k)^2 - \Delta t_k.
$$

Since $\sum_{k=1}^n h_{xx}(t_{k-1}, X_{k-1})\Delta t_k \to \int_a^b h_{xx}(s, X(s))ds$ in probability, it is enough to show that

$$
S_n := \sum_{k=1}^n h_{xx}(t_{k-1}, X_{k-1})H_{nk} \to 0 \quad \text{in probability.} \tag{6.42}
$$

S_n may have no moments, but they exist for its truncation

$$
S_n^{(N)} := \sum_{k=1}^n h_{xx}(t_{k-1}, X_{k-1})H_{nk}I_{N,k}(\omega)
$$

with $N > 0$ and

$$
I_{N,k}(\omega) := \begin{cases} 1 & \text{if } |X(t_{n,i}, \omega)| \le N \text{ for all the } i < k \\ 0 & \text{otherwise.} \end{cases}
$$

Since $\mathbb{E}[H_{nk}] = 0$, $\mathbb{E}[H_{nk}^2] = 2(\Delta t_k)^2$ and the H_{nk} $(k = 1, 2, \ldots, n)$ are independent, we have $\mathbb{E}[S_n^{(N)}] = 0$ and

$$\mathbb{E}[(S_n^{(N)})^2] = \sum_{k=1}^{n} \mathbb{E}[(h_{xx}(t_{k-1}, X_{k-1})I_{N,k})^2]\mathbb{E}[H_{nk}^2]$$

$$\leq 2(\max_{a \leq t \leq b, |x| \leq N} |h_{xx}(t,x)|) \sum_{k=1}^{n} (\Delta t_k)^2 \to 0$$

when $\delta_n \to 0$. For any fixed N, $S_n^{(N)}$ converges to zero in m.s. and therefore also in probability. The truncation error is $P[S_n \neq S_n^{(N)}] = P[\max_{a \leq t \leq b} |X(t)| > N]$. Since $\max_{a \leq t \leq b} |X(t)| = \max_{a \leq t \leq b} |X(a) + (t - a)F + (W(t) - W(a))G| \leq |X(a)| + (b - a)|F| + |G|\max_{a \leq t \leq b} |W(t) - W(a)|$ is an a.s. finite r.v., $P[S_n \neq S_n^{(N)}]$ can become as small as desired for every sufficiently large N. Since $P[|S_n| > \varepsilon] = P[|S_n^{(N)}| > \varepsilon] + P[S_m \neq S_n^{(N)}]$, one recognizes that $P[|S_n| > \varepsilon] \to 0$ and therefore S_n converges in probability to zero, as required. ∎

6.6 The calculi of Itô and Stratonovich

As mentioned in Observation 6.3, the difference between Itô formula and the corresponding formula under normal calculus rules is that in the latter the second-order term with respect to the state variable x in the Taylor expansion is negligible (is of lower order). So, it disappears when we go to the limit to obtain the differential form, i.e. for ordinary calculus $(dX(t))^2 = 0$. This is not so under Itô calculus because of the irregularity (non-differentiability) of the Wiener process. Let us look to this matter with more detail. Consider a small time interval $]t, t + \Delta t]$ (with $\Delta t \to 0$) and the increment $\Delta W(t) = W(t + \Delta t) - W(t)$. We have $E[(\Delta W(t))^2] = \Delta t$, which is not negligible (it is of the same order of Δt, not of lower order) and suggests the rule $(dW(t))^2 = dt$ that we have used. To see that this is due to the irregularity of the Wiener process sample paths, let us imagine approximating $W(t)$ by a smoother differentiable process $\tilde{W}(t)$. We would have $(\Delta \tilde{W}(t))^2 = \left(\frac{d\tilde{W}(t)}{dt}\Delta t + o(\Delta t)\right)^2 = o(\Delta t)$, which is negligible (smaller order than Δt), and the ordinary calculus rules would apply.

This is the reason the *Stratonovich calculus*, which verifies the ordinary calculus rules, is recommended under certain circumstances as the appropriate stochastic calculus. In an SDE, the random perturbations are driven by a white noise, or, looking at their cumulative effect or integral, are driven by a Wiener process. If this is just a mathematically convenient approximation and 'reality' is better modelled by a random differential equation (which means that time varies continuously) with perturbations driven by a coloured noise, the integral of which is a differentiable process, then, when solving the SDE, Stratonovich calculus may give a better approximation to the 'realistic' model. In fact, under appropriate conditions (see Arnold (1974), Clark (1966), Gray (1964), Khashminsky (1969), Wong and Zakai (1965)), if we consider smooth (or even polygonal) approximations $\tilde{W}_n(t)$ of $W(t)$ converging to $W(t)$ as $n \to +\infty$, the solutions of the 'realistic' random differential equations driven by $\tilde{W}_n(t)$ (solutions that are not Markov processes) converge to the Stratonovich calculus solution of the SDE (driven by the Wiener process $W(t)$).

Stratonovich calculus does not possess the nice probabilistic properties of the Itô integral, for example the martingale property (as a function of the upper limit) and the zero expectation property may fail. The example given in Section 6.1 shows this clearly. By (6.7) and (6.10), which give, respectively, the Itô and Stratonovich integrals, one can check that the Itô integral has zero expectation and is a martingale (see part (b) of Property 6.1 and Property 6.2), while the Stratonovich integral has expectation $\mathbb{E}\left[(S)\int_0^t W(s)dW(s)\right] = \mathbb{E}\left[\frac{1}{2}W^2(t)\right] = \frac{t}{2}$ and is clearly not a martingale (remember that martingales have a constant expectation). Furthermore, the Stratonovich integral is defined for a much more restricted class of integrand functions.

However, as we have mentioned, it may be the more appropriate calculus when one uses an SDE (driven by continuous-time white noise $\varepsilon(t)$) as a convenient approximation of the behaviour of a dynamical phenomenon occurring in continuous time when perturbations are driven by a slightly coloured continuous-time noise $\tilde{\varepsilon}(t)$.

However, Itô calculus seems to be the more appropriate calculus when one uses an SDE as a convenient approximation of the behaviour of a dynamical phenomenon occurring in discrete time (so the more 'realistic' model would be a stochastic difference equation) perturbed by a discrete-time white noise $\bar{\varepsilon}(t)$. In fact, in parallel to what happens in the very construction of the Itô integral and under appropriate conditions, the solutions of stochastic difference equations $X(t_k + \Delta t) - X(t_k) = f(t_k, X(t_k))\Delta t + g(t_k, X(t_k))\bar{\varepsilon}(t_k)\Delta t$ (with $t_k = k\Delta t$ and $\Delta t > 0$, $k = 0, 1, 2, \ldots$) converge as $\Delta t \to 0$ to the Itô calculus solution of the approximating SDE $dX(t) = f(t, X(t))dt + g(t, X(t))dW(t)$.

Why should we use an SDE even in the case where 'reality' is better modelled by a random differential equation with coloured noise or by a stochastic difference equation? Because these more 'realistic' models are usually not mathematically tractable. Coloured noises need the knowledge of auto-correlations for all time interval durations, while white noise has zero correlations for all positive time interval durations, and correlations make calculus with coloured noise almost impossible to handle except in simple cases. Difference equations are usually much harder to handle than differential equations even in the deterministic case and the situation does not improve when going to the stochastic case.

Since Itô and Stratonovich calculi give usually different results, it is important for applications to see whether the conditions recommend the use of one or other calculus as a better approximation. Even assuming that conditions are appropriate to the use of an SDE as an approximation, we are faced with the decision whether the phenomenon under study intrinsically occurs in discrete or in continuous time, and that is quite often hard to say. Later we will see that, for vast classes of models, such a decision is not necessary and that both calculi give identical results if we properly interpret the type of averages we use (which differ in the two calculi).

We will now look at the *Stratonovich calculus* in a more general context that the example of Section 6.1. We will define *Stratonovich stochastic integrals* just for the case of integrand functions $G(s, \omega)$ in which time dependence is not direct and arbitrary but rather an indirect dependence through the solution of an SDE. Since we have not yet studied SDEs in a mathematically structured way (that will be done in the next chapter), we will deal with SDEs in an informal manner.

Like we did in (6.1) and (6.2) for the Itô calculus, we say that by Stratonovich calculus solution of the SDE

$$(S) \quad dX_t = f(t, X(t))dt + g(t, X(t))dW(t), \qquad X(0) = X_0, \tag{6.43}$$

we mean the solution of the corresponding integral equation

$$X(t) = X_0 + \int_0^t f(s, X(s))ds + (S) \int_0^t g(s, X(s))dW(s) \tag{6.44}$$

when we use the Stratonovich integral $(S) \int_0^t g(s, X(s))dW(s)$. We will also say that the SDE (6.43) is a *Stratonovich stochastic differential equation*. The use of (S) signals that we are not using Itô calculus but rather Stratonovich calculus. Alternatively, one can also write $\int_0^t g(s, X(s)) \circ dW(s)$ and $dX = fdt + g \circ dW(t)$. In order to define the *Stratonovich integral* $(S) \int_0^t g(s, X(s))dW(s)$, we are going to demand that $G(s, \omega) = g(s, X(s, \omega))$ (where $X(s, \omega)$ is the solution of (6.43)) be continuous in m.s., so that we can use a definition of Stratonovich integral similar to the one used in the example given in Section 6.1. We also demand that the SDE (6.43) has a unique solution (an issue that we will study later). Then, we define the Stratonovich integral by

$$(S) \int_0^t g(s, X(s))dW(s) = \int_0^t g(s, X(s)) \circ dW(s) :=$$
$$\text{l.i.m.} \sum_{k=1}^n g\left(t_{n,k-1}, \frac{X(t_{n,k-1}) + X(t_{n,k})}{2}\right)(W(t_{n,k}) - W(t_{n,k-1})), \tag{6.45}$$

where $0 = t_{n,0} < t_{n,1} < \ldots < t_{n,n-1} < t_{n,n} = t$ $(n = 1, 2, \ldots)$ is a sequence of partitions of $[0, t]$ with mesh $\delta_n = \max_{k=1,\ldots,n}(t_{n,k} - t_{n,k-1}) \to 0$ as $n \to +\infty$.

Please notice the difference to what would result from the definition of Itô integral under similar conditions:

$$\int_0^t g(s, X(s))dW(s) =$$

$$\text{l.i.m.} \sum_{k=1}^n g(t_{n,k-1}, X(t_{n,k-1}))(W(t_{n,k}) - W(t_{n,k-1}));$$

this integral appears in the integral form of the Itô SDE

$$dX_t = f(t, X(t))dt + g(t, X(t))dW(t), \qquad X(0) = X_0. \qquad (6.46)$$

Comparing the expression of the Itô integral with the expression (6.45) of the Stratonovich integral, one sees that:

- If the integrand function $g(t, x)$ does not depend on x, i.e. $g(t, x) \equiv g(t)$, then the Stratonovich and Itô integrals are identical.
- Otherwise, the two integrals differ.

For $g(t, x)$ continuous in t and with continuous partial derivative in x, using a first-order Taylor expansion about $x = X(t_{n,k-1})$ in (6.45) and comparing with the expression for the Itô integral, one concludes that the Stratonovich SDE (6.43) is equivalent, i.e. has the same solution, as the Itô SDE

$$dX(t) = \left(f(t, X(t)) + \frac{1}{2}\frac{\partial g(t, X(t))}{\partial x}g(t, X(t)) \right) dt + g(t, X(t))dW(t), \qquad (6.47)$$
$$X(0) = X_0.$$

The interested reader may try this and the exercise is trivial if one ignores some fine technical issues, particularly ensuring that the approximation error involved in stopping the Taylor expansion at first order is not a problem. Of course, a proper proof requires the technical issues to be addressed, which requires some techniques used in Section 6.5 to prove Itô theorem (fortunately, since here we use m.s. convergence, without the complications in that proof caused by the use of convergence in probability). For a complete proof, we refer the reader to Wong and Zakai (1969) (see also Arnold (1974)). A relevant work by Stratonovich on the integral he invented is Stratonovich (1966).

This relation between Itô and Stratonovich calculi, when applied to SDEs, can be used in the opposite direction. Therefore, the Itô SDE (6.46) is equivalent (has the same solution) as the Stratonovich SDE

$$(S) \quad dX(t) = \left(f(t, X(t)) - \frac{1}{2}\frac{\partial g(t, X(t))}{\partial x}g(t, X(t)) \right) dt + g(t, X(t))dW(t), \qquad (6.48)$$
$$X(0) = X_0.$$

This relation can be very useful because it allows us to avoid the more complex Itô calculus rules and use instead the ordinary rules of calculus with which we are well acquainted. In fact, we can solve the Itô SDE (6.46) directly using Itô calculus rules, but we get the same result if we first convert (6.46) to the equivalent Stratonovich SDE (6.48) and then solve this using ordinary calculus rules (which are also the Stratonovich rules).

From the conversion formulas between the two calculi, one immediately sees that, *if $g(t,x)$ does not depend on x, there is no difference between the two calculi* (because in this case $\frac{\partial g(t,X(t))}{\partial x} = 0$).

6.7 The multidimensional integral

The Itô integral can be easily generalized to a multidimensional setting. In this setting, the integrator process is a standard m-dimensional Wiener process $\mathbf{W}(s,\omega) = [W_1(s,\omega), W_2(s,\omega), \dots, W_m(s,\omega)]^T$ and the *integrand* is a $n \times m$-*matrix-valued function* $\mathbf{G}(s,\omega) = [G_{ij}(s,\omega)]_{i=1,\dots,n;j=1,\dots,m}$, i.e. its values are in $\mathbb{R}^{n \times m}$. Note that the coordinates of the Wiener process are independent but that is not really a restriction since we can introduce a correlation structure by pre-multiplying $\mathbf{W}(s)$ by an appropriate matrix, which can be absorbed by $\mathbf{G}(s)$. The *multidimensional Itô integral* is

$$\int_0^t \mathbf{G}(s,\omega)d\mathbf{W}(s,\omega) = \left[\sum_{j=1}^m \int_0^t G_{1j}dW_j(s), \dots, \sum_{j=1}^m \int_0^t G_{nj}dW_j(s) \right]^T$$

and, as can be seen, is written in terms of unidimensional Itô integrals. Everything is similar to the unidimensional case. The spaces $H^2[0,t]$ and $M^2[0,t]$ for the integrand functions $\mathbf{G}$ are defined as in the unidimensional case, with the $L^{2*}[0,t]$ *norm* defined as in (6.12) but with $|G(s)|^2$ (which was equal to $G^2(s)$) replaced by $|\mathbf{G}|^2 := \left(\sum_{i=1}^n \sum_{j=1}^m G_{ij}^2 \right) = \text{trace}(\mathbf{GG}^T)$. The *inner product* in $H^2[0,t]$ is now defined by

$$\langle \mathbf{G}, \mathbf{H} \rangle_* := \left(\int_0^t \mathbb{E}[\text{trace}(\mathbf{G}(s)\mathbf{H}^T(s))]ds \right).$$

We now have the conditions to deal with a *system of stochastic differential equations* or a *multidimensional stochastic differential equation*. Consider an n-dimensional stochastic process $\mathbf{X}(s,\omega) = [X_1(s,\omega), X_2(s,\omega), \dots, X_n(s,\omega)]^T$, an n-dimensional function $\mathbf{f}(s,\mathbf{x}) = [f_1(s,\mathbf{x}), f_2(s,\mathbf{x}), \dots, f_n(s,\mathbf{x})]^T$ defined for $(s,\mathbf{x}) \in [0,t] \times \mathbb{R}^n$, and an $n \times m$-matrix-valued function

$$\mathbf{g}(s,\mathbf{x}) = [g_{ij}(s,\mathbf{x})]_{i=1,\dots,n;j=1,\dots,m}$$

defined for $(s,\mathbf{x}) \in [0,t] \times \mathbb{R}^n$. We can now consider a *system of stochastic differential equations* (omitting, as usual, ω to simplify the notation)

$$d\mathbf{X}(t) = \mathbf{f}(t,\mathbf{X}(t))dt + \mathbf{g}(t,\mathbf{X}(t))d\mathbf{W}(t), \qquad \mathbf{X}(0) = \mathbf{X}_0,$$

where $\mathbf{X}_0$ is a random vector $\mathcal{A}_0$-measurable (therefore, independent of the Wiener process $\mathbf{W}(t)$). The system of SDEs is equivalent to the *system of stochastic integral equations*

$$\mathbf{X}(t) = \mathbf{X}_0 + \int_0^t \mathbf{f}(s,\mathbf{X}(s))ds + \int_0^t \mathbf{g}(s,\mathbf{X}(s))d\mathbf{W}(s).$$

The *multidimensional Itô processes* $\mathbf{X(t)}$ ($t \in [0, T]$) now takes the differential form

$$d\mathbf{X}(t, \omega) = \mathbf{F}(t, \omega)dt + \mathbf{G}(t, \omega)d\mathbf{W}(t, \omega)$$

or the integral form

$$\mathbf{X}(t, \omega) = \mathbf{X}_0(\omega) + \int_0^t \mathbf{F}(s, \omega)ds + \int_0^t \mathbf{G}(s, \omega)d\mathbf{W}(s, \omega),$$

where

- $\mathbf{X}_0$ is an n-dimensional random vector $\mathcal{A}_0$-measurable, therefore is independent of the Wiener process. It can, in particular, be an n-dimensional deterministic constant.
- $\mathbf{F}$ is a jointly measurable n-dimensional-valued function adapted to the filtration $\mathcal{A}_s$ and such that $\int_0^t |\mathbf{F}(s)|ds < +\infty$ a.s., where $|\mathbf{F}(s)|^2 = \sum_{i=1}^n (F_i(s))^2$.
- $\mathbf{G} \in M^2[0, t]$ is a $n \times m$-matrix-valued function.

In the multidimensional version of Itô theorem and Itô formula, $\mathbf{h}(t, \mathbf{x}) = [h_1(t, \mathbf{x}), \dots, h_k(t, \mathbf{x})]^T$ is now a k-dimensional column vector of continuous real functions defined on $[0, T] \times \mathbb{R}^n$ with continuous first-order partial derivatives in t, with continuous first-order partial derivatives in the x_i ($i = 1, \dots, n$) and continuous second-order partial derivatives in the x_i and x_j ($i, j = 1, \dots, n$). Let $\mathbf{Y}(t) = \mathbf{h}(t, \mathbf{X}(t))$, where $\mathbf{X}(t)$ is the n-dimensional Itô process. Let us represent by $\frac{\partial \mathbf{h}}{\partial t}$ the k-dimensional column vector of partial derivatives $\frac{\partial h_r}{\partial t}$, by $\frac{\partial \mathbf{h}}{\partial \mathbf{x}}$ the $k \times n$-matrix of partial derivatives $\frac{\partial h_r}{\partial x_i}$, by $\frac{\partial^2 \mathbf{h}}{\partial x_i \partial x_j}$ the k-dimensional column vector of partial derivatives $\frac{\partial^2 h_r}{\partial x_i \partial x_j}$, and by $\frac{\partial^2 \mathbf{h}}{\partial \mathbf{x} \partial \mathbf{x}}$ the $n \times n$-matrix whose elements are vectors, namely the vectors $\frac{\partial^2 \mathbf{h}}{\partial x_i \partial x_j}$. The Itô formula becomes now

$$d\mathbf{Y}(t) = \frac{\partial \mathbf{h}}{\partial t}dt + \frac{\partial \mathbf{h}}{\partial \mathbf{x}}d\mathbf{X}(t) + \frac{1}{2}\,\text{trace}\left(\mathbf{G}\mathbf{G}^T \frac{\partial^2 \mathbf{h}}{\partial \mathbf{x} \partial \mathbf{x}}\right)dt,$$

with all derivatives computed at the point $(t, \mathbf{X}(t))$.

Exercise 6.12 *Show that, if $X_1(t)$ and $X_2(t)$ are Itô processes, then $Y(t) = X_1(t)X_2(t)$ is an Itô process and $dY(t) = X_1(t)dX_2(t) + X_2(t)dX_1(t) + dX_1(t)dX_2(t)$.*

Suggestion: In the computation of $dX_1(t)dX_2(t)$ you can use the multiplication table $(dt)^2 = 0$, $dtdW_i = 0$, $(dW_i)^2 = dt$, $dW_i dW_j = 0$ ($i \neq j$).

7

Stochastic differential equations

7.1 Existence and uniqueness theorem and main proprieties of the solution

Let $(\Omega, \mathcal{F}, P)$ be a complete probability space and $W(t)$ a standard Wiener process defined on it. Let $(\mathcal{A}_s)_{s \in [0,T]}$ be a non-anticipative filtration. Let X_0 be a $\mathcal{A}_0$-measurable r.v. (therefore independent of the Wiener process); in particular, it can be a deterministic constant. Having defined in Chapter 6 the Itô stochastic integrals, the stochastic integral equation

$$X(t, \omega) = X_0(\omega) + \int_0^t f(s, X(s, \omega))ds + \int_0^t g(s, X(s, \omega))dW(s, \omega) \qquad (7.1)$$

does have a meaning for $t \in [0, T]$ if, as in Definition 6.4 of an Itô process,

$$F(s, \omega) = f(s, X(s, \omega)) \in M^1[0, T]$$
$$G(s, \omega) = g(s, X(s, \omega)) \in M^2[0, T].$$

Of course we need to impose appropriate conditions on f and g to insure that $F \in M^1[0, T]$ and $G \in M^2[0, T]$, so that the integral form (7.1) has meaning and so the corresponding (Itô) *stochastic differential equation* (SDE)

$$dX(t, \omega) = f(t, X(t, \omega))dt + g(t, X(t, \omega))dW(t, \omega) \qquad X(0, \omega) = X_0(\omega)$$
$$(7.2)$$

also has a meaning for $t \in [0, T]$.

Besides having a meaning, we also want the SDE (7.2) to have a unique solution and this may require appropriate conditions for f and g. Like ordinary differential equations (ODEs), a restriction on the growth of these functions will avoid explosions of the solution (i.e. avoid the solution to diverge to ∞ in finite time) and a Lipschitz condition will insure uniqueness of the solution.

Introduction to Stochastic Differential Equations with Applications to Modelling in Biology and Finance, First Edition. Carlos A. Braumann.
© 2019 John Wiley & Sons Ltd. Published 2019 by John Wiley & Sons Ltd.
Companion Website: www.wiley.com/go/braumann/stochastic-differential-equations

We will denote by $\mathcal{E}$ the set of real-valued functions $h(t, x)$ with domain $[0, T] \times \mathbb{R}$ that are Borel-measurable jointly in t and x^1 and, for some $K > 0$ and all $t \in [0, T]$ and $x, y \in \mathbb{R}$, verify the two properties:

- $|h(t, x) - h(t, y)| \leq K|x - y|$ (*Lipschitz condition*)
- $|h(t, x)| \leq K(1 + |x|^2)^{1/2}$ (*restriction on growth*).

We will follow the usual convention of not making explicit the dependence of random variables and stochastic processes on chance ω, but one should keep always in mind that such dependence exists and so the Wiener process and the solutions of SDE do depend on ω.

We now state an existence and uniqueness theorem for SDEs which is, with small nuances, the one commonly shown in the literature. Note that the conditions stated in the theorem are sufficient (not necessary), so there may be some improvements (i.e. the class $\mathcal{E}$ of functions for which existence and uniqueness is insured may be enlarged), particularly in special cases. We will talk about that later.

Of course the starting time was labelled 0 and we will work on a time interval $[0, T]$ just for convenience, but nothing prevents one from working on intervals like $[t_0, T]$.

Although further complications arise in the stochastic case, the theorem and the proof are inspired on an analogous existence and uniqueness theorem for ODEs. Both use the technique of proving uniqueness by assuming two solutions and showing they must be identical. Both use the constructive Picard's iterative method of approximating the solution by feeding one approximation into the integral equation to get the next approximation:

$$X_0(t) \equiv X_0$$
$$X_{n+1}(t) = X_0 + \int_0^t f(s, X_n(s))ds + \int_0^t g(s, X_n(s))dW(s) \quad (n = 0, 1, \dots).$$

This method, which is very convenient for the proof, can be used in practice to approximate the solution, although it is quite slow compared to other available methods.

Monte Carlo simulation of SDEs

Although a more complete treatment will be presented in Section 12.2, we mention here some ideas on how to simulate trajectories of the solution of an SDE.

If an explicit expression for the solution of the SDE can be obtained, it can be used to simulate trajectories of $X(t)$. If the explicit expression only involves the Wiener process directly at the present time, we can simulate the Wiener process

---

1 This means that h is measurable with respect to the product σ-algebra $\mathcal{B}_{[0,T]} \times \mathcal{B}$, where $\mathcal{B}_{[0,T]}$ is the Borel σ-algebra on $[0, T]$ and $\mathcal{B}$ is the Borel σ-algebra on $\mathbb{R}$, so inverse images by h of Borel sets in $\mathbb{R}$ will be in $\mathcal{B}_{[0,T]} \times \mathcal{B}$.

by the method of Section 4.1 and plug the values into the expression, as we will do to simulate the solution of the Black–Scholes model in Section 8.1. Monte Carlo simulation for more complicated explicit expressions will be addressed in Section 12.2.

Unfortunately, in many cases we are unable to obtain an explicit expression, in which case the trajectories can be simulated using an approximation of $X(t)$ obtained by discretizing time. One uses a partition $0 = t_0 < t_1 < t_1 < \cdots < t_n = T$ of $[0, T]$ with a sufficiently large n to reduce the numerical error. Typically, but not necessarily, the partition points are equidistant, with $t_k = kT/n$. A trajectory is simulated iteratively at the partition points. One starts (step 0) simulating $X(t_0) = X_0$ by using the distribution of X_0; if $X_0 = x_0$ is a deterministic value, then $X(t_0) = x_0$ and no simulation is required. At each iteration (step k with $k = 1, 2, \ldots, n$), one uses the previously simulated value $X_{k-1} = X(t_{k-1})$ to simulate the next value $X_k = X(t_k)$. The simplest method to do that is the *Euler method*, the same as is used for ODEs. In the context of SDEs, it is also known as the *Euler–Maruyama method*. In this method, $f(s, X(s))$ and $g(s, X(s))$ are approximated in the interval $[t_{k-1}, t_k[$ by constants, namely by the values they take at the beginning point of the interval (to ensure the non-antecipative property of the Itô calculus). This leads to the first-order approximation scheme

$$X_k = X_{k-1} + \int_{t_{k-1}}^{t_k} f(s, X(s))ds + \int_{t_{k-1}}^{t_k} g(s, X(s))dW(s)$$
$$\simeq X_{k-1} + f(t_{k-1}, X_{k-1})(t_k - t_{k-1}) + g(t_{k-1}, X_{k-1})(W(t_k) - W(t_{k-1})).$$

Since the approximate value of $X(t_{k-1})$ is known from the previous iteration, the only thing random in the above scheme is the increment $W(t_k) - W(t_{k-1})$. The increments $W(t_k) - W(t_{k-1})$ $(k = 1, \ldots, n)$ are easily simulated (we have done this in Section 4.1 when simulating trajectories of the Wiener process), since they are independent random variables having a normal distribution with mean zero and variance $t_k - t_{k-1}$. At the end of the iterations, we have the approximate values at the time partition points of a simulated trajectory. Of course, this iterative procedure can be repeated to produce other trajectories.

There are faster methods (i.e. requiring not so large values of n), like the *Milstein method*. On the numerical resolution/simulation of SDE, we refer the reader to Kloeden and Platen (1992), Kloeden *et al.* (1994), Bouleau and Lépingle (1994), and Iacus (2008).

SDE existence and uniqueness theorem

Theorem 7.1 (SDE existence and uniqueness theorem)   *Let $X_0 \in L^2$ (i.e. X_0 has finite variance) be a $\mathcal{A}_0$-measurable r.v. (therefore independent of the Wiener process) and consider on the interval $[0, T]$ $(T > 0)$ an SDE*

$$dX(t) = f(t, X(t))dt + g(t, X(t))dW(t), \qquad X(0) = X_0, \tag{7.3}$$

or the corresponding integral form, the stochastic integral equation[2]

$$X(t) = X_0 + \int_0^t f(s, X(s))ds + \int_0^t g(s, X(s))dW(s). \tag{7.4}$$

Suppose that $f, g \in \mathcal{E}$.
  Then:

(a) *[Existence] There is an a.s. continuous stochastic process $X(t) = X(t, \omega)$ ($t \in [0, T]$) that is a solution of (7.3), i.e. $X(t)$ satisfies a.s. (7.4) for all $t \in [0, T]$.*
(b) *[Uniqueness] The solution is a.s. unique (we will simply say 'unique') in the sense that, given two a.s. continuous solutions $X(t)$ and $X^*(t)$, we have a.s.*

$$\sup_{0 \le t \le T} |X(t) - X^*(t)| = 0.$$

(c) *We have:*[3]

$$\mathbb{E}[(X(t))^2] \le (1 + \mathbb{E}[(X_0)^2]) \exp(K(K+2)t) - 1$$
$$\mathbb{E}[(X(t) - X_0)^2] \le 2K^2(T+1)(1 + \mathbb{E}[(X_0)^2])t \exp(K(K+2)t). \tag{7.5}$$

(d) *The solution $X(t) \in H^2[0, T]$ and is m.s continuous.*
(e) *[Semigroup property] For $0 \le s \le t \le T$, $X(t) = X_{s,X(s)}(t)$.*
    We have used the notation $X_{s,x}(t)$ ($t \in [s, T]$) to represent the a.s. continuous unique solution of $dX(t) = f(t, X(t))dt + g(t, X(t))dW(t)$ with initial condition $X(s) = x$, i.e. the solution of $X(t) = x + \int_s^t f(u, X(u))du + \int_s^t g(u, X(u))dW(u)$.
(f) *[Markov property] The solution $X(t)$ is a Markov process. The initial distribution coincides with the distribution of X_0 and the transition probabilities verify $P(t, B|s, x) = P[X_{s,x}(t) \in B]$ ($s \le t$).*
(g) *[Diffusion] If f and g are also continuous in t, then the solution $X(t)$ is a diffusion process with drift coefficient $a(s, x) = f(s, x)$ and diffusion coefficient $b(s, x) = |g(s, x)|^2$.*

Observation 7.1 *Assume $g(t, x)$ is continuous in t and has continuous partial derivatives in x. Assume also that $f(t, x) + \frac{1}{2} \frac{\partial g(t,x)}{\partial x} g(t, x)$ and $g(t, x)$ are in $\mathcal{E}$ and are continuous functions of t. Consider the Stratonovich SDE*

  (S) $dX_t = f(t, X(t))dt + g(t, X(t))dW(t).$

  From the fist set of assumptions, we know from (6.47) in Section 6.6 that it is equivalent to the Itô SDE

$$dX_t = \left(f(t, X(t)) + \frac{1}{2} \frac{\partial g(t, X(t))}{\partial x} g(t, X(t)) \right) dt + g(t, X(t))dW(t).$$

---

2 It is understood without saying that the second integral in the integral form (7.4) is the Itô integral and so the corresponding SDE (7.3) is to be interpreted in the Itô sense.
3 Some other possible bounds are available in the literature.

Applying Theorem 7.1 to this SDE, we know, from the second set of assumptions, that the solution exists, is unique, and, from part (f), that it is a diffusion process.

Therefore, the solution of the Stratonovich SDE exists, is unique, and is a diffusion process with drift coefficient $a(s,x) = f(s,x) + \frac{1}{2}\frac{\partial g(s,x)}{\partial x}g(s,x) = f(s,x) + \frac{1}{4}\frac{\partial b(s,x)}{\partial x}$ and diffusion coefficient $b(s,x) = |g(s,x)|^2$.

7.2 Proof of the existence and uniqueness theorem

We now present a proof of the existence and uniqueness Theorem 7.1, which can be skipped by non-interested readers. Certain parts of the proof are presented in a sketchy way; for more details, the reader can consult, for instance, Arnold (1974), Øksendal (2003), Gard (1988), Schuss (1980) or Wong and Hajek (1985).

(a) Proof of uniqueness

Let $X(t)$ and $X^*(t)$ be two a.s. continuous solutions. They satisfy (7.4) and so are non-anticipative (only depend on past values of themselves and of the Wiener process). Since $f(t,x)$ and $g(t,x)$ are Borel measurable functions, $f(t,X(t))$, $g(t,X(t))$, $f(t,X^*(t))$, and $g(t,X^*(t))$ are non-anticipative. We could work with $\mathbb{E}[|X(t) - X^*(t)|^2]$, but we have not yet proved that such second-order moment indeed exists. So we replace it by the truncated moment $\mathbb{E}[|X(t) - X^*(t)|^2 I_N(t)]$, with $I_N(t) = 1$ if $|X(u)| \leq N$ and $|X^*(u)| \leq N$ for all $0 \leq u \leq t$ and $I_N(t) = 0$ otherwise. Note that $I_N(t)$ is non-anticipative, that $I_N(t) \in H^2[0,T]$, and that $I_N(t) = I_N(t)I_N(s)$ for $s \leq t$. Since $X(t)$ and $X^*(t)$ satisfy (7.4), using the inequality $(a + b)^2 \leq 2a^2 + 2b^2$, one gets

$$\mathbb{E}[|X(t) - X^*(t)|^2 I_N(t)]$$
$$= \mathbb{E}\left[I_N(t)\left(\int_0^t I_N(s)(f(s,X(s)) - f(s,X^*(s)))ds\right. \right.$$
$$\left.\left. + \int_0^t I_N(s)(g(s,X(s)) - g(s,X^*(s)))dW(s)\right)^2\right]$$
$$\leq 2\mathbb{E}\left[\left(\int_0^t I_N(s)(f(s,X(s)) - f(s,X^*(s)))ds\right)^2\right]$$
$$+ 2\mathbb{E}\left[\left(\int_0^t I_N(s)(g(s,X(s)) - g(s,X^*(s)))dW(s)\right)^2\right].$$

Using the Schwarz inequality (which, in particular, implies $(\int_0^t \varphi(s)ds)^2 \leq \int_0^t 1^2 ds \int_0^t \varphi^2(s)ds$), the fact that $t \leq T$, and the norm preservation property for Itô integrals of $H^2[0,T]$ functions, one gets

$$\mathbb{E}[|X(t) - X^*(t)|^2 I_N(t)]$$
$$\leq 2T \int_0^t \mathbb{E}[I_N(s)(f(s,X(s)) - f(s,X^*(s)))^2]ds$$
$$+ 2 \int_0^t \mathbb{E}[I_N(s)(g(s,X(s)) - g(s,X^*(s)))^2]ds.$$

Putting $L = 2(T + 1)K^2$ and using the Lipschitz condition, one has

$$\mathbb{E}[|X(t) - X^*(t)|^2 I_N(t)] \leq L \int_0^t \mathbb{E}[|X(s) - X^*(s)|^2 I_N(s)] ds. \tag{7.6}$$

Using the *Bellman–Gronwall lemma*,[4] also known as *Gronwall's inequality*, one gets $\mathbb{E}[|X(t) - X^*(t)|^2 I_N(t)] = 0$, which implies $I_N(t)X(t) = I_N(t)X^*(t)$ a.s. But

$$P[I_N(t) \not\equiv 1 \text{ in } [0, T]] \leq P\left[\sup_{0 \leq t \leq T} |X(t)| > N\right] + P\left[\sup_{0 \leq t \leq T} |X^*(t)| > N\right]$$

and since $X(t)$ are $X^*(t)$ are a.s. bounded in $[0, T]$ (due to their a.s. continuity), both probabilities on the right-hand side become arbitrarily small for every sufficiently large N. So, $X(t) = X^*(t)$ a.s. for t, and so also for all $t \in \mathbb{Q} \cap [0, T]$ (since the set of rational numbers $\mathbb{Q}$ is countable). Due to the continuity, we have a.s. $X(t) = X^*(t)$ for all $t \in [0, T]$ and therefore $\sup_{0 \leq t \leq T} |X(t) - X^*(t)| = 0$ a.s.

(b) Proof of existence

The proof of existence is based on the same Picard's method that is used on a similar proof for ODEs. It is an iterative method of successive approximations, starting with

$$X_0(t) \equiv X_0 \tag{7.7}$$

and using the iteration

$$X_{n+1}(t) = X_0 + \int_0^t f(s, X_n(s)) ds + \int_0^t g(s, X_n(s)) dW(s) \quad (n = 0, 1, 2, \ldots). \tag{7.8}$$

We just need to prove that $X_n(t)$ ($n = 0, 1, 2, \ldots$) are a.s. continuous functions and that this sequence uniformly converges a.s. The limit will then be a.s. a continuous function, which, as we will show, is the solution of the SDE.

Since $X_0 \in L^2$, we have $\sup_{0 \leq t \leq T} \mathbb{E}[(X_0(t))^2] < +\infty$ and will see by induction that $X_n(t) \in L^2$ and $\sup_{0 \leq t \leq T} \mathbb{E}[(X_n(t))^2] < +\infty$ for all n. In fact, assuming this is true for $n - 1$, we show it is true for n because (due to the restriction on growth, the norm preservation of the Itô integral, the inequality $(a + b + c)^2 \leq$

---

4 The *Bellman–Gronwall lemma* states that, given $\varphi(t) \geq 0$, $h(t)$ integrable in $[a, b]$, and a positive constant L, if $\varphi(t) \leq L \int_0^t \varphi(s) ds + h(t)$ for all $t \in [a, b]$, then $\varphi(t) \leq L \int_0^t \exp(L(t - s)) h(s) ds + h(t)$ for all $t \in [a, b]$. A proof can be seen, for instance, in Gikhman and Skorohod (1969). Here, we have applied the lemma to the functions $\varphi(t) = \mathbb{E}[|X(t) - X^*(t)|^2 I_N(t)]$ and $h(t) \equiv 0$.

$3a^2 + 3b^2 + 3c^2$, and the Schwarz's inequality), with $B = 3K^2(T + 1)$,

$$
\begin{aligned}
\mathbb{E}[(X_n(t))^2] &\leq 3\mathbb{E}[(X_0)^2] + 3K^2T \int_0^t (1 + \mathbb{E}[(X_{n-1}(s))^2])ds \\
&\quad + 3K^2 \int_0^t (1 + \mathbb{E}[(X_{n-1}(s))^2])ds \\
&= 3\mathbb{E}[(X_0)^2] + B \int_0^t (1 + \mathbb{E}[(X_{n-1}(s))^2])ds \\
&\leq 3\mathbb{E}[(X_0)^2] + BT \left(1 + \sup_{0 \leq t \leq T} \mathbb{E}[(X_{n-1}(t))^2]\right).
\end{aligned}
\tag{7.9}
$$

Using a reasoning similar to the one used in the proof of (7.6), but now with no need to used truncated moments (since the non-truncated moments exist), one obtains

$$
\mathbb{E}[|X_{n+1}(t) - X_n(t)|^2] \leq L \int_0^t \mathbb{E}[|X_n(s) - X_{n-1}(s)|^2]ds.
\tag{7.10}
$$

Iterating (7.10), one obtains by induction

$$
\mathbb{E}[|X_{n+1}(t) - X_n(t)|^2] \leq L^n \int_0^t \frac{(t-s)^{n-1}}{(n-1)!} \mathbb{E}[|X_1(s) - X_0(s)|^2]ds.
$$

Using the restriction on growth, one obtains

$$
\begin{aligned}
\mathbb{E}[|X_1(s) - X_0(s)|^2] &= \mathbb{E}\left[|\int_0^t f(s, X_0)ds + \int_0^t g(s, X_0)dW(s)|^2\right] \\
&\leq 2K^2(T+1) \int_0^t (1 + \mathbb{E}[(X_0)^2])ds \leq 2K^2(T+1)T(1 + \mathbb{E}[(X_0)^2]) =: M,
\end{aligned}
$$

and so

$$
\sup_{0 \leq t \leq T} \mathbb{E}[|X_{n+1}(t) - X_n(t)|^2] \leq M(LT)^n/n!.
$$

Since

$$
\begin{aligned}
d_n &:= \sup_{0 \leq t \leq T} |X_{n+1}(t) - X_n(t)| \\
&\leq \int_0^T |f(s, X_n(s)) - f(s, X_{n-1}(s))|ds \\
&\quad + \sup_{0 \leq t \leq T} \left|\int_0^t (g(s, X_n(s)) - g(s, X_{n-1}(s)))dW(s)\right|,
\end{aligned}
$$

using (6.27) and the Lipschitz condition, one gets

$$
\begin{aligned}
\mathbb{E}[(d_n)^2] &\leq 2K^2T \int_0^T \mathbb{E}[|X_n(s) - X_{n-1}(s)|^2]ds \\
&\quad + 8K^2 \int_0^T \mathbb{E}[|X_n(s) - X_{n-1}(s)|^2]ds \\
&\leq (2K^2T + 8K^2)TM(LT)^{n-1}/(n-1)! = C(LT)^{n-1}/(n-1)!.
\end{aligned}
$$

By Thebyshev inequality,

$$\sum_{n=1}^{+\infty} P[d_n > 1/2^{n-1}] \le \sum_n 2^{2(n-1)} \mathbb{E}[(d_n)^2]$$

$$\le C \sum_n (4LT)^{n-1}/(n-1)! < +\infty.$$

By the Borel–Cantelli lemma,

$$P[d_n > 1/2^{n-1} \text{ for an infinite number of values of } n] = 0,$$

and so, for every sufficiently large n, $d_n \le 1/2^{n-1}$ a.s. Since the series $\sum_{k=0}^{+\infty} d_k$ is convergent, the series $\sum_{k=0}^{+\infty}(X_{k+1}(t) - X_k(t))$ converges uniformly a.s. on $[0, T]$ (notice that the terms are bounded by d_k). Therefore

$$X_n(t) = X_0(t) + \sum_{k=0}^{n-1}(X_{k+1}(t) - X_k(t))$$

converges uniformly a.s. on $[0, T]$.

This shows the a.s. uniform convergence of $X_n(t)$ on $[0, T]$ as $n \to +\infty$. Denote the limit by $X(t)$. Since $X_n(t)$ are obviously non-anticipative, the same happens to $X(t)$. The a.s. continuity of the $X_n(t)$ and the uniform convergence imply the a.s. continuity of $X(t)$. Obviously, from the restriction on growth and a.s. continuity of $X(t)$, we have $\int_0^t |f(s, X(s))|ds \le \int_0^t (1 + |X(s)|^2)^{1/2}ds < +\infty$ a.s. and $\int_0^t |g(s, X(s))|^2 ds \le \int_0^t (1 + |X(s)|^2)ds < +\infty$ a.s. Consequently, $X(t)$ is an Itô process and the integrals in (7.4) make sense.

The only thing missing is to show that $X(t)$ is indeed a solution, i.e. that $X(t)$ satisfies (7.4) for $t \in [0, T]$. Apply limits in probability to both sides of (7.8). The left side $X_n(t)$ converges in probability to $X(t)$ (the convergence is even a stronger a.s. uniform convergence). On the right-hand side, we have:

- The integrals $\int_0^t f(s, X_n(s))ds$ converge a.s., and so converge in probability to $\int_0^t f(s, X(s))ds$. In fact, $\left| \int_0^t f(s, X_n(s))ds - \int_0^t f(s, X(s))ds \right| \le K \int_0^t |X_n(s) - X(s)|ds \to 0$ a.s. (due to the Lipschitz condition).
- The integrals $\int_0^t g(s, X_n(s))dW(s)$ converge to $\int_0^t g(s, X(s))dW(s)$ in probability. In fact, the Lipschitz condition implies $\int_0^t |g(s, X_n(s)) - g(s, X(s))|^2 ds \le K^2 \int_0^t |X_n(s) - X(s)|^2 ds \to 0$ a.s. and we can use the result of Exercise 6.9 in Section 6.4.
- So, the right-hand side $X_0 + \int_0^t f(s, X_n(s))ds + \int_0^t g(s, X_n(s))dW(s)$ converges in probability to $X_0 + \int_0^t f(s, X(s))ds + \int_0^t g(s, X(s))dW(s)$.

Since the limits in probability are a.s. unique, we have

$$X(t) = X_0 + \int_0^t f(s, X(s))ds + \int_0^t g(s, X(s))dW(s) \text{ a.s.},$$

i.e. $X(t)$ satisfies (7.4).

Proof that $X(t) \in H^2[0, T]$ and $\sup_{0 \le t \le T} \mathbb{E}[(X(t))^2] < +\infty$

From (7.9), putting $h_n(t) = 1 + \mathbb{E}[(X_n(t))^2]$ gives $h_n(t) \le 3h_0 + B \int_0^t h_{n-1}(s)ds$. Iterating, we have

$$h_n(t) \le 3 \left(h_0 + h_0 Bt + h_0 \frac{(Bt)^2}{2!} + \cdots + h_0 \frac{(Bt)^{n-1}}{(n-1)!} \right)$$

$$+ B^n \int_0^t \frac{(t-s)^{n-1}}{(n-1)!} h_0 ds$$

$$\le 3h_0 \left(1 + Bt + \frac{(Bt)^2}{2!} + \dots + \frac{(Bt)^n}{n!} \right) \le 3h_0 e^{Bt}$$

and therefore $\mathbb{E}[(X_n(t))^2] \le 3(1 + \mathbb{E}[(X_0)^2])e^{Bt}$. Letting $n \to \infty$, by the dominated convergence theorem, we get

$$\mathbb{E}[(X(t))^2] \le 3(1 + \mathbb{E}[(X_0)^2])e^{Bt}. \tag{7.11}$$

Consequently, $X(t) \in H^2[0, T]$ and $\sup_{0 \le t \le T} \mathbb{E}[(X(t))^2] < +\infty$.

Proof that $F(s, \omega) = f(s, X(s, \omega))$ and $G(s, \omega) = g(s, X(s, \omega))$ are in $H^2[0, T]$

From the previous result, $\int_0^t \mathbb{E}[f^2(s, X(s))]ds \le K^2 \int_0^t (1 + \mathbb{E}[X^2(s)])ds < +\infty$ and $\int_0^t \mathbb{E}[g^2(s, X(s))]ds \le K^2 \int_0^t (1 + \mathbb{E}[X^2(s)])ds < +\infty$, and so F and G are in $H^2[0, T]$.

As a consequence, the stochastic integral $\int_0^t g(s, X(s))dW(s)$ is a martingale.

Proof of (7.5)

By the Itô formula (6.37),

$$(X(t))^2 = (X_0)^2 + \int_0^t (2X(s)f(s, X(s)) + g^2(s, X(s)))ds \tag{7.12}$$

$$+ \int_0^t 2X(s)g(s, X(s))dW(s).$$

Applying mathematical expectations, one gets

$$\mathbb{E}[(X(t))^2] = \mathbb{E}[(X_0)^2] + \int_0^t \mathbb{E}[2X(s)f(s, X(s)) + g^2(s, X(s))]ds.$$

We have assumed that the stochastic integral has a null expected value, although we have not shown that the integrand function was in $H^2[0, T]$. Since this may fail, we should, in good rigour, have truncated $|X(t)|$ by N (to ensure the nullity of expectation of the stochastic integral) and then go to the limit as $N \to +\infty$.

Using the restriction on growth and the inequality $|x| \leq (1 + x^2)^{1/2}$, one gets

$$\mathbb{E}[(X(t))^2]$$
$$\leq \mathbb{E}[(X_0)^2] + \int_0^t \mathbb{E}[2K|X(s)|(1 + |X(s)|^2)^{1/2} + K^2(1 + |X(s)|^2)]\,ds$$
$$\leq \mathbb{E}[(X_0)^2] + K(K + 2)\int_0^t \mathbb{E}[(1 + \mathbb{E}[|X(s)|^2])]ds.$$

Put $\varphi(s) = 1 + \mathbb{E}[|X(s)|^2]$ and $h(t) \equiv h = 1 + \mathbb{E}[|X_0|^2]$, note that $\varphi(t) \leq h + K(K + 2)\int_0^t \varphi(s)ds$ and apply Gronwall's inequality to obtain $1 + \mathbb{E}[|X(t)|^2] \leq h + K(K + 2)e^{K(K+2)t}\int_0^t e^{-K(K+2)s}h\,ds$, and so the first inequality of (7.5).

From (7.4), squaring and applying mathematical expectations, one gets

$$\mathbb{E}[(X(t) - X_0)^2] \leq 2\mathbb{E}\left[\left|\int_0^t f(s, X(s))ds\right|^2\right] + 2\mathbb{E}\left[\left|\int_0^t g(s, X(s))dW(s)\right|^2\right]$$
$$\leq 2t\int_0^t \mathbb{E}[|f(s, X(s))|^2]ds + 2\int_0^t \mathbb{E}[|g(s, X(s))|^2]ds$$
$$\leq 2(T + 1)K^2\int_0^t(1 + \mathbb{E}[|X(s)|^2])ds.$$

Applying the first expression of (7.5) and bounding the $e^{K(K+2)s}$ that shows up in the integral by $e^{K(K+2)t}$, one gets the second expression in (7.5).

Proof that the solution is m.s. continuous

For $t \geq s$, $X(t)$ is also solution of $X(t) = X(s) + \int_s^t f(u, X(u))du + \int_s^t g(u, X(u))\,dW(u)$. Since now the initial condition is the value $X(s)$ at time s and the length of the time interval is $t - s$, the second inequality in (7.5) now reads

$$\mathbb{E}[(X(t) - X(s))^2]$$
$$\leq 2K^2(T + 1)(1 + \mathbb{E}[(X(s))^2])(t - s)\exp(K(K + 2)(t - s)) \tag{7.13}$$

and we can now use the first inequality of (7.5) to bound $1 + \mathbb{E}[(X(s))^2]$ by $(1 + \mathbb{E}[(X(0))^2])e^{K(K+2)T}$. We conclude that $\mathbb{E}[(X(t) - X(s))^2] \to 0$ as $t - s \to 0$, which proves the m.s. continuity.

Proof of the semigroup property: For $0 \leq s \leq t \leq T$, $X(t) = X_{s,X(s)}(t)$

This is an easy consequence of the trivial property of integrals (if one splits the integration interval into two subintervals, the integral over the interval is the sum of the integrals over the subintervals), which gives $X(t) = X_0 + \int_0^t f(u, X(u))du + \int_0^t g(u, X(u))dW(u) = X_0 + \int_0^s f(u, X(u))du + \int_0^s g(u, X(u))dW(u) + \int_s^t f(u, X(u))du + \int_s^t g(u, X(u))dW(u) = X(s) + \int_s^t f(u, X(u))du + \int_s^t g(u, X(u))dW(u)$.

Proof that the solution is a Markov process

Let $0 \leq s \leq t \leq T$. The intuitive justification that $X(t)$ is a Markov process comes from the semigroup property $X(t) = X_{s,X(s)}(t)$. This means that $X(t)$ can be obtained in the interval $[s, t]$ as solution of $X(t) = X(s) + \int_s^t f(u, X(u))du + \int_s^t g(u, X(u))dW(u)$. So, given $X(s)$, $X(t)$ is defined in terms of $X(s)$ and

$W(u) - W(s)$ ($u \in [s, t]$), and so is measurable with respect to $\sigma(X(s), W(u) - W(s) : s \leq u \leq t)$. Since $X(s)$ is $\mathcal{A}_s$-measurable, and so is independent of $\sigma(W(u) - W(s) : s \leq u \leq t)$, we conclude that, given $X(s)$, $X(t)$ only depends on $\sigma(W(u) - W(s) : s \leq u \leq t)$. Since $\sigma(X(u) : 0 \leq u \leq s) \subset \mathcal{A}_s$ is indepen-dent of $\sigma(W(u) - W(s) : s \leq u \leq t)$, also $X(t)$ (future value) given $X(s)$ (present value) is independent of $\sigma(X(u) : 0 \leq u \leq s)$ (past values). Thus, it is a Markov process.

A formal proof can been seen, for instance, in Gihman and Skorohod (1972), Arnold (1974), and Gard (1988).

Proof that, if f and g are also continuous in t, then $X(t)$ is a diffusion process

Let $0 \leq s \leq t \leq T$ and $\Delta = t - s$. Conditioning on $X(s) = x$ (deterministic ini-tial condition), from (7.13) we obtain

$$\mathbb{E}_{s,x}[(X(t) - X(s))^2] \leq 2K^2(T + 1)(1 + x^2)(t - s)\exp(K(K + 2)(t - s)). \tag{7.14}$$

First of all, let us note that expressions similar to (7.5) can be obtained for higher even $2n$-order moments if $X_0 \in L^{2n}$. Therefore, since now the initial condition $X(s) = x$ is deterministic and so has moments of any order $2n$, we can use such expressions to obtain expressions of higher order similar to (7.14). We get

$$\mathbb{E}_{s,x}[(X(t) - X(s))^{2n}] \leq C(1 + x^{2n})(t - s)^n \exp(D(t - s)), \tag{7.15}$$

with C and D appropriate positive constants. Given $X(s) = x$, since $X(t) = X_{s,x}(t)$, we get

$$\mathbb{E}_{s,x}[(X(t) - X(s))^k] := \mathbb{E}[(X(t) - x)^k | X(s) = x] = \mathbb{E}[(X_{s,x}(t) - x)^k]$$

Due to (7.15), these moments exist for all $k = 1, 2, 3, 4, \ldots$ (even for odd k, the moment exists since the moment of order $k + 1$ exists). Since $X(t)$ is a Markov process with a.s. continuous trajectories, we just need to show that (5.1), (5.2), and (5.3) hold.

Due to the Lipschitz condition, notice that f and g are also continuous func-tions of x.

From (7.15) with $n = 2$ we get $\mathbb{E}_{s,x}[(X(s + \Delta) - x)^4] \leq A\Delta^2$, with $A = C(1 + x^{2n})\exp(DT)$ constant, so that $\frac{1}{\Delta}\mathbb{E}_{s,x}[(X(s + \Delta) - x)^4] \to 0$ as $\Delta \to 0$. Therefore

$$\frac{1}{\Delta}P_{s,x}[|X(s + \Delta) - x| \geq \varepsilon] \leq \frac{1}{\Delta}\frac{1}{\varepsilon^4}\mathbb{E}_{s,x}[(X(s + \Delta) - x)^4] \to 0.$$

So, (5.1) holds.

Starting from (7.4), since the Itô integral has zero expectation, we get

$$\mathbb{E}_{s,x}[X(s + \Delta) - x] = \int_s^{s+\Delta} \mathbb{E}_{s,x}[f(u, X_{s,x}(u))]du = \int_s^{s+\Delta} f(u, x)du + \int_s^{s+\Delta} \mathbb{E}_{s,x}[(f(u, X_{s,x}(u)) - f(u, x))]du. \tag{7.16}$$

We also have, using the Lipschitz condition, Schwarz's inequality, and (7.14),

$$
\begin{aligned}
&\left| \int_s^{s+\Delta} \mathbb{E}_{s,x}[(f(u, X_{s,x}(u)) - f(u,x))]du \right| \\
&\leq \Delta^{1/2} \left(\int_s^{s+\Delta} \mathbb{E}_{s,x}[|f(u, X_{s,x}(u)) - f(u,x)|^2]du \right)^{1/2} \\
&\leq K\Delta^{1/2} \left(\int_s^{s+\Delta} \mathbb{E}_{s,x}[|X_{s,x}(u) - x|^2]du \right)^{1/2} \\
&\leq K\Delta^{1/2} \, (2K^2(T+1)(1+x^2)\Delta \exp(K(K+2)T))^{1/2} \, \Delta^{1/2} = B\Delta^{3/2},
\end{aligned}
\tag{7.17}
$$

where B is a positive constant that does not depend on s. The continuity of f in t holds on the close interval $[0, T]$ and is uniform. Therefore, given an arbitrary $\varepsilon > 0$, there is a $\Delta > 0$ independent of s such that

$$
\begin{aligned}
&\left| \int_s^{s+\Delta} f(u,x)du - f(s,x)\Delta \right| \\
&= \left| \int_s^{s+\Delta} (f(u,x) - f(s,x))du \right| \leq \int_0^t \varepsilon ds = \varepsilon\Delta.
\end{aligned}
\tag{7.18}
$$

From (7.16), (7.17), and (7.18), we obtain (5.2) with $a(s,x) = f(s,x)$.

To obtain (5.3) with $b(s,x) = g^2(s,x)$, one starts from (7.12) instead of (7.4), using also as initial condition $X(s) = x$ and applying similar techniques.

This concludes the proof of Theorem 7.1. ∎

7.3 Observations and extensions to the existence and uniqueness theorem

The condition $X_0 \in L^2$ (i.e. X_0 has finite variance) is really not required and we could prove Theorem 7.1, with the exception of parts (c) and (d), without making that assumption. Of course, for parts (c) and (d) we would need the assumption, since otherwise the required second-order moments of $X(t)$ might not exist.

In fact, except for parts (c) and (d), we could easily adapt the proof presented in Section 7.2 in order to wave that assumption. We would just replace X_0 by its truncation to an interval $[-N, N]$. Since the truncated r.v. is in L^2, the proof would stand for the truncated X_0, and then we would go to the limit as $N \to +\infty$.

The restriction to growth and the Lipschitz condition for $f(t,x)$ and $g(t,x)$ do not always need to hold for all points $x \in \mathbb{R}$, or $x \in \mathbb{R}$ and $y \in \mathbb{R}$. In the case where the solution $X(t)$ of the SDE has values that always belong to a set $D \subset \mathbb{R}$, then it is sufficient that the restriction on growth and the Lipschitz conditions are valid on D. In fact, in that case nothing changes if we replace f and g by other functions that coincide with them on D and have zero values out of D, and these other functions satisfy the restriction on growth and the Lipschitz condition.

For existence and uniqueness, we could use a weaker *local Lipschitz condition* for f and g instead of the global Lipschitz condition we have assumed in Theorem 7.1.

A function $h(s, x)$ with domain $[0, T] \times \mathbb{R}$ satisfies a *local Lipschitz condition* if, for any $N > 0$, there is a $K_N > 0$ such that, for all $t \in [0, T]$ and $|x| \leq N$, $|y| \leq N$, we have

$$|h(t, x) - h(t, y)| \leq K_N |x - y|. \tag{7.19}$$

The proof of existence uses a truncation of $X(s)$ to the interval $[-N, N]$ and ends by taking limits as $N \to +\infty$.

Consider f and g fixed. We can say that the SDE $dX(t) = f dt + g dW(t)$, $X(0) = X_0$, or the corresponding stochastic integral equation $X(t) = X_0 + \int_0^t f(s, X(s)) ds + \int_0^t g(s, X(s)) dW(s)$, is a map (or transformation) that, given a r.v. X_0 and a Wiener process $\{W(t)\}_{t \in [0,T]}$ (on a certain given probability space endowed with a non-anticipative filtration $\mathcal{A}_s$), transforms them into the a.s. unique solution $\{X(t)\}_{t \in [0,T]}$ of the SDE, which is adapted to the filtration. If we choose a different Wiener process, the solution changes. This type of solution, which is the one we have studied so far, is the one that is understood by default (i.e. if nothing in contrary is said). It is called a *strong solution*. Of course, once the Wiener process is chosen, the solution is unique a.s. and to each $\omega \in \Omega$ there corresponds the value of $X_0(\omega)$ and the whole trajectory $\{W(t, \omega)\}_{t \in [0,T]}$ of the Wiener process, which the SDE (or the corresponding stochastic integral equation) transforms into the trajectory $\{X(t, \omega)\}_{t \in [0,T]}$ of the SDE solution. In summary, X_0 and the Wiener process are both given and we seek the associated unique solution of the SDE.

There are also *weak solutions*. Again, consider f and g fixed. Now, however, only X_0 is given, not the Wiener process. What we seek now is to find a probability space and a pair of processes $\{\overline{X}(t)\}_{t \in [0,T]}$ and $\{\overline{W}(t)\}_{t \in [0,T]}$ on that space such that $\overline{X}(t) = X_0 + \int_0^t f(s, \overline{X}(s)) ds + \int_0^t g(s, \overline{X}(s)) d\overline{W}(s)$; of course, these are not unique. The difference is that, in the strong solution, the Wiener process is given *a priori* and chosen freely, the solution being dependent on the chosen Wiener process, while in the weak solution, the Wiener process is obtained *a posteriori* and is part of the solution.[5] Of course, a strong solution is also a weak solution, but the reverse may fail. One can see a counter-example in Øksendal (2003), in which the SDE has no strong solutions but does have weak solutions.

The uniqueness considered in Theorem 7.1 is the so-called *strong uniqueness*, i.e. given two solutions, their sample paths coincide for all $t \in [0, T]$ with probability one. We also have *weak uniqueness*, which means that, given two

---

5 Of course, one has to be careful to consider a filtration $\overline{\mathcal{A}}_s$ to which $\overline{X}(t)$ is adapted and for which the Wiener process $\overline{W}(t)$ is an $\overline{\mathcal{A}}_t$-martingale. It is possible to extend the definition of stochastic integral $\int_0^t g(s, \overline{X}(s)) d\overline{W}(s)$ to this situation, even when $\overline{X}(t)$ is not adapted to the filtration generated by the Wiener process and the initial condition.

solutions (no matter if they are weak or strong solutions), they have the same finite-dimensional distributions. Of course, strong uniqueness implies weak uniqueness, but the reverse may fail. Under the conditions assumed in the existence and uniqueness Theorem 7.1, two weak or strong solutions are weakly unique. In fact, given two solutions, either strong or weak, they are also weak solutions. Let them be $(\overline{X}(t), \overline{W}(t))$ and $(\overline{X}^*(t), \overline{W}^*(t))$ (remember that a weak solution is a pair of the 'solution itself' and a Wiener process). Then, since by the theorem there are strong solutions, let $X(t)$ and $Y(t)$ be the strong solutions corresponding to the Wiener process choices $\overline{W}(t)$ and $\overline{W}^*(t)$. By Picard's method of successive approximations given by (7.7)–(7.8), the approximating sequences have the same finite-dimensional distributions, so the same happens to their a.s. limits $X(t)$ and $Y(t)$.

It is important to stress that, under the conditions assumed in the existence and uniqueness Theorem 7.1, from the probabilistic point of view, i.e. from the point of view of finite-dimensional distributions, there is no difference between weak and strong solutions nor between the different possible weak solutions. This may be convenient since, to determine the probabilistic properties of the strong solution, we may work with weak solutions and get the same results.

Sometimes the Lipschitz condition or the restriction on growth are not valid and so the existence of strong solution is not guaranteed. In that case, one can see if there are weak solutions. Conditions for the existence of weak solutions can be seen in Stroock and Varadhan (2006) and Karatzas and Shreve (1991). Another interesting result is (see Karatzas and Shreve (1991)) that, if there is a weak solution and strong uniqueness holds, then there is a strong solution.

We have seen that, when f and g, besides satisfying the other assumptions of the existence and uniqueness Theorem 7.1, were also continuous functions of time, the solution of the SDE was a diffusion process with drift coefficient $f(t, x)$ and diffusion coefficient $g^2(t, x)$.

The reciprocal problem is also interesting. Given a diffusion process $X(t)$ ($t \in [0, T]$) in a complete probability space $(\Omega, \mathcal{F}, P)$ with drift coefficient $a(s, x)$ and diffusion coefficient $b(s, x)$, is there an SDE which has such a process as a weak solution? Under appropriate regularity conditions, the answer is positive. Some results on this issue can be seen in Stroock and Varadhan (2006), Karatzas and Shreve (1991), and Gihman and Skorohod (1972). So, in a way, there is a correspondence between solutions of SDE and diffusion processes.

In the particular case that f and g do not depend on time (and satisfy the assumptions of the existence and uniqueness Theorem 7.1), they are automatically continuous functions of time and therefore the solution of the SDE will be a homogeneous diffusion process with drift coefficient $f(x)$ and diffusion coefficient $g^2(x)$. In this case, the SDE is an *autonomous stochastic differential equation* and its solution is also called an *Itô diffusion*. In this case, one does not need to verify the restriction on growth assumption since this is a direct consequence of the Lipschitz condition (this latter condition, of course, needs to be checked). In the autonomous case, one can work in the interval $t \in [0, +\infty[$ since f and g do not depend on time.

In the autonomous case, under the assumptions of the existence and uniqueness Theorem 7.1, one can even conclude that the solution of the SDE is a *strong Markov process* (see, for instance, Gihman and Skorohod (1972) or Øksendal (2003)).

In the autonomous case, if we are working in one dimension (so this is not generalizable to multidimensional SDEs), we can get results even when the Lipschitz condition fails but $f(x)$ and $g(x)$ are continuously differentiable (i.e. are of class C^1). Note that, if they are of class C^1, they may or may not verify a Lipschitz condition; if they have bounded derivatives, they satisfy a Lipschitz condition, but this is not the case if the derivatives are unbounded.

The result by McKean (1969) for autonomous one-dimensional SDEs is that, if f and g are of class C^1, then there is a unique (strong) solution up to a possible explosion time $T_\infty(\omega)$. By explosion, we mean the solution becoming ∞. If f and g also satisfy a Lipschitz condition, that is sufficient to prevent explosions (i.e. one has $T_\infty = +\infty$ a.s.) and the solution exists for all times and is unique. If f and g are of class C^1 but fail to satisfy a Lipschitz condition, one cannot exclude the possibility of an explosion, but there are cases in which one can show that an explosion is not possible (or, more precisely, has a zero probability of occurring) and therefore the solution exists for all times and is unique. We will see later some examples of such cases and of the simple methods used to show that in such cases there is no explosion. There is also a test to determine whether an explosion will or will not occur, the *Feller test* (see McKean (1969)).

Let us now consider *multidimensional stochastic differential equations*. The setting is similar to that of the multidimensional Itô processes of Section 6.7. The difference is that now one uses $\mathbf{G}(s, \omega) = \mathbf{g}(s, \mathbf{X}(s, \omega))$ and $\mathbf{F}(s, \omega) = \mathbf{f}(s, \mathbf{X}(s, \omega))$. The Lipschitz condition and the restriction on growth of $\mathbf{f}$ and $\mathbf{g}$ in the existence and uniqueness Theorem 7.1 are identical with $|\mathbf{x} - \mathbf{y}|$ representing the euclidean distance. The results of Theorem 7.1 remain valid. When the solution is a diffusion process, the drift coefficient is the vector $\mathbf{f}$ and the diffusion coefficient is the matrix $\mathbf{g}\mathbf{g}^T$.

Notice, however, that, given a multidimensional diffusion process with drift coefficient $\mathbf{a}$ and diffusion coefficient $\mathbf{b}$, there are several matrices $\mathbf{g}$ for which $\mathbf{g}\mathbf{g}^T = \mathbf{b}$, but all of them result in the same probabilistic properties for the solution of the SDE. So, from the point of view of probabilistic properties or of weak solutions, it is irrelevant which $\mathbf{g}$ one chooses.

The treatment in this chapter allows the inclusion of SDE of the type

$$dY(t) = f(t, Y(t), W(t))dt + g(t, Y(t), W(t))dW(t),$$

by adding the equation $dW(t) = dW(t)$ and working in two dimensions with the vector $\mathbf{X}(t) = [Y(t), W(t)]^T$ and the SDE

$$dX(t) = \begin{bmatrix} f \\ 0 \end{bmatrix} dt + \begin{bmatrix} g \\ 1 \end{bmatrix} dW(t).$$

The initial condition takes the form $\mathbf{X}(0) = [Y_0, 0]^T$. Note that $f(t, Y(t), W(t)) = f(t, \mathbf{X}(t))$ and $g(t, Y(t), W(t)) = g(t, \mathbf{X}(t))$.

We can also have *higher order stochastic differential equations* of the form (here the superscript (n) represents the derivative of order n)

$$Y^{(n)}(t) = f(t, Y(t), ..., Y^{(n-1)}(t)) + g(t, Y(t), ..., Y^{(n-1)}(t))\varepsilon(t)$$

with initial conditions $Y^{(i)}(0) = a_i$ ($i = 0, 1, ..., n - 1$). For that, we use the same technique that is used for higher order ODEs, working with the vector

$$\mathbf{X}(t) = [Y(t), Y^{(1)}(t), ..., Y^{(n-1)}(t)]^T$$

and the SDE

$$dX(t) = \begin{bmatrix} Y^{(1)}(t) \\ Y^{(2)}(t) \\ ... \\ f(t, \mathbf{X}(t)) \end{bmatrix} dt + \begin{bmatrix} 0 \\ 0 \\ 0 \\ g(t, \mathbf{X}(t)) \end{bmatrix} dW(t).$$

We can also have functions f and g that depend on chance ω in a more general direct way (thus far they did not depend directly on chance, but they did so indirectly through $X(s)$). This poses no problem as long as F and G are kept non-anticipative.

8

Study of geometric Brownian motion (the stochastic Malthusian model or Black–Scholes model)

8.1 Study using Itô calculus

The *Black–Scholes model* or *stochastic Malthusian model*, which was introduced in Chapter 3, can, among other applications, be used to model the price of a stock in the stock market or the growth of a population of living beings with abundant resources under *environmental stochasticity*. By *environmental stochasticity* it is meant that the population lives in an environment subject to random fluctuations that affect the *per capita* growth rate. You should be aware that this model is not appropriate to deal with *demographic stochasticity*, i.e. the random sampling variations in the number of births and deaths that occur even when the environment is steady and therefore birth and death rates are not influenced by environmental conditions. The model is described by the SDE

$$dX(t) = RX(t)dt + \sigma X(t)dW(t), \qquad X(0) = x_0 \tag{8.1}$$

or by the corresponding integral form

$$X(t) = x_0 + \int_0^t RX(s)ds + \int_0^t \sigma X(s)dW(s). \tag{8.2}$$

Here, R represents the 'average' return or growth rate. It is always a per unit (*per capita* in populations, per unit of capital in stocks) rate of growth, but we will simply call it average return rate in case of stocks and average growth rate in case of populations. As for σ, it measures the strength of the effect on the return/growth rate of environmental fluctuations in the market/environment. It gives you an idea of the standard deviation of fluctuations about the average value of the rates and is known in the financial literature by *volatility*. We will assume that the initial value x_0 is known, so it can be treated as deterministic, and that $x_0 > 0$. Please keep in mind that $W(t) = W(t, \omega)$ and $X(t) = X(t, \omega)$ depend on chance ω (market scenario or state of nature in the examples given) even when we do not mention that dependence explicitly.

Introduction to Stochastic Differential Equations with Applications to Modelling in Biology and Finance, First Edition. Carlos A. Braumann.
© 2019 John Wiley & Sons Ltd. Published 2019 by John Wiley & Sons Ltd.
Companion Website: www.wiley.com/go/braumann/stochastic-differential-equations

In this section we will adopt the Itô calculus, as is the default choice when nothing is said in contrary. So, the second integral in (8.2) is to be interpreted as the *Itô integral*.

Let us look for the solution of the SDE (8.1), which is the solution of its integral form (8.2).

Here, $f(t, x) = f(x) = Rx$ and $g(t, x) = g(x) = \sigma x$. We have an autonomous SDE. Therefore, since f and g satisfy the Lipschitz condition (no need to check it directly, since this results from the fact that they have continuous bounded derivatives), they automatically satisfy the restriction on growth condition and we can apply the existence and uniqueness Theorem 7.1. So, there is an a.s. unique continuous solution $X(t)$, which is a Markov process and even an Itô diffusion, i.e. a homogeneous diffusion process. The drift coefficient is $a(x) = f(x) = Rx$ and the diffusion coefficient is $b(x) = g^2(x) = \sigma^2 x^2$.

Since the phenomenon is of a multiplicative nature, it can be useful to work in logarithmic scale, using the new variable

$$Z(t) = \ln(X(t)/x_0). \tag{8.3}$$

In order to explore various concepts and techniques, we are going to obtain the solution of the SDE by three different methods.

Solution via use of the Itô formula

Let $h(t, x) = \ln(x/x_0)$, so that $Z(t) = h(t, X(t))$. Applying the *Itô formula* (6.36) or the mnemonic in Observation 6.3, we obtain

$$dZ(t) = rdt + \sigma dW(t), \quad Z(0) = 0, \tag{8.4}$$

with $r = R - \sigma^2/2$. In the integral form, we have $Z(t) = \int_0^t rds + \int_0^t \sigma dW(s)$. Using (6.13), $\int_0^t \sigma dW(t) = \sigma(W(t) - W(0)) = \sigma W(t)$, since the integrand is a step function with a single step (see also Exercise 6.5b). So, the desired solution is

$$Z(t) = rt + \sigma W(t). \tag{8.5}$$

If we prefer to write the result in the original scale, it will be

$$X(t) = x_0 \exp(rt + \sigma W(t)),$$

which is known as *geometric Brownian notion*.

A brief study of the solution

We see that

$$Z(t) \frown \mathcal{N}(rt, \sigma^2 t). \tag{8.6}$$

Notice that the mean (expected value) rt of $Z(t)$ differs from the solution of the deterministic model $dX = RXdt$ (the model corresponding to $\sigma = 0$), which is $Z(t) = Rt$.

We conclude that $X(t)$ has a lognormal distribution with expectation[1]

$$\mathbb{E}[X(t)] = x_0 \exp(rt + \sigma^2 t/2) = x_0 \exp(Rt),\tag{8.7}$$

which coincides with the deterministic solution $X(t) = x_0 \exp(Rt)$.

Since (see Section 4.3) $W(t)/t \to 0$ a.s. as $t \to +\infty$, we have asymptotically $Z(t) = t(r + \sigma W(t)/t) \sim rt$ a.s. Therefore, as $t \to +\infty$, we get almost surely

$$X(t) \to \begin{cases} +\infty & \text{if } r > 0, \text{ i.e. if } R > \sigma^2/2 \\ 0 & \text{if } r < 0, \text{ i.e. if } R < \sigma^2/2. \end{cases}\tag{8.8}$$

This behaviour apparently differs from the deterministic model, for which $R > 0$ or $R < 0$ determines a $+\infty$ limit or a zero limit, respectively. This different qualitative behaviour allows, in the population growth case, that the population will, with probability one, become extinct even when the 'average' growth rate R has a positive value, as long as such value is smaller than $\sigma^2/2$.

From the normal distribution of $Z(t)$, one easily obtains the pdf $p(t, y) = f_{X(t)}(y)$ of $X(t)$:

$$p(t, y) = \frac{1}{y\sqrt{2\pi\sigma^2 t}} \exp\left(-\frac{(\ln(y/x_0) - rt)^2}{2\sigma^2 t}\right) \qquad (y > 0).\tag{8.9}$$

For $s > t$, we can obtain the distribution of $X(t)$ conditioned on the value of $X(s)$. Since $X(t) = X(s) + \int_s^t RX(u)du + \int_s^t \sigma X(u)dW(u)$ or, if one prefers to work in log scale, $Z(t) = Z(s) + \int_s^t rdu + \int_s^t \sigma dW(u)$, we get

$$(Z(t) \mid Z(s) = c) \frown \mathcal{N}(c + r(t - s), \sigma^2(t - s)).\tag{8.10}$$

So, with $c = \ln(x/x_0)$, the transition pdf for $X(t)$ is, for x, $y > 0$ and $t > s$,

$$p(t - s, y|x) = \frac{1}{y\sqrt{2\pi\sigma^2(t - s)}} \exp\left(-\frac{(\ln(y/x) - r(t - s))^2}{2\sigma^2(t - s)}\right).\tag{8.11}$$

This is the pdf $f_{X(t)|X(s)=x}(y)$ of $X(t)$ conditioned on having $X(s) = x$.

Solution via Kolmogorov equations

The expression (8.11) can also be obtained resolving directly the *forward Kolmogorov equation* (we put $\tau = t - s$)

$$\frac{\partial p(\tau, y|x)}{\partial \tau} + \frac{\partial}{\partial y}(a(y)p(\tau, y|x)) - \frac{1}{2}\frac{\partial^2}{\partial y^2}(b(y)p(\tau, y|x)) = 0$$

with the condition $\lim_{\tau \downarrow 0} p(\tau, y|x) = \delta(y - x)$ and with $a(y) = Ry$ and $b(y) = \sigma^2 y^2$. To solve this equation, it helps to perform the change of variable $z = \ln(y/x_0)$.

---

1 Remember that, for a Gaussian random variable Z, we have $\mathbb{E}[e^Z] = \exp\left(\mathbb{E}[Z] + \frac{1}{2}VAR[Z]\right)$.

The expression (8.9) can also be obtained by solving directly the forward Kolmogorov equation

$$\frac{\partial p(t, y)}{\partial t} + \frac{\partial}{\partial y}(a(y)p(t, y)) - \frac{1}{2}\frac{\partial^2}{\partial y^2}(b(y)p(t, y)) = 0$$

with the condition $\lim_{\tau \downarrow 0} p(\tau, y) = \delta(y - x_0)$ and with $a(y) = Ry$ and $b(y) = \sigma^2 y^2$. This is, however, unnecessary since we had already obtained (8.11) and, since our initial condition is deterministic ($X_0 = x_0$), we just need to attend to the fact that $p(t, y) = p(t, y|x_0)$.

Note that the Kolmogorov equations (either the forward or the backward equation) allow us to determine the transition densities. Since in this case of a deterministic known initial condition $X(0) = x_0$ we have the initial (degenerate) probability distribution, we can determine the finite-dimensional distributions of the Markov process $X(t)$ and this fully characterizes $X(t)$ from the probabilistic point of view; from here we can easily find a weak solution of the SDE. This method, besides being more complicated, does not provide us with a strong solution and so, contrary to the method used initially in this section, does not allow us to express the solution $X(t)$ as a function of the initial condition and of the concrete trajectory of the Wiener process $W(t)$. However, in most cases the trajectory of $W(t)$ is not known and our only interest is to compute the probabilities of certain events concerning the process $X(t)$; if so, then the knowledge of the finite-dimensional distributions is sufficient.

Solution via conversion to a Stratonovich equation

Another way of solving the Itô SDE (8.1) is to convert it to an equivalent Stratonovich SDE, which has the same solution, by using the equivalence of (6.46) to (6.47). The latter is to be solved using Stratonovich calculus, i.e. using the ordinary calculus rules with which we are well acquainted. So, the Itô SDE (8.1) is equivalent (has the same solution) to the Stratonovich SDE

$$\begin{aligned} (S) \, dX(t) &= \left(RX(t) - \tfrac{1}{2}\sigma\sigma X(t)\right) dt + \sigma X(t)dW(t) \\ &= rX(t)dt + \sigma X(t)dW(t), \quad X(0) = x_0. \end{aligned} \tag{8.12}$$

Using the Stratonovich rules of calculus (i.e. the usual rules) to differentiate a composite function, one gets from (8.12)

$$(S) \, dZ(t) = \frac{1}{X(t)} dX(t) = rdt + \sigma dW(t),$$

which takes the integral form $Z(t) = \int_0^t rdt + (S)\int_0^t \sigma dW(t)$. Since, in this Stratonovich integral, the integrand does not depend on x, the integral, as seen in Section 6.6, does not differ from the Itô integral $\int_0^t \sigma dW(t) = \sigma W(t)$.

We arrive at $Z(t) = rt + \sigma W(t)$. So, we end up having the same result $Z(t) = rt + \sigma W(t)$ as solving directly the Itô SDE (8.1).

We shall stress that we have used Stratonovich calculus just as an auxiliary, since the purpose was to solve the original SDE (8.1) according to Itô calculus. To do this, we took advantage of this being equivalent to solving the modified equation (8.12) using Stratonovich calculus.

Application and discussion

Whatever the method used, the solution of the Black–Scholes model is

$$X(t) = x_0 \exp(rt + \sigma W(t)), \quad \text{with } r = R - \sigma^2/2.$$

If we have observed data values $X_k := X(t_k)$ $(k = 0, 1, 2, \ldots, n)$ at times $0 = t_0 < t_1 < t_2 \cdots < t_n$, putting $\Delta_k = t_k - t_{k-1}$ the *log-returns* are

$$L_k := \ln \frac{X_k}{X_{k-1}} = r\Delta_k + \sigma(W(t_k) - W(t_{k-1}) \frown \mathcal{N}(r\Delta_k, \sigma^2\Delta_k)$$

We invite the reader to look at Figure 2.3, which shows in logarithmic scale the real trajectory (corresponding to the market scenario that has effectively occurred) of the price $X(t)$ of a stock (Renault) at the Euronext Paris during the period of 2 January 2012 ($t = 0$ with a stock price $X_0 = x_0 = €27.595$) to 31 December 2015 (we used daily closing prices).

We estimated the parameters of model (8.1) based on these observations using the *maximum likelihood method*, as described in detail at the end of this section. We then obtained the following estimates: $\hat{r} = 0.303/\text{year}$ (with $r = R - \sigma^2/2$) and $\hat{\sigma}^2 = 0.149/\text{year}$ (i.e. $\hat{\sigma} = 0.387/\sqrt{\text{year}}$).

We now invite the reader to look at Figure 2.4, which shows, also in logarithmic scale, two simulated trajectories of geometric Brownian motion using as parameters the values that have been estimated for the Renault stock, i.e. using $r = R - \sigma^2/2 = 0.303/\text{year}$ and $\sigma = 0.387/\sqrt{\text{year}}$ (i.e. $\sigma^2 = 0.149/\text{year}$). Neither of these two trajectories corresponds to the market scenario that has effectively occurred, but, assuming the model is correct and the estimated values of the parameters were exactly estimated (not quite true since there is always an estimation error), these trajectories correspond to two other values of ω, i.e. to two other simulated theoretically possible market scenarios that might have occurred.

In this case, the simulations were performed exactly since we have an explicit expression for the solution $X(t) = x_0 \exp(rt + \sigma W(t))$ in terms of the Wiener process and this process appears in that expression only at the present time t. So, we can just simulate trajectories of the Wiener process $W(t)$ by the technique used in Section 4.1 and plug the Wiener process values into the expression of $X(t)$ to obtain trajectories of our process.

Of course, we do not always have such a favourable situation. We refer to Sections 7.1 and 12.2 for Monte Carlo simulation of solutions of SDE under less favourable conditions.

The reader should know that Figure 2.3 refers to the real data and Figure 2.4 to two simulated trajectories, but, if the reader did not know this it would be hard to guess which of the trajectories is real and which are simulated. This indicates that the Black–Scholes model does give a reasonable representation of the behaviour of stock prices. Deeper studies indicate that the behaviour of stock prices seems to differ, in some finer aspects, from the behaviour of geometric Brownian motion. For example, while for geometric Brownian motion the *log-returns* $L_k = \ln(X(t_k)/X(t_{k-1}))$ are independent Gaussian random variables with mean $r(t_k - t_{k-1})$ and variance $\sigma^2(t_k - t_{k-1})$, the real log-returns seem to have a slightly asymmetric distribution with heavier tails than the normal distributions, suggesting the use of a Lévy process instead of a Wiener process. Some improved models have even been proposed in which the volatility σ, instead of being a deterministic and constant value, can vary randomly with dynamics governed by another stochastic differential equation; these are called stochastic volatility models. Still, the Black–Scholes model seems to be a reasonable approximation and is widely used (it is indeed the reference model in the markets).

In Figure 8.1 one can see the effect of the volatility parameter σ. We have used the parameters of the Renault stock estimated from the real data and simulated one trajectory (corresponding to one simulated trajectory of the Wiener process $W(t)$, i.e. corresponding to a given simulated market scenario ω). That gives the black jumpier trajectory. Using exactly the same market scenario ω, i.e. the same values of the Wiener trajectory $W(t)$, and using the same values of x_0 and R, we have obtained the much less jumpier trajectory in grey. The

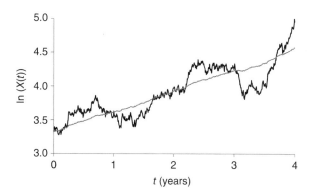

Figure 8.1 Simulated trajectory of the logarithm of geometric Brownian motion $X(t)$, with $X(0) = 27.595$, $r = R - \sigma^2/2 = 0.303$/year and with $\sigma = 0.387/\sqrt{\text{year}}$ ('jumpier' dark curve), $\sigma^2 = 0.149$/year. The trajectory was simulated but the parameters were estimated using prices (in euros) of a real stock (Renault) from 2 January 2012 ($t = 0$) to 31 December 2015 at Euronext Paris. The lighter less 'jumpy' curve uses exactly the same ω, i.e. the same trajectory of the perturbing noise, and the same parameters with the exception of σ, which is now $\sigma = 0.0387/\sqrt{\text{year}}$, ten times smaller.

difference is that we have used a value of the volatility σ ten times smaller that the value estimated from observations. Of course, if we keep on reducing the volatility until it reaches zero, we would have the deterministic model solution and the figure would show a perfect straight line deterministic trajectory for the $\ln X(t)$.

Exercise 8.1 *Let $X(t)$ be the value of a stock at time $t \geq 0$ and assume it satisfies the Black-Scholes model (8.1) with $X(0) = €27.595$, $R = 0.378/year$ and $\sigma^2 = 0.149/year$. Determine:*

(a) $\mathbb{E}[X(1\ year)]$
(b) $VAR\,[X(1\ year)]$
    Suggestion: $X^2(t) = x_0^2\ e^{2Z(t)}$ and, since $2Z(t)$ is Gaussian, $\mathbb{E}[e^{2Z(t)}] = \exp\left(\mathbb{E}[2Z(t)] + \frac{1}{2} VAR\,[2Z(t)]\right).$
(c) $P[€30 \leq X(1\ year) \leq €45]$
(d) $\mathbb{E}[X(1.5\ years) \mid X(1\ year) = €40.200]$
(e) $VAR\,[X(1.5\ years) \mid X(1\ year) = €40.200]$
(f) $\mathbb{E}[X(1.5\ years) \mid X(0.5\ years) = €30.275\ and\ X(1\ year) = €40.200]$
(g) $P[X(1.5\ years) > €50 \mid X(1\ year) = €40.200]$
(h) $P[X(1.5\ years) > €50 \mid X(0.5\ years) = €30.275\ and\ X(1\ year) = €40.200].$

Parameter estimation

Let us see how can one perform parameter estimation in this case. Assume we have observed data values $X_k := X(t_k)$ ($k = 0, 1, 2, \ldots, n$) of the stock prices at times $0 = t_0 < t_1 < t_2 \cdots < t_n$, with $X(0) = x_0$, and put $\Delta_k = t_k - t_{k-1}$.

In this case we have an explicit solution and know the transition probabilities, so parameter estimation can be performed by the method of maximum likelihood, using the Markov property of the solution. In fact, the transition distribution has in this case a known pdf $p(t_k, y|t_{k-1}, x)$, and so the joint pdf of the observations $X_1, X_2, \ldots, X_n$, which depends on the parameters r and σ^2, is just the product $\prod_{k=1}^{n} p(t_k, X_k|t_{k-1}, X_{k-1})$. This, considered as a function of the parameters, is called the *likelihood function* and its logarithm is the *log-likelihood* function. Maximizing it w.r.t. the parameters gives the maximum likelihood estimators. Under certain conditions, one knows that asymptotically (as $n \to +\infty$), the maximum likelihood estimators are unbiased and normally distributed with a variance-covariance matrix equal to the inverse of the Fisher information matrix. This can be used to obtain approximate confidence intervals for the parameters.

In the special case of the Black–Scholes model, since the information of the log-returns is equivalent to the sample data information, an equivalent and simpler procedure is to maximize instead the log-likelihood of the log-returns. This is very easy to obtain due to the independence of the log-returns in the Black–Scholes model. The likelihood of the log-returns $L_k := \ln \frac{X_k}{X_{k-1}} \frown \mathcal{N}(r\Delta_k, \sigma^2\Delta_k)$ $(k = 1, 2, \ldots, n)$ is then just the product of their Gaussian pdfs. So, the log-likelihood function of the log-returns is

$$L(r, \sigma^2) = \sum_{k=1}^{n} \ln\left(\frac{1}{\sqrt{2\pi\sigma^2\Delta_k}} \exp\left(-\frac{(L_k - r\Delta_k)^2}{2\sigma^2\Delta_k}\right)\right)$$

$$= -\frac{n}{2}\ln(2\pi) - \frac{n}{2}\ln\sigma^2 - \frac{1}{2}\sum_{k=1}^{n}\ln(\Delta_k) - \frac{1}{2\sigma^2}\sum_{k=1}^{n}\frac{(L_k - r\Delta_k)^2}{\Delta_k}.$$

If we have equidistant times of observation, then $\Delta_k \equiv \Delta$ and $t_k = k\Delta$, and this expression simplifies to:

$$L(r, \sigma^2) = -\frac{n}{2}\ln(2\pi) - \frac{n}{2}\ln\sigma^2 - \frac{n}{2}\ln\Delta - \frac{1}{2\sigma^2\Delta}\sum_{k=1}^{n}(L_k - r\Delta)^2.$$

Maximizing it leads to the maximum likelihood estimators:

$$\hat{r} = \frac{1}{n\Delta}\sum_{k=1}^{n}L_k = \frac{\ln(X_n/x_0)}{n\Delta} \quad \text{and} \quad \hat{\sigma}^2 = \frac{1}{n\Delta}\sum_{k=1}^{n}(L_k - \hat{r}\Delta)^2.$$

We have used this simplification in estimating the parameters of the Renault stock. In fact, we have daily observations, but only on weekdays. Since on weekends and holidays there are no transactions, we decided to ignore such days and treat weekdays as being consecutive. Then, with N being the (average) number of weekdays in a year (around 250), $\Delta = 1/N$. With this simplification, we have obtained $\hat{r} = 0.303/\text{year}$ and $\hat{\sigma}^2 = 0.149/\text{year}$. The *Fisher information matrix* has, with $V = \sigma^2$, entries $F_{11} = \mathbb{E}\left[-\frac{\partial^2 L(r,V)}{\partial r^2}\right] = \frac{n\Delta}{V}$, $F_{12} = F_{12} = \mathbb{E}\left[-\frac{\partial^2 L(r,V)}{\partial r\partial V}\right] = 0$, and $F_{22} = \mathbb{E}\left[-\frac{\partial^2 L(r,V)}{\partial V^2}\right] = \frac{n}{2V^2}$. In some cases (not here), the expectations are difficult to obtain, and one can replace them by the sample values as an approximation. The inverse matrix is diagonal with entries $\frac{\sigma^2}{n\Delta}$ and $\frac{2\sigma^4}{n}$. So, approximate 95% confidence intervals for r and σ^2 are, respectively, $\hat{r} \pm 1.96\sqrt{\sigma^2/(n\Delta)} \simeq 0.303 \pm 0.379/\text{year}$ and $\hat{\sigma}^2 \pm 1.96\sqrt{2\sigma^4/n} \simeq 0.149 \pm 0.013/\text{year}$ (we have approximated σ^2 by $\hat{\sigma}^2$ to compute these expressions). Notice the low degree of precision in estimating the return rate; in fact, its precision does not depend on the number n of observed data values, but on the time interval $n\Delta = t_n$ between the first and last observations, which in this case is only 4 years. The estimate of the volatility σ depends mostly on the number of observed data values, which in this case is around a thousand, so we have good precision.

Prediction

Having observed the data of the 4-year period ending on 31 December 2015, we may wish at this last day of 2015 to predict the then unknown stock price at 31 March 2016, i.e. $\tau = 0.25$ years later. The last known stock price is precisely $X_n = X(t_n) = €92.63$, where t_n is 31 December 2015. Due to the Markov property, conditional on knowing that $X(t_n) = €92.63$, past values are irrelevant for the probability distribution of $X(t_n + \tau)$, the value we want to predict. In this case, it is easier to work with logarithms, and the conditional distribution of $\ln X(t_n + \tau)$ is $\mathcal{N}(\ln X_n + r\tau, \sigma^2\tau)$. If we knew exactly the parameters r and σ, a *point prediction* for $\ln X(t_n + \tau)$ would be $\ln X_n + r\tau$ and a 95% prediction interval would be $\ln X_n + r\tau \pm 1.96\sigma\sqrt{\tau}$.

Since we do not know the parameters, the point prediction will use the maximum likelihood estimator of r, and will therefore be

$$\ln \hat{X}(t_n + \tau) = \ln X_n + \hat{r}\tau = \ln 92.63 + 0.303 \times 0.25 = 4.6044,$$

so that $\hat{X}(t_n + 0.25) = e^{4.6044} = €99.92$. We can compare this with the true value of the stock that was later observed at 31 March 2016, $X(t_n + 0.25) = €87.32$.

For a confidence interval, we have to look at the prediction error. Working with logarithms, which is more convenient due to their normal distribution, the prediction error is

$$\ln \hat{X}(t_n + \tau) - \ln X(t_n + \tau)$$
$$= (\ln X_n + \hat{r}\tau) - (\ln X_n + r\tau + \sigma(W(t_n + \tau) - W(t_n)))$$
$$= (\hat{r} - r)\tau - \sigma(W(t_n + \tau) - W(t_n)).$$

Since $\hat{r}$ uses only information up to time t_n, it is independent of $W(t_n + \tau) - W(t_n)$. Therefore, since $\hat{r} \frown \mathcal{N}(r, \frac{\sigma^2}{n\Delta})$ and $W(t_n + \tau) - W(t_n) \frown \mathcal{N}(0, \tau)$, the prediction error is $\frown \mathcal{N}(0, \frac{\sigma^2}{n\Delta}\tau^2 + \sigma^2\tau) = \mathcal{N}(0, \sigma^2\tau(1 + \tau/t_n))$. A 95% confidence prediction interval for $\ln X(t_n + \tau)$ would be $\ln \hat{X}(t_n + \tau) \pm 1.96\sqrt{\sigma^2\tau(1 + \tau/t_n)}$. Since we do not know the true value of σ, we can approximate it by the maximum likelihood estimator (which is quite precise), leading to the approximate 95% confidence prediction interval for $\ln X(t_n + \tau)$

$$\ln \hat{X}(t_n + \tau) \pm 1.96\sqrt{\hat{\sigma}^2\tau(1 + \tau/t_n)} = 4.6044 \pm 0.3899$$
$$= [4.2144, 4.9943].$$

Applying exponentials to the extremes, the corresponding prediction interval for $X(t_n + \tau)$ is approximately $[€67.66, €147.57]$.

Instead of predicting 3 months ahead, one can do step-by-step prediction, just predicting the next day at a time. Of course, in each day, the basis for the prediction of next day would be the current day stock price and the parameter estimates could also be updated every day. This way, the prediction errors would be smaller since $\tau = 1$ day (around 0.004 years).

For more information on estimation, testing and prediction, including comparison tests among average return/growth rates of different stocks or populations, for this type of Black–Scholes model, we refer to Braumann (1993) and Braumann (1999a).

Quite often, though, one does not have an explicit solution nor the transition densities for the likelihood function. Even in such cases, one would like to estimate parameters and make predictions. We will discuss these issues in Section 12.2.

8.2   Study using Stratonovich calculus

Here we are going to solve an SDE looking exactly like (8.1) but using Stratonovich calculus, i.e. our model is now the *Stratonovich stochastic differential equation*

$$(S)\, dX(t) = R_S X(t)dt + \sigma X(t)dW(t), \quad X(0) = x_0. \tag{8.13}$$

This equation is *not equivalent* to the Itô SDE (8.1). For that reason, we have used R_S instead of R to facilitate the distinction between this model and the Itô calculus model (8.1); as for σ, no distinction is necessary.

The solution of (8.13) can be easily obtained by using Stratonovich ordinary rules of calculus. So, for the change of variable (8.3), we now get

$$(S)\, dZ(t) = \frac{1}{X(t)} dX(t) = R_S dt + \sigma dW(t), \tag{8.14}$$

or, in the integral form, $Z(t) = \int_0^t R_S dt + (S) \int_0^t \sigma dW(t)$, from which we immediately get the desired solution

$$Z(t) = R_S t + \sigma W(t) \frown \mathcal{N}(R_S t, \sigma^2 t). \tag{8.15}$$

This looks different from the result (8.5)–(8.6) obtained under Itô calculus. Now, the expected value of $Z(t)$ coincides with the deterministic model solution, but the expected value of $X(t)$, $\mathbb{E}[X(t)] = x_0 \exp(R_S t + \sigma^2 t/2)$ does not coincide with the solution of the deterministic model, quite the opposite of what happened under Itô calculus.

Now we have $Z(t) = t(R_S + W(t)/t) \sim R_S t$ a.s. So, as $t \to +\infty$, we have a.s.

$$X(t) \to \begin{cases} +\infty & \text{if } R_S > 0 \\ 0 & \text{if } R_S < 0. \end{cases} \tag{8.16}$$

This behaviour agrees with the deterministic model, but disagrees with the behaviour of the Itô calculus model (8.1) used in Section 8.1. Now, in the application to population growth, whenever the 'average' growth rate R_S is positive, the population grows without bound and, when that rate is negative, the population, with probability one, becomes extinct.

These differences in the qualitative behaviour when one uses different stochastic calculi pose no mathematical problem, since the integrals and the SDE are differently defined and shown to follow different rules. So, it comes as no surprise that the results differ. The differences pose, however, a delicate problem in applications, particularly since this has an effect on issues as important as predicting whether a wildlife population will or will not become extinct, for example. The question in such applications is: Which calculus should one use? Which gives the correct qualitative behaviour and the correct predictions?

We will examine these modelling issues in Chapter 9.

9

The issue of the Itô and Stratonovich calculi

9.1 Controversy

For the Black–Scholes model, we have shown in Chapter 8 that, using Itô or Stratonovich calculus in the resolution of stochastic differential equations (SDEs) (i.e. using Itô or Stratonovich integrals in the integral form of such SDEs) usually leads to different results and the associated models (they are indeed different models) usually have qualitatively different behaviours. This is just an example of a much more general feature of SDEs, with the exception of SDEs having $g(s, x)$ functions not depending on x (i.e. $g(s, x) = g(s)$) since, as mentioned in Section 6.6, in this exceptional case the two calculi coincide.

In the example, when applied to population growth in a randomly varying environment (in which case the model is known as the stochastic Malthusian model), the two calculi give different behaviours with respect to population extinction. This also happens for other more realistic models of population growth, like the stochastic logistic model. There are situations where the Stratonovich calculus model predicts non-extinction of the population (and even the existence of a stationary density, which corresponds to a stochastic equilibrium), while the Itô calculus model predicts extinction with probability one. This fact produced some controversies in the literature on what calculus is more appropriate in the applications. There was even a theory put forward by May and MacArthur in 1972 on niche overlap, stating that there is a bound on the amount of niche overlap that can sustain coexistence of all the species in a community of competing species (see May and MacArthur (1972) and May (1973,1974)). Besides other issues, the theory was sustained on an Itô calculus model and would completely fail if one were to use instead the Stratonovich calculus model (see Feldman and Roughgarden (1975) and Turelli (1978)).

Coming back to the stochastic Malthusian model, the key parameter is R, interpreted as 'average' growth rate; to better distinguish between the Itô calculus model and the Stratonovich calculus model we have used R in the first

Introduction to Stochastic Differential Equations with Applications to Modelling in Biology and Finance,
First Edition. Carlos A. Braumann.
© 2019 John Wiley & Sons Ltd. Published 2019 by John Wiley & Sons Ltd.
Companion Website: www.wiley.com/go/braumann/stochastic-differential-equations

case and R_S in the second case. For the Itô calculus model, extinction occurs with probability one if that parameter is smaller than $\sigma^2/2$, even if it is positive. For the Stratonovich calculus model, extinction will have zero probability of occurring whenever the parameter is positive. The reader may imagine what would happen if a biologist were to ask the modeller the following question: 'I have a population living in a randomly varying environment and the stochastic Malthusian model seems to be appropriate. The average growth rate is positive but smaller than $\sigma^2/2$. Will the population go extinct or not?' A conscientious modeller might reply: 'The answer is YES if you use Itô calculus and is NO if you use Stratonovich calculus.' The same situation applies to the stochastic logistic model and other stochastic population models.

This indeed poses a problem on the practical use of SDEs and the credibility of model qualitative predictions. As mentioned in Section 6.6 and suggested by several textbooks, Itô calculus seems to be the better approximation when the phenomenon intrinsically occurs in discrete time with perturbations induced by a discrete-time white noise. If the phenomenon intrinsically occurs in continuous time with perturbations induced by a slightly coloured continuous-time noise, Stratonovich calculus seems to be the better approximation. But, in real life, it is quite difficult to distinguish between these two situations, as in the case of population growth. At first sight, if generations overlapped with reproduction continuously occurring over time, the indication could be to use Stratonovich calculus. If, however, there are short and well-defined reproductive seasons and non-overlapping generations, the indication could be to use a discrete time model and therefore use an Itô calculus SDE as a better approximation. These generic indications are for the most part useless and hide the real issue. One can argue (and indeed it has been argued) that one should use the calculus that better approaches a given real-life situation. We should remember that births and deaths are discrete events but may occur at any time in a continuous time interval and, even in the case where there is a reproductive season, births are spread along the season. Another example of use of the Black–Scholes is the price of a stock. Trading seems to occur in continuous time with trading orders processed by computers, but, of course, the trading computer works with machine cycles; the cycles are very short, giving the appearance of continuity, but actually there is no trading in between two computer cycles.

This problem of which calculus is the most appropriate in a given application has been much discussed and even been the object of controversy, but it is really just a semantic problem. Let us go back to the Black–Scholes/stochastic Malthusian growth models (8.1) or (8.13), according to whether one uses Itô calculus or Stratonovich calculus. Will the parameter R, which is interpreted as the 'average' return/growth rate, indeed represent the same 'average' rate? Or does it represent two different types of 'average'? We will see that this is indeed the case and the whole controversy on which calculus to use in applications is based on a semantic confusion. The problem lies in the literature implicitly assuming that the letter R used for both calculi meant the same thing in both, namely the 'average' return/growth rate. Implicitly, the presumption was made that the term 'average', without any specification of the type of average, was unequivocal and, therefore, the same 'average' would be used in both calculi. As we shall see, this is not true and is the source of the controversy.

9.2   Resolution of the controversy for the particular model

Let us take a closer look at the meaning of return rate (or growth rate) and of 'average' return rate (or growth rate). We will, to have a more concrete understanding, consider $X(t)$ to be the price of a stock (which is the capital of its owner) or the size of a population and assume they follow the Black–Scholes/stochastic Malthusian growth model.

For the deterministic model $dX(t)/dt = RX(t)$, if the stock price (the capital of the owner of the stock) at time t is x, the return rate is, by definition, the instantaneous rate of capital increase at time t per unit of capital. In the case of a population, the growth rate is similarly the instantaneous rate of growth *per capita*. So we have for the return/growth rate

$$\frac{1}{X(t)}\frac{dX(t)}{dt} = \frac{1}{x}\lim_{\Delta t \downarrow 0}\frac{1}{\Delta t}(X(t+\Delta t) - x),$$

and this is simply equal to R. So, in the deterministic case, the interpretation of the parameter is clear, it is the return/growth rate at time t when the stock price/population size at that time is x.

In the stochastic case, $X(t + \Delta t)$ is a r.v. and we recur to some average. But there are several types of averages. If we use the arithmetic average, we will obtain the *arithmetic average return/growth rate* at time t when the stock price/population size at that time is x. This is given by:[1]

$$R_a = \frac{1}{x}\lim_{\Delta t \downarrow 0}\frac{1}{\Delta t}(\mathbb{E}_{t,x}[X(t+\Delta t)] - x). \tag{9.1}$$

This average is equal to R if we use the Itô calculus model (8.1) and is equal to $R_S + \sigma^2/2$ if we use the Stratonovich calculus model (8.13).

Exercise 9.1   *Use the results obtained in Chapter 8 to prove this statement.*

---

1 Since we are dealing with autonomous equations, this rate does not depend on t, just on the value x that the process takes at time t. For this Black–Scholes/stochastic Malthusian growth model, which is very simple, it also does not depend on x, but the situation is different for other autonomous models (models where the return/growth rate in the deterministic model depends on x).

We can also use the *geometric average return/growth rate* at time t when the stock price/population size at that time is x.[2] This is given by:

$$R_g = \frac{1}{x} \lim_{\Delta t \downarrow 0} \frac{1}{\Delta t} (\exp(\mathbb{E}_{t,x}[\ln X(t + \Delta t)]) - x). \tag{9.2}$$

This average is equal to $r = R - \sigma^2/2$ if we use the Itô calculus model (8.1) and is equal to R_S if we use the Stratonovich calculus model (8.13).

Exercise 9.2 *Use the results obtained in Chapter 8 to prove this statement.*

So, when the authors used the same letter R in both calculi thinking they were referring to the same 'average' return/growth rate, they were mistaken. That is the reason why we have used R for the Itô calculus and R_S for the Stratonovich calculus, in order to avoid committing the same mistake. *As we have seen, in Itô calculus, R means the arithmetic average return/growth rate R_a.*[3] *In Stratonovich calculus, R, which we have denoted by R_S, means the geometric average growth rate R_g.*[4]

Therefore, if we specify (as we should) what is the average we are using, the two calculi produce exactly the same quantitative and qualitative results. Namely, in terms of the geometric average return/growth rate, we have for both calculi

$$
\begin{aligned}
Z(t) &= R_g t + \sigma W(t) \\
X(t) &= x_0 \exp(R_g t + \sigma W(t))
\end{aligned}
\tag{9.3}
$$

and so almost surely as $t \to +\infty$

$$X(t) \to \begin{cases} +\infty & \text{if } R_g > 0 \\ 0 & \text{if } R_g < 0. \end{cases} \tag{9.4}$$

Therefore, *for population growth and with probability one, there is extinction $(X(t) \to 0$ as $t \to +\infty)$ or growth without bound $(X(t) \to +\infty)$ according to whether the geometric average growth rate R_g is negative or positive, respectively.*

Since this a multiplicative phenomenon, it is natural that expressing results in terms of the geometric average makes them look qualitatively analogous to the deterministic model. The reader can see what would be the *same* results written in terms of the arithmetic average growth rate, which do not look so similar to the deterministic case. But that is just an aesthetic question. What is important to stress here is that the results are identical for the two calculi.

The apparent difference between the two calculi, which generated controversy in the literature over which would be the most appropriate calculus in different applications, was just the result of a semantic confusion. Once we

---

2 Remember that the geometric average is the exponential of the arithmetic average of the logarithms.

3 The geometric average growth rate would be given by $R_g = r = R - \sigma^2/2$.

4 The arithmetic average growth rate would be given by $R_a = R_S + \sigma^2/2$.

have eliminated the confusion, we see that the two calculi can be used indifferently, producing exactly the same results. We only need to use for the value of R the value of the appropriate average of the calculus we are using: the arithmetic average if we are using Itô calculus, the geometric average if we are using Stratonovich calculus. More details can be found in Braumann (2003). If you are using an estimate based on observations, be careful to use the appropriate sample average.[5]

Is this conclusion valid just for the Black–Scholes/stochastic Malthusian growth model? In Braumann (2007a) we have shown that the same result applies to autonomous SDE models of the form

$$dX(t) = R(X(t))X(t)dt + \sigma X(t)dW(t)$$

(with R satisfying some regularity conditions), where again the two calculi give exactly the same result if we use for $R(x)$ the expression of the arithmetic (for Itô calculus) or the geometric (for Stratonovich calculus) average return/growth rate at time t when the population size at that time is x. This is the correct interpretation of R under the corresponding calculus. The definitions are obviously identical to the ones given above by (9.1) and (9.2), except that now the averages $R_a(x)$ and $R_g(x)$ depend on x.

In the next section we will extend this type of result to a general autonomous SDE model (with some regularity conditions), showing that the same problem of a semantic confusion between two types of averages exists and, once eliminated, the two calculi give exactly the same results. We will, however, work not with the *per capita* growth rate but with the total return/growth rate because, in the general case, this is more convenient; the appropriate type of average for Itô calculus is still the arithmetic average, but for Stratonovich calculus the correct type of average depends on the expression of the diffusion coefficient.

9.3 Resolution of the controversy for general autonomous models

The issue raised in the previous section is not specific to the example discussed there. Considering the fact that the two calculi, Itô or Stratonovich, usually give apparently different results, even qualitative results that lead to different fates

---

5 For example, suppose you have data $X(t_k)$ $(k = 0, 1, \ldots, n)$ from a single sample-path at times $t_k = k\Delta$ $(k = 0, 1, \ldots, n)$. Consider the log-returns $L_k = \ln \frac{X(t_k)}{X(t_{k-1})}$ $(k = 1, 2, \ldots, n)$ and let $\bar{L} = \frac{1}{n} \sum_{k=1}^{n} L_k$. The geometric average return/growth rate, to be used for R_S if you choose Stratonovich calculus, can be estimated by its maximum likelihood estimator $\hat{R}_g = \frac{1}{\Delta}\bar{L}$. The arithmetic average growth rate, to be used for R if you choose Itô calculus, can be estimated by its maximum likelihood estimator $\hat{R}_a = \hat{R}_g + \hat{\sigma}^2/2$. Here, $\hat{\sigma}^2 = \frac{1}{\Delta}\frac{1}{n}\sum_{k=1}^{n}(L_k - \bar{L})^2$ is the maximum likelihood estimator of σ^2 for both calculi.

for the stochastic process under study (population size or some other process), the issue of which calculus is more appropriate for a given application can show up and lead to similar controversies. For quite general autonomous SDE models having a unique solution, we will show again, following Braumann (2007b), that we are faced with an apparent difference due to a semantic confusion that, once cleared, shows that the results of the two calculi are indeed completely coincidental.

Consider, for $t \geq 0$, an autonomous SDE model of the form

$$dX(t) = f(X(t))dt + g(X(t))dW(t) \tag{9.5}$$

describing the time evolution of a certain quantity. To exemplify, we will talk about the size of an animal population, but it could be any other variable and phenomenon. We assume that $f(x)$ is of class C^1 and $g(x)$ is of class C^2 and is positive for all interior points of the state space.

For convenience, we will now look at the total growth rate of the population (meaning the rate at which the population size is growing) instead of the *per capita* growth rate (the one we abbreviate to 'growth rate'), but, if we prefer to work with the latter, just divide the total growth rate by the population size.

In the applications, $f(x)$ is interpreted as the 'average' total growth rate of the population when its size is equal to x. Unless $g(x)$ is constant, the two calculi, Itô and Stratonovich, will give apparently different solutions and even apparently different qualitative behaviours concerning the fate of the population. Again one suspects that there is a semantic confusion of implicitly assuming it is the same type of average for both calculi, when indeed one should care not about the letters used but rather about their true meaning in terms of the evolution of population trajectories. Therefore, to avoid such semantic confusion upon which may rest a controversy in the literature, we will use f_I when using Itô calculus and f_S when using Stratonovich calculus. So we have

$$dX(t) = f_I(X(t))dt + g(X(t))dW(t) \tag{9.6}$$

$$(S) \quad dX(t) = f_S(X(t))dt + g(X(t))dW(t) \tag{9.7}$$

and assume that, in both cases, there exists a unique continuous solution for all $t \geq 0$. The diffusion and drift coefficients are now

$$b(x) = g^2(x) \tag{9.8}$$

$$a_I(x) = f_I(x) \tag{9.9}$$

$$a_S(x) = f_S(x) + \frac{1}{4}\frac{db(x)}{dx} = f_S(x) + \frac{1}{2}g(x)\frac{dg(x)}{dx}. \tag{9.10}$$

Of course, for the deterministic model $\frac{dX}{dt} = f(X)$, the total growth rate when the population size is x at time t is by definition

$$\lim_{\Delta t \downarrow 0} \frac{X(t + \Delta t) - x}{\Delta t} = f(x).$$

For the stochastic models, the arithmetic average total growth rate when the population size is x at time t is, by definition,

$$f_a(x) := \lim_{\Delta t \downarrow 0} \frac{\mathbb{E}_{t,x}[X(t + \Delta t)] - x}{\Delta t}, \tag{9.11}$$

which is exactly the drift coefficient (see (5.2)). So, for the Itô SDE (9.6), the arithmetic average total growth rate is

$$f_a(x) = a_I(x) = f_I(x) \tag{9.12}$$

and, for the Stratonovich SDE (9.7), the arithmetic average total growth rate is

$$f_a(x) = a_S(x) = f_S(x) + \frac{1}{2} g(x) \frac{dg(x)}{dx}. \tag{9.13}$$

In conclusion, the arithmetic average total growth rate is $f_I(x)$ for the Itô SDE but $f_S(x)$ is not the arithmetic average total growth rate for the Stratonovich SDE.

One may ask if, for the Stratonovich SDE, $f_S(x)$ is the geometric average total growth rate $f_g(x) := \lim_{\Delta t \downarrow 0} \frac{\exp(\mathbb{E}_{t,x}[\ln X(t + \Delta t)]) - x}{\Delta t}$, as has happened in the previous section for the stochastic Malthusian model. Unless $g(x)$ is of the form σx for some $\sigma > 0$ as in the previous section, the answer is 'No'. So, the geometric average, based on using a logarithmic transformation, does not work for general $g(x)$. We shall try a more general transformation, leading therefore to a different type of average. Let us see what type of average $f_S(x)$ is for the Stratonovich SDE.

Instead of a logarithm transformation, we will use the transformation by the function

$$\phi(y) = \int_c^y \frac{1}{g(z)} dz, \tag{9.14}$$

where c is a fixed arbitrary point in the interior of the state space (the choice of c is irrelevant since different choices lead to the same results). Since $\phi(y)$ is strictly increasing, it has an inverse ϕ^{-1}. We now define the *ϕ-average growth rate*

$$f_\phi(x) := \lim_{\Delta t \downarrow 0} \frac{\phi^{-1}(\mathbb{E}_{t,x}[\phi(X(t + \Delta t))]) - x}{\Delta t}. \tag{9.15}$$

Notice that, in the particular case of the stochastic Malthusian model of the previous section, we have $g(x) = \sigma x$ (with $\sigma > 0$) and $]0, +\infty[$ as the sate space, so that the assumption of $g(x) > 0$ in the interior of the state space holds true; in this particular case, $\phi(x) = \frac{1}{\sigma} \ln x +$ constant and so the ϕ-average coincides with the geometric average.

Let us compute, for the Stratonovich SDE (9.7), the ϕ-average total growth rate at time t if the population has size x at that time, using the transformation

$Y = \phi(X)$ so that $Y(t) = \phi(X(t))$. From (9.7), we get, using Stratonovich calculus rules (which are the ordinary calculus rules) and the fact that $\frac{d\phi(x)}{dx} = \frac{1}{g(x)}$:

$$(S) \quad dY(t) = \frac{f_S(X(t))}{g(X(t))} dt + dW(t) = \frac{f_S(\phi^{-1}(Y(t)))}{g(\phi^{-1}(Y(t)))} dt + 1 dW(t).$$

The drift coefficient for Y is therefore $a_Y(y) = \frac{f_S(\phi^{-1}(y))}{g(\phi^{-1}(y))} + \frac{1}{2} \times 1 \times \frac{d1}{dy} = \frac{f_S(\phi^{-1}(y))}{g(\phi^{-1}(y))}$
(since the second term in the sum is zero), and this must be equal to the expression of the definition of drift coefficient of Y, which is $\lim_{\Delta t \downarrow 0} \frac{\mathbb{E}_{t,x}[Y(t+\Delta t)] - y}{\Delta t}$.
Therefore, with $y = \phi(x)$, $\mathbb{E}_{t,x}[Y(t + \Delta t)] := \mathbb{E}[Y(t + \Delta t)|X(t) = x] =$
$\mathbb{E}[Y(t + \Delta t)|Y(t) = y] = y + \frac{f_S(\phi^{-1}(y))}{g(\phi^{-1}(y))} \Delta t + o(\Delta t)$. Apply ϕ^{-1} to both sides, make
a Taylor first-order expansion about y and notice that $\frac{d\phi^{-1}(y)}{dy} = \frac{1}{d\phi(x)/dx} = g(x)$
to obtain $\phi^{-1}(\mathbb{E}_{t,x}[Y(t + \Delta t)]) = x + f_S(x)\Delta t + o(\Delta t)$. Replacing this in (9.15), we conclude that, for the Stratonovich SDE (9.7), the ϕ-average total growth rate at time t if the population has size x at that time is given by

$$f_\phi(x) = f_S(x). \tag{9.16}$$

In conclusion, contrary to what has been often presumed in the literature, $f(x)$ means two different 'average' total growth rates under the two calculi. It is the arithmetic average total growth rate under Itô calculus and the ϕ-average total growth rate under Stratonovich calculus (which coincides with the geometric average for the case of $g(x) = \sigma x$). Taking into account the difference between the two averages, the two calculi give completely coincidental results.

10

Study of some functionals

10.1 Dynkin's formula

Considerer an *autonomous stochastic differential equation* (SDE)

$$dX(t) = f(X(t))dt + g(X(t))dW(t), \quad t \geq s, \quad X(s) = x \tag{10.1}$$

with $f(x)$ and $g(x)$ satisfying a Lipschitz condition. Remembering that, in this autonomous case, the Lipschitz condition implies the restriction on growth condition, we conclude that the assumptions of the existence and uniqueness Theorem 7.1 are verified. So, the solution exists and is unique and it is an a.s. continuous homogeneous diffusion process with drift coefficient

$$a(x) = f(x)$$

and diffusion coefficient

$$b(x) = g^2(x),$$

often called an *Itô diffusion*. Among the continuous homogeneous diffusion processes, we have also the Feller diffusions, which come with different definitions (i.e. different additional assumptions), some very similar to what we have just called Itô diffusions.

The homogeneity implies that $X_{s,x}(s + \tau)$ and $X_{0,x}(\tau)$, as processes in $\tau \geq 0$, have the same initial distribution (with all probability mass concentrated at the deterministic value x) and the same transition distributions, so they have the same finite-dimensional distributions. Therefore, to compute probabilities and expected values, it is indifferent to use one or the other. This does not mean that they have equal values at the same instants τ; usually, they do not.[1]

---

1 It is like the r.v. X and Y corresponding to the height of two people chosen at random from some population. The two r.v. have the same probability distribution but, if you pick two people at random, it is very unlikely that the r.v. are equal, i.e. that the two people will have the same height.

Introduction to Stochastic Differential Equations with Applications to Modelling in Biology and Finance,
First Edition. Carlos A. Braumann.
© 2019 John Wiley & Sons Ltd. Published 2019 by John Wiley & Sons Ltd.
Companion Website: www.wiley.com/go/braumann/stochastic-differential-equations

When working with conditional expectations, where the condition is the value of the solution at a given time (like $\mathbb{E}_{s,x}$), only the transition probabilities matter and we can even forget what the initial condition is, so that we may denote simply by $X(t)$ a generic solution of the SDE $dX(t) = f(X(t))dt + g(X(t))dW(t)$ without specifying *a priori* what is the initial condition at time 0. As we know, when such an initial condition is fixed, the solution is a.s. unique.

Since we are only going to work with the probability distributions, keep in mind that everything that follows in this chapter can indeed be applied to any homogeneous diffusion process with drift coefficient $a(x) = f(x)$ and diffusion coefficient $b(x) = g^2(x)$ that are continuous functions of x, even if f and g do not satisfy a Lipschitz condition, as long as there is a weak solution of the corresponding SDE and weak uniqueness holds. We know that, in the particular case that f and g have a continuous derivative, there is even a unique strong solution up to a possible explosion time and we only need to verify that explosion does not occur a.s. (i.e. the explosion time is a.s. infinite).

Let

$$u(t,x) := \mathbb{E}_x[h(X_t)], \tag{10.2}$$

where we have abbreviated $\mathbb{E}[\dots|X(0) = x] = \mathbb{E}_{0,x}[\dots]$ to $\mathbb{E}_x[\dots]$ and where h is a bounded continuous Borel measurable function.

If u is a bounded continuous function with bounded continuous first- and second-order partial derivatives in x, then, from (5.12), (5.13), and (5.14), it satisfies the *backward Kolmogorov equation* (BKE)

$$\frac{\partial u}{\partial t} = Du \tag{10.3}$$

with terminal condition

$$\lim_{t \downarrow 0} u(t,x) = h(x),$$

where

$$D = a(x)\frac{\partial}{\partial x} + \frac{1}{2}b(x)\frac{\partial^2}{\partial x^2}$$

is the *diffusion operator*.

We now define the *infinitesimal operator* of the Itô diffusion $X(t)$:

$$\mathcal{A}h(x) := \lim_{t \downarrow 0} \frac{\mathbb{E}_x[h(X(t)] - h(x)}{t}. \tag{10.4}$$

The set of real functions $h(x)$ for which the limit exists for all x is the *domain* of $\mathcal{A}$. Note that, if $h(x)$ is of class C^2 and has compact support,[2] then h is in the

---

2 This means that there is a compact set such that the function has zero values outside that set.

domain of $\mathcal{A}$ and $\mathcal{A} = D$. In fact, we can apply Itô Theorem 6.1 to $Y(t) = h(X(t))$ and obtain

$$dY(t) = \frac{\partial h(X(t))}{\partial x}dX(t) + \frac{1}{2}\frac{\partial^2 h(X(t))}{\partial x^2}(dX(t))^2$$

$$= \left(f(X(t))\frac{\partial h(X(t))}{\partial x} + \frac{1}{2}g^2(X(t))\frac{\partial^2 h(X(t))}{\partial x^2} \right) dt$$

$$+ \left(g(X(t))\frac{\partial h(X(t))}{\partial x} \right) dW(t),$$

which can be written in integral form

$$Y(t) = h(X(t))$$

$$= h(x) + \int_0^t \left(f(X(s))\frac{\partial h(X(s))}{\partial x} + \frac{1}{2}g^2(X(s))\frac{\partial^2 h(X(s))}{\partial x^2} \right) ds$$

$$+ \int_0^t g(X(s))\frac{\partial h(X(s))}{\partial x}dW(s); \tag{10.5}$$

we get the desired result after some computations by applying mathematical expectations $\mathbb{E}_x$, noting that the expectation of the second integral is zero. It can also be proved that, if h is of class C^2 (even if it does not have compact support) and if h is in the domain of $\mathcal{A}$, then $\mathcal{A} = D$ (see Øksendal (2003)).

Property 10.1 (Dynkin's formula) *Let $X(t)$ be an Itô diffusion and $h(x)$ be a function of class C^2 with compact support.*

(a) Then we have Dynkin's formula

$$\mathbb{E}_x[h(X(t))] = h(x) + \mathbb{E}_x \left[\int_0^t \mathcal{A}h(X(s))ds \right]$$

$$= h(x) + \mathbb{E}_x \left[\int_0^t Dh(X(s))ds \right]. \tag{10.6}$$

(b) If τ is a Markov time such that $\mathbb{E}_x[\tau] < +\infty$, then we have the more general Dynkin's formula

$$\mathbb{E}_x[h(X(\tau))] = h(x) + \mathbb{E}_x \left[\int_0^\tau Dh(X(s))ds \right]. \tag{10.7}$$

To obtain (a) we just need to use (10.5) and apply $\mathbb{E}_x$. For a proof of (b) we refer the reader, for instance, to Øksendal (2003).

Observation 10.1 *Consider the case where τ is the first exit time of a bounded interval $I =]a, b[$, i.e. $\tau = \inf\{t \geq 0 : X(t) \notin I\} = \inf\{t \geq 0 : X(t) = a$ or $X(t) = b\}$. If $x \in I$ and if $\mathbb{E}_x[\tau] < +\infty$, then Dynkin's formula (10.7) holds for any function $h(x)$ of class C^2, even if it does not have compact support.*

In fact, in this case, the values of $X(s)$ that show up in the integral are confined to the interval I, so h can be replaced by a function with compact support coinciding with h on that interval.

From (10.6), one easily obtains, for h of class C^2 with compact support, that $u(t, x) := \mathbb{E}_x[h(X(t))]$ satisfies

$$\frac{\partial u}{\partial t} = \mathbb{E}_x[Dh(X(t))]. \tag{10.8}$$

We now show, following Øksendal (2003), that u satisfies the BKE.

Property 10.2   *Let $h(x)$ be a function of class C^2 with compact support. Then the function $u(t, x) := \mathbb{E}_x[h(X(t))]$ satisfies the backward Kolmogorov equation*

$$\frac{\partial u}{\partial t} = Du \tag{10.9}$$

with terminal condition $u(0^+, x) = h(x)$.

Observation 10.2   *Note that u is bounded of class C^{12}. In Øksendal (2003) it is even shown that the solution of (10.9) is unique in the sense that, if $\tilde{u}(t, x)$ is a bounded solution of class C^{12} of the BKE (10.9) satisfying the terminal condition $\tilde{u}(0^+, x) = h(x)$, then $\tilde{u}(t, x) = u(t, x)$.*

Proof of (10.9) (sketch) To prove (10.9), note that u is differentiable in t (see (10.8)) and that, with $r > 0$,

$$\frac{1}{r}(\mathbb{E}_x[u(t, X(r))] - u(t, x)) = \frac{1}{r}\mathbb{E}_x[\mathbb{E}_{X(r)}[h(X(t))] - \mathbb{E}_x[h(X(t))]]$$

$$= \frac{1}{r}\mathbb{E}_x[\mathbb{E}_x[h(X(t+r))|\mathcal{A}_r] - \mathbb{E}_x[h(X(t))|\mathcal{A}_r]]$$

$$= \frac{1}{r}\mathbb{E}_x[h(X(t+r)) - h(X(t))] = \frac{u(t+r, x) - u(t, x)}{r},$$

and let $r \downarrow 0$. On the left-hand side we get $\mathcal{A}u = Du$ and on the right-hand side we get $\frac{\partial u}{\partial t}$. ∎

The results of this section, which will be quite useful in Chapter 11, are easily generalized to the multidimensional case.

10.2   Feynman–Kac formula

The BKE can be generalized to the *Feynman–Kac formula*:

Property 10.3 (Feynman–Kac formula) *Let $h(x)$ be a function of class C^2 with compact support and $q(x)$ a continuous lower bounded function. Then, the function*

$$v(t,x) := \mathbb{E}_x \left[\exp \left(- \int_0^t q(X(s))ds \right) h(X(t)) \right] \qquad (10.10)$$

satisfies the partial differential equation

$$\frac{\partial v}{\partial t} = Dv - qv \qquad (10.11)$$

with the terminal condition $v(0^+, x) = h(x)$.

Observation 10.3 *Note that v is of class $C^{1,2}$ and is bounded on sets of the form $K \times \mathbb{R}$, for compact real sets K. In Øksendal (2003) it is even shown that a solution of (10.11) with these properties is unique in the sense that, if $\tilde{v}(t,x)$ is another solution of (10.11) with those properties and satisfying the terminal condition $\tilde{v}(0^+, x) = h(x)$, then $\tilde{v}(t,x) = v(t,x)$.*

Proof of Feynman–Kac formula (sketch) For more details, see, for instance, Øksendal (2003). Let $Y(t) = h(X(t))$ and $Z(t) = \exp\left(- \int_0^t q(X(s))ds \right)$. We have already applied Itô Theorem 6.1 to $Y(t)$, obtaining (10.5). Applying it to $Z(t)$ and $Y(t)Z(t)$, one obtains $dZ(t) = -Z(t)q(X(t))dt$ and $d(Y(t)Z(t)) = Y(t)dZ(t) + Z(t)dY(t) + dY(t)dZ(t) = Y(t)dZ(t) + Z(t)dY(t) + 0$. The function $v(t,x) = \mathbb{E}_x[Y(t)Z(t)]$ is differentiable in t (one can use a similar reasoning as used for $u(t,x)$). Then

$$\frac{1}{r}(\mathbb{E}_x[v(t,X(r))] - v(t,x))$$

$$= \frac{1}{r}\mathbb{E}_x[\mathbb{E}_{X(r)}[Z(t)h(X(t))] - \mathbb{E}_x[Z(t)h(X(t))]]$$

$$= \frac{1}{r}\mathbb{E}_x \left[\mathbb{E}_x \left[h(X(t+r))e^{-\int_0^t q(X(s+r))ds}|\mathcal{A}_r \right] - \mathbb{E}_x[Z(t)h(X(t))|\mathcal{A}_r] \right]$$

$$= \frac{1}{r}\mathbb{E}_x[Z(t+r) \exp\left(\int_0^r q(X(\theta))d\theta \right) h(X(t+r)) - Z(t)h(X(t))]$$

$$= \frac{1}{r}\mathbb{E}_x[h(X(t+r)Z(t+r) - h(X(t))Z(t)]$$

$$+ \frac{1}{r}\mathbb{E}_x \left[h(X(t+r))Z(t+r) \left(\exp\left(\int_0^r q(X(s))ds \right) - 1 \right) \right].$$

Letting $r \downarrow 0$ and noting that $\frac{1}{r}h(X(t+r))Z(t+r)\left(\exp\left(\int_0^r q(X(s))ds\right) - 1\right)$ is bounded and converges to $h(X(t))Z(t)q(X(0))$, one gets $\mathcal{A}v(t,x) = \frac{\partial}{\partial t}v(t,x) + q(x)v(t,x)$ and we have again $\mathcal{A} = \mathcal{D}$. ∎

There are generalizations of the Feynman–Kac formula to more complex functionals (see, for instance, Karatzas and Shreve (1991)). The results of this section, which will be quite useful in Chapter 11, are easily generalized to the multidimensional case.

11

Introduction to the study of unidimensional Itô diffusions

11.1 The Ornstein–Uhlenbeck process and the Vasicek model

In Chapter 3 we saw that the projection into a coordinate axis of the *Brownian motion* of a particle suspended on a fluid was modelled by Einstein as a Wiener process of the form $X(t) = x_0 + \sigma W(t)$, where x_0 is the initial position of the particle and σ^2 is the diffusion coefficient that gives the speed of change in the variance of the particle's position. This model, however, does not take into account that the friction is finite and so the particle, after a collision with a molecule of the fluid, does not change instantly its position and stop, but rather moves continuously with decreasing speed. In Einstein's model, the particle does not even have a defined speed (since the Wiener process does not have a derivative in the ordinary sense).

As an improvement to Einstein's model, the *Ornstein–Uhlenbeck model* (see Uhlenbeck and Ornstein (1930)) appeared in 1930. So far as I know, it was the first SDE to appear in the literature. In this model, it is assumed that the particle is subjected to two forces, the friction force and the force due to the random collisions with the fluid molecules. Let $V(t)$ be the particle's speed (in the direction of the coordinate axis) and m the mass of the particle. The force $mdV(t)/dt$ acting on the particle is then given by the SDE

$$m\frac{dV(t)}{dt} = -fV(t) + \beta\varepsilon(t), \tag{11.1}$$

where f is the friction coefficient. Putting $\alpha = f/m$ and $\sigma = \beta/m$, we have

$$dV(t) = -\alpha V(t)dt + \sigma dW(t). \tag{11.2}$$

Introduction to Stochastic Differential Equations with Applications to Modelling in Biology and Finance, First Edition. Carlos A. Braumann.
© 2019 John Wiley & Sons Ltd. Published 2019 by John Wiley & Sons Ltd.
Companion Website: www.wiley.com/go/braumann/stochastic-differential-equations

Given the initial speed, we can solve this SDE to obtain the particle's speed $V(t)$ (now well defined), which is known as the *Ornstein–Uhlenbeck process*.[1] If we want the particle's position, it will be given by $x_0 + \int_0^t V(s)ds$, where x_0 is the initial position.

A variant of this model is used to model the dynamics of interest rates and also of exchange rates between two currencies (e.g. between the euro and the US dollar). It is known as the *Vasicek model*. This model considers that there is a kind of reference rate $R > 0$ and proposes that the *interest rate* (or the *exchange rate*) $X(t)$ at time t is described by the SDE

$$dX(t) = -\alpha(X(t) - R)dt + \sigma dW(t), \quad X(0) = x_0, \tag{11.3}$$

with $\alpha > 0$ and $\sigma > 0$. Note that, on average, there is a tendency for the rate to revert to the reference rate R, a phenomenon known as *reversion to the mean* or *mean reversion*. In fact, if we ignore market fluctuations (put $\sigma = 0$) in order to look at the average behaviour, we see that $X(t)$ would converge to R as $t \to +\infty$.[2] Of course, with market fluctuations ($\sigma > 0$), the tendency to revert to the mean is still there but is perturbed by the market fluctuations.

This is an autonomous SDE with $f(x) = -\alpha(x - R)$ and $g(x) \equiv \sigma$. These functions have bounded continuous derivatives and so they satisfy a Lipschitz condition. So, by the existence and uniqueness Theorem 7.1, the solution $X(t)$ exists and is a.s. unique. This *Itô diffusion* $X(t)$ is a homogeneous diffusion process with drift coefficient $a(x) = -\alpha(x - R)$ and diffusion coefficient $b(x) \equiv \sigma^2$. Since $g(x)$ does not depend on x, the calculi of Itô and Stratonovich will in this case give coincidental results and so one may use indifferently the Itô rules of calculus or the ordinary rules of calculus.

Putting $V(t) = X(t) - R$, one sees that $V(t)$ satisfies the Ornstein–Uhlenbeck model $dV(t) = dX(t) = -\alpha V(t)dt + \sigma dW(t)$ (which is just a particular case of the Vasicek model when $R = 0$). So, $e^{\alpha t}dV(t) + \alpha e^{\alpha t}V(t)dt = \sigma e^{\alpha t}dW(t)$. Since in this case we can apply ordinary rules of calculus, this can be written as $d(e^{\alpha t}V(t)) = \sigma e^{\alpha t}dW(t)$. So, putting $Z(t) = e^{\alpha t}V(t)$ and noting that $V(0) = x_0 - R$, we have $Z(t) = (x_0 - R) + \int_0^t \sigma e^{\alpha s}dW(s)$. Therefore

$$X(t) = V(t) + R = e^{-\alpha t}Z(t) + R$$

$$= e^{-\alpha t}\left((x_0 - R) + \int_0^t \sigma e^{\alpha s}dW(s)\right) + R$$

$$= R + e^{-\alpha t}(x_0 - R) + \sigma e^{-\alpha t}\int_0^t e^{\alpha s}dW(s).$$

---

1 In some literature (not here), the term Ornstein–Uhlenbeck process is reserved to a stationary solution of (11.2) (i.e. a solution $V(t)$ that is a stationary process), which requires that the initial speed, the r.v. $V(0)$, has the same probability distribution as the other $V(t)$.

2 If $X(t) > R$, then $\frac{dX(t)}{dt} = -\alpha(X(t) - R) < 0$ and $X(t)$ will decrease, approaching R. If $X(t) < R$, then $\frac{dX(t)}{dt} = -\alpha(X(t) - R) > 0$ and $X(t)$ will increase, again approaching R.

Notice that the integrand in the stochastic integral $\int_0^t e^{\alpha s} dW(s)$ is a deterministic function. So, by Property 6.1f, the stochastic integral has distribution

$$\mathcal{N}\left(0, \int_0^t (e^{\alpha s})^2 ds\right) = \mathcal{N}\left(0, \frac{1}{2\alpha}(e^{2\alpha t} - 1)\right).$$

We conclude that

$$X(t) \frown \mathcal{N}\left(R + e^{-\alpha t}(x_0 - R), \frac{\sigma^2}{2\alpha}(1 - e^{-2\alpha t})\right).$$

This *transient distribution* (i.e. the distribution of the process at a finite time) has therefore the pdf

$$p(t, y) = f_{X(t)}(y)$$

$$= \frac{1}{\sqrt{2\pi \frac{\sigma^2}{2\alpha}(1 - e^{-2\alpha t})}} \exp\left(-\frac{(y - (R + e^{-\alpha t}(x_0 - R)))^2}{2\frac{\sigma^2}{2\alpha}(1 - e^{-2\alpha t})}\right)$$

$$(-\infty < y < +\infty).$$

The transient distribution converges as $t \to +\infty$ to a limiting distribution, called the *stationary distribution*, namely to

$$\mathcal{N}\left(R, \frac{\sigma^2}{2\alpha}\right).$$

If we denote by X_∞ a r.v. having the stationary distribution, we may say that $X(t)$ converges in distribution to X_∞. The *stationary density* is the pdf of the stationary distribution, in this case given by

$$p(y) = f_{X_\infty}(y) = \frac{1}{\sqrt{2\pi \frac{\sigma^2}{2\alpha}}} \exp\left(-\frac{(y - R)^2}{2\frac{\sigma^2}{2\alpha}}\right) \quad (-\infty < y < +\infty).$$

In the deterministic ($\sigma = 0$) model $dX(t) = -\alpha(X(t) - R)dt$, we saw that $X(t)$ converges to the stable equilibrium point R. Instead, in the stochastic ($\sigma > 0$) model (11.3), the interest rate (or exchange rate) $X(t)$ does not converge to an equilibrium point, but rather keeps fluctuating randomly for ever. However, the probability distribution of $X(t)$ does converge to an equilibrium distribution, the stationary distribution, and we can speak of a *stochastic equilibrium*. We will later see that $X(t)$ is even an *ergodic process* and we can determine estimates of the asymptotic mean R and the asymptotic variance $\frac{\sigma^2}{2\alpha}$ by using sample time averages and variances across a single trajectory of $X(t)$. This is quite advantageous, since in financial and other applications there is usually only one available trajectory, the single observed trajectory corresponding to the market scenario ω that has effectively occurred.

The application of the Vasicek model (11.3) to interest rates or exchange rates faces the problem that negative values of $X(t)$ are theoretically possible, since the distribution of $X(t)$ is Gaussian and so can take negative values. For typical values of the parameters in these applications, the probability of negative values is positive but so negligible (practically zero) that this model can be safely used (remember that models are usually just approximations, preferably good ones, of reality). Anyway, some years ago, when negative interest rates were not observed in the markets (the situation nowadays changed and negative interest rates are indeed observed in some markets), this was still a theoretical (although not a practical) inconvenience. It is still a theoretical inconvenience for exchange rates. This led to the alternative model in which the rates cannot be negative at all, the *Cox–Ingersoll–Ross model (CIR model)*

$$dX(t) = -\alpha(X(t) - R)dt + \sigma\sqrt{X(t)}dW(t), \quad X(0) = x_0 > 0. \tag{11.4}$$

Coming back to the Vasicek model (11.3), we were fortunate and have obtained an explicit solution $X(t)$ for any finite time t (transient solution) and its pdf in explicit form. This allowed us, by letting $t \to +\infty$, to verify the convergence to a stationary distribution and to determine the stationary density. For most models, however, we are not so fortunate and cannot determine an explicit expression for the transient pdf because we do not have an explicit expression for $X(t)$ and the alternative method of solving the forward Kolmogorov equation to determine the transient pdf does not provide an explicit expression either. So, we are unable to determine the stationary density (assuming it exists) by letting $t \to +\infty$ in the expression of the transient density. Even so, if there is a stationary density, we can obtain its expression. In fact, the stationary density must be time invariant and so must be a solution of (5.20), which is an ordinary differential equation (ODE) that is very easy to solve, as we will see later. However, having a solution of (5.20) that is also a pdf is not by itself a guarantee that the d.f. of $X(t)$ converges as $t \to +\infty$ to a distribution with that pdf. This will have to be deduced by other methods, namely, in the unidimensional case, by studying the behaviour of $X(t)$ at the boundaries of its state space.

An introduction to the study of one-dimensional Itô diffusions will be the object of next sections. Most of the methods we will use cannot unfortunately be extended to multidimensional Itô diffusions. The main reason is that such methods are based on the fact that, in one dimension, the continuity of the trajectories of an Itô diffusion $X(t)$ implies that, given $z \in]a, b[$ (where $[a, b]$ is an interval contained in the state space), if the process passes by a at time s and by b at time $t > s$ (or vice versa), then it must pass by z at some time $u \in]s, t[$.

11.2 First exit time from an interval

Consider a one-dimensional homogeneous diffusion process $X(t)$ ($t \in [0, +\infty[$) with drift coefficient and diffusion coefficients

$$a(x) = f(x) \quad \text{and} \quad b(x) = g^2(x),$$

both continuous functions of x. Let the state space be an interval (which may be open, closed or mixed) with lower extreme r_1 and upper extreme r_2, with $-\infty \le r_1 < r_2 \le +\infty$. Any of the boundaries r_1 and r_2 may or may not belong to the state space. We will assume that $b(x) > 0$ for $x \in]r_1, r_2[$ and that the process $X(t)$ is a weak solution of the SDE $dX(t) = f(X(t))dt + g(X(t))dW(t)$ with initial condition $X(0)$ and assume that this solution is weakly unique. We will further assume that the process is *regular*, i.e.

$$P_x[T_y < +\infty] := P[T_y < +\infty | X(0) = x] > 0 \quad (r_1 < x, y < r_2),$$

where

$$T_y = \inf\{t \ge 0 : X(t) = y\} \tag{11.5}$$

is the *first passage time* by y. If for a given trajectory the process never passes by y, we have $T_y = +\infty$ (by convention, the infimum of an empty set of real numbers is $+\infty$). So, a regular process is a process for which every point in the interior of the state space communicates with any other interior point (i.e. there is a positive probability of eventually moving from one state to the other).

   Assume that, at the present time (which we may label as time 0 for convenience), the process has value x, i.e. $X(0) = x$, with $r_1 < x < r_2$. Let a and b, with $r_1 < a < x < b < r_2$, be two interior points of the state space and denote by

$$T_{a,b} = \min\{T_a, T_b\} \tag{11.6}$$

the moment in which the process $X(t)$ reaches a or b for the first time, i.e. the *first exit time* from the interval $]a, b[$. Remember that we use the abbreviated notation $\mathbb{E}_x[\dots] := \mathbb{E}[\dots | X(0) = x]$.

Property 11.1 (First exit time) *Let h be a bounded continuous function and*

$$v_h(x) := \mathbb{E}_x\left[\int_0^{T_{a,b}} h(X(s))ds\right]. \tag{11.7}$$

Let $v(x)$ ($x \in [a, b]$) be the solution of the ODE

$$Dv(x) = -h(x) \tag{11.8}$$

such that $v(a) = v(b) = 0$. Then we have $v_h(x) < +\infty$ and $v_h(x) = v(x)$.

In particular, when $h(x) \equiv 1$, we conclude that $\mathbb{E}_x[T_{a,b}] < +\infty$ and is given by the solution of the ODE $Dv(x) = -1$ such that $v(a) = v(b) = 0$. The finiteness of the expectation implies that $T_{a,b}$ is a.s. finite.

Proof: We do not know *a priori* that $\mathbb{E}_x[T_{a,b}] < +\infty$, so let us use the truncation $\tau = min\{T, T_{a,b}\}$ with fixed $T > 0$. Obviously $\mathbb{E}_x[\tau] < +\infty$ and we can apply Dynkin's formula (10.7) to v, obtaining

$$\mathbb{E}_x[v(X(\tau))] = v(x) + \mathbb{E}_x\left[\int_0^\tau Dv(X(s))ds\right].$$

For $s \le \tau$, we have $X(s) \in [a, b]$ and so $Dv(X(s)) = -h(X(s))$. Therefore,

$$\mathbb{E}_x[v(X(\tau))] = v(x) - \mathbb{E}_x\left[\int_0^\tau h(X(s))ds\right]. \tag{11.9}$$

As $T \uparrow +\infty$, we get $\tau \uparrow T_{a,b}$. Since $\mathbb{E}_x\left[\int_0^\tau h(X(s))ds\right] = v(x) - \mathbb{E}_x[v(X(\tau))]$ is uniformly bounded in T, we get $\mathbb{E}_x\left[\int_0^\tau h(X(s))ds\right]$ converging to the finite limit $\mathbb{E}_x\left[\int_0^{T_{a,b}} h(X(s))ds\right]$. We also have $X(\tau) \to X(T_{a,b}) = a$ or b and, therefore, $v(X(\tau)) \to 0$. So, from (11.9), we get $\mathbb{E}_x\left[\int_0^{T_{a,b}} h(X(s))ds\right] = v(x)$, as required. ∎

Property 11.2 (Order of exit) *Let $u(x)$ ($x \in [a, b]$) be the solution of the ODE*

$$Du(x) = 0 \tag{11.10}$$

such that $u(a) = 0$ and $u(b) = 1$. Then

$$P_x[T_b < T_a] := P[T_b < T_a | X(0) = x] = u(x). \tag{11.11}$$

Proof: Let τ be as in the previous proof and apply Dynkin's formula (10.7) to u. We have

$$\mathbb{E}_x[u(X(\tau))] = u(x) + \mathbb{E}_x\left[\int_0^\tau Du(X(s))ds\right]. \tag{11.12}$$

For $s \le \tau$, $X(s) \in [a, b]$ and so $Du(X(s)) = 0$. Therefore, $\mathbb{E}_x[u(X(\tau))] = u(x)$. As $T \uparrow +\infty$, we get $\mathbb{E}_x[u(X(T_{a,b}))] = u(x)$. Since $X(T_{a,b})$ can only take the values a (when $T_a < T_b$) or b (when $T_b < T_a$), we get $\mathbb{E}_x[u(X(T_{a,b}))] = u(a)(1 - P_x[T_b < T_a]) + u(b)P_x[T_b < T_a] = P_x[T_b < T_a] = u(x)$ as required. ∎

In order to obtain moments of $T_{a,b}$ of higher order, we need the following generalization of Proposition 11.1:

Property 11.3 *Let h be a continuous bounded function, q a function of class C^2, and*

$$U_{q,h}(x) := \mathbb{E}_x\left[q\left(\int_0^{T_{a,b}} h(X(s))ds\right)\right]. \tag{11.13}$$

Let $U(x)$ $(x \in [a, b])$ be the solution of the ODE

$$DU(x) = -h(x)V(x) \tag{11.14}$$

such that $U(a) = U(b) = q(0)$, with $V(x) := \mathbb{E}_x \left[q' \left(\int_0^{T_{a,b}} h(X(s))ds \right) \right]$ (where q' denotes the derivative of q). Then $U_{q,h}(x) < +\infty$ and $U_{q,h}(x) = U(x)$.

In particular, when $h(x) \equiv 1$, $q(x) = x^n$ and $U_n(x) := \mathbb{E}_x[(T_{a,b})^n]$ $(n = 1, 2, \ldots)$, we conclude that $\mathbb{E}_x[(T_{a,b})^n] < +\infty$ is given by the solution of the ODE $DU_n(x) = -nU_{n-1}(x)$ such that $U_n(a) = U_n(b) = 0$.

Proof: The technique is similar to the previous proofs, so we will skip the steps of starting with the truncation $\tau = min\{T, T_{a,b}\}$ and letting $T \uparrow +\infty$ afterwards, as well as the step of showing that the desired expectation is finite.

Applying Dynkin's formula (10.7) to U, we get

$$\mathbb{E}_x[U(X(T_{a,b}))] = U(x) + \mathbb{E}_x \left[\int_0^{T_{a,b}} DU(X(s))ds \right]. \tag{11.15}$$

Attending to (11.14) and denoting by $T_{a,b}(y)$ the first exit time from $]a, b[$ when the process $X(t)$ has the initial condition $X(0) = y,$[3] we have

$$
\begin{aligned}
\mathbb{E}_x \left[\int_0^{T_{a,b}} DU(X(s))ds \right] &= -\mathbb{E}_x \left[\int_0^{T_{a,b}} h(X(s))V(X(s))ds \right] \\
&= -\mathbb{E}_x \left[\int_0^{T_{a,b}} h(X(s))\mathbb{E}_{X(s)} \left[q' \left(\int_0^{T_{a,b}(X(s))} h(X(u))du \right) \right] ds \right] \\
&= -\mathbb{E}_x \left[\int_0^{T_{a,b}} h(X(s))q' \left(\int_s^{T_{a,b}(X(s))+s} h(X(u))du \right) ds \right] \\
&= \mathbb{E}_x \left[-\int_0^{T_{a,b}} h(X(s))q' \left(\int_s^{T_{a,b}} h(X(u))du \right) ds \right] \\
&= \mathbb{E}_x \left[\int_0^{T_{a,b}} \frac{dq \left(\int_s^{T_{a,b}} h(X(u))du \right)}{ds} ds \right] \\
&= q(0) - \mathbb{E}_x \left[q \left(\int_0^{T_{a,b}} h(X(u))du \right) \right]
\end{aligned}
$$

Replacing in (11.15) and attending to $\mathbb{E}_x[U(X(T_{a,b}))] = q(0)$ (since $U(a) = U(b) = q(0)$), we obtain the desired result

$$\mathbb{E}_x \left[q \left(\int_0^{T_{a,b}} h(X(u))du \right) \right] = U(x). \qquad \blacksquare$$

Of course, explicit expressions involve solving the ODEs of the previous propositions.

---

3 When $y = x$, we omit the reference to the initial condition and simply write, as we have done so far, $T_{a,b}$.

With that purpose, we introduce two measures, the *scale measure S* and the *speed measure M*.[4] Following Karlin and Taylor (1981), these measures can be defined by their densities. These densities are functions defined for $\xi \in]r_1, r_2[$. The *scale density* is defined by

$$s(\xi) = \exp\left(-\int_{\xi_0}^{\xi} \frac{2a(\eta)}{b(\eta)} d\eta \right), \tag{11.16}$$

where $\xi_0 \in]r_1, r_2[$ is fixed but arbitrarily chosen. So, this density is defined up to a multiplicative constant, which is irrelevant for our purposes. The *speed density* is defined by

$$m(\xi) = \frac{1}{b(\xi)s(\xi)} \tag{11.17}$$

and so, again, since it involves $s(\xi)$, is defined up to a multiplicative constant irrelevant to our purposes. Although these measures are usually not probabilities, we may define on $]r_1, r_2[$ something somewhat similar to distribution functions. Namely, the *scale function* is defined by

$$S(x) = \int_{]x_0, x]} s(\xi)d\xi = \int_{x_0}^{x} s(\xi)d\xi \tag{11.18}$$

(it is irrelevant whether the integration interval is open, semi-open or closed since s is a continuous function) and the *speed function* by (note that m is also continuous)

$$M(x) = \int_{x_0}^{x} m(\xi)d\xi, \tag{11.19}$$

where $x_0 \in]r_1, r_2[$ is fixed but arbitrarily chosen. So, these functions are defined up to an additive constant in terms of the respective densities, which were defined up to a multiplicative constant; these constants are irrelevant for our purposes. We can even define $S(r_1^+) = \lim_{x \downarrow r_1} S(x)$ and $S(r_2^-) = \lim_{x \uparrow r_2} S(x)$ and similarly for M.

Given $a, b \in]r_1, r_2[$, the scale and speed measures of the interval $]a, b]$ are given by

$$S(]a, b]) = \int_{]a, b]} s(\xi)d\xi = S(b) - S(a) \tag{11.20}$$

---

4 As mentioned before, a measure is 'almost' a probability, since it is non-negative and σ-additive, but the measure of the universal set (in this case $]r_1, r_2[$) does not need to be equal to one and may be finite or even $+\infty$. For those more knowledgeable in measure theory, the scale and speed measures can be defined for Lebesgue sets A of $]r_1, r_2[$; in fact, these measures are absolutely continuous with respect to the Lebesgue measure λ of $]r_1, r_2[$. This means that they have Radon–Nikodym derivatives with respect to the Lebesgue measure, also called densities, which we denote by $s = \frac{dS}{d\lambda}$ and $m = \frac{dM}{d\lambda}$, respectively, so that the scale measure of A is given by $S(A) = \int_A s(\xi)d\lambda(\xi) = \int_A s(\xi)d\xi$ and the speed measure of A is given by $M(A) = \int_A m(\xi)d\lambda(\xi) = \int_A m(\xi)d\xi$.

and

$$M(]a, b]) = M(b) - M(a) \tag{11.21}$$

and are defined up to a multiplicative constant, which is irrelevant for our purposes. Due to the continuity of s and m, the scale and speed measures are not affected if we replace the interval $]a, b]$ by $]a, b[$ or $[a, b[$ or $[a, b]$. More generally, the scale and speed measures of other Borel sets (and even Lebesgue sets) $A \subset]r_1, r_2[$ are defined by $S(A) = \int_A s(x) dx$ and $M(A) = \int_A m(x) dx$.

For a function $h(x)$ of class C^2, easy computations show that

$$Dh(x) = \frac{1}{2} \left(\frac{1}{m(x)} \right) \frac{d}{dx} \left(\frac{1}{s(x)} \frac{dh(x)}{dx} \right),$$

and therefore

$$Dh(x) = \frac{1}{2} \frac{d}{dM(x)} \left(\frac{dh(x)}{dS(x)} \right). \tag{11.22}$$

The advantage of using these functions is that they convert the diffusion operator into a consecutive application of two derivatives.

This immediately allows us to obtain an expression for $u(x)$. In fact, (11.10) can now be written as $\frac{d}{dM(x)} \left(\frac{du(x)}{dS(x)} \right) = 0$. Integrating first with respect to M and then with respect to S, we get $u(x) = B + AS(x)$, with A and B integrating constants. Using $u(a) = 0$ and $u(b) = 1$, we can determine A and B and obtain

$$u(x) := P_x[T_b < T_a] = \frac{S(x) - S(a)}{S(b) - S(a)}. \tag{11.23}$$

Exercise 11.1 *Let $S(x)$ be the scale function. Since it is defined up to additive and multiplicative constants, choose those constants at will (i.e. choose ξ_0 in (11.16) and x_0 in (11.18) at will). Let $Y(t) = S(X(t))$ be the transformed process that results from $X(t)$ by applying the transformation S. Use Itô Theorem 6.1 and determine the drift and diffusion coefficients of $Y(t)$. Note that the transformation changed the drift coefficient to zero (in Chapter 13 we will see another method of changing the drift coefficient to zero). Therefore, the scale function of $Y(t)$ is $S_Y(y) = y$ (do not confuse this with the scale function of X, which is $S(x)$). To be more precise, it has more generally the form $= y - y_0$, but we can choose the additive constant $y_0 = 0$.*

Therefore, if $Y(0) = y$ with $c < y < d$, the probability that $Y(t)$ passes by d before passing by c is $= \frac{S_Y(y) - S_Y(c)}{S_Y(d) - S_Y(c)} = \frac{y-c}{d-c}$, a proportionality to the distances rule. When this proportionality occurs, we say that the process is in natural scale. The reason why S is called the 'scale function' is that it transforms the original process $X(t)$ into a process $Y(t) = S(X(t))$ that is in natural scale.

To obtain an expression for $v_h(x)$, which is a solution of (11.8), we can write this equation in the form $\frac{d}{dM(x)} \left(\frac{dv_h(x)}{dS(x)} \right) = -2h(x)$. Integrating with respect to

M, we obtain $\frac{dv_h(\eta)}{dS(\eta)} = -2\int_a^\eta h(\xi)dM(\xi) + A = -2\int_a^\eta h(\xi)m(\xi)d\xi + A$ (where A is an integration constant). Integrating this with respect to S, we obtain

$$v_h(x) = -2\int_a^x\int_a^\eta h(\xi)m(\xi)d\xi dS(\eta) + A(S(x) - S(a)) + B$$

with A and B constants. Using $v_h(a) = 0$ and $v_h(b) = 0$, we get $B = 0$ and $A = \frac{2}{S(b)-S(a)}\int_a^b\int_a^\eta h(\xi)m(\xi)d\xi dS(\eta)$. Therefore,

$$v_h(x) = 2\left(u(x)\int_a^b\int_a^\eta h(\xi)m(\xi)d\xi dS(\eta) - \int_a^x\int_a^\eta h(\xi)m(\xi)d\xi dS(\eta)\right).$$

A change of variable and some computations leads to:

$$v_h(x) := \mathbb{E}_x\left[\int_0^{T_{a,b}} h(X(s))ds\right]$$

$$= 2u(x)\int_x^b (S(b) - S(\xi))m(\xi)h(\xi)d\xi$$

$$+ 2(1 - u(x))\int_a^x (S(\xi) - S(a))m(\xi)h(\xi)d\xi. \tag{11.24}$$

As already mentioned, in the particular case $h(x) \equiv 1$ we get an expression for $\mathbb{E}_x[T_{a,b}]$.

Following the methods in Braumann *et al.* (2009) and Carlos *et al.* (2013), we can use the particular case mentioned in Property 11.3 to obtain recursively moments $U_n(x) := \mathbb{E}_x[(T_{ab})^n]$ ($n = 1, 2, \ldots$) of the first passage time T_{ab}, taking into account that $U_0(x) \equiv 1$. We know that $U_n(x)$ satisfies, for $n = 1, 2, \ldots$, the ODE

$$DU_n(x) = -nU_{n-1}(x),$$

with the initial conditions $U_n(a) = U_n(b) = 0$. From (11.22), the ODE can be written in the form

$$\frac{d}{dM(x)}\left(\frac{dU_n(x)}{dS(x)}\right) = -2nU_{n-1}(x) \quad (n = 1, 2, \ldots). \tag{11.25}$$

Integrating first with respect to $dM(x) = m(x)dx$, we get

$$\frac{dU_n(x)}{dS(x)} = 2n\left(-\int_a^x U_{n-1}(\zeta)m(\zeta)d\zeta + A\right),$$

where A is an integrating constant. Integrating now with respect to $dS(x) = s(x)dx$, we get

$$U_n(x) = 2n\left(-\int_a^x\left(\int_a^\xi U_{n-1}(\zeta)m(\zeta)d\zeta\right)s(\xi)d\xi + A\int_a^x s(\xi)d\xi + B\right),$$

where B is another integrating constant. Since $U_n(a) = 0$, this implies $0 = 2n(0 + 0 + B)$ and so $B = 0$. Since $U_n(b) = 0$, this implies

$$0 = 2n \left(-\int_a^b \left(\int_a^\xi U_{n-1}(\zeta)m(\zeta)d\zeta \right) s(\xi)d\xi + A \int_a^b s(\xi)d\xi \right)$$

and so

$$A = \frac{\int_a^b \left(\int_a^\xi U_{n-1}(\zeta)m(\zeta)d\zeta \right) s(\xi)d\xi}{\int_a^b s(\xi)d\xi}.$$

Replacing the values of A and B and attending to (11.23), one gets

$$U_n(x) := \mathbb{E}_x[(T_{ab})^n] = 2n \left(-\int_a^x \int_a^\xi U_{n-1}(\zeta)m(\zeta)d\zeta \, s(\xi)d\xi \right.$$
$$\left. + u(x) \int_a^b \int_a^\xi U_{n-1}(\zeta)m(\zeta)d\zeta \, s(\xi)d\xi \right). \quad (11.26)$$

Since $U_0(x) = \mathbb{E}_x[(T_{ab})^0] = 1$, we can use expression (11.26), together with expression (11.23), to iteratively obtain the successive moments $U_1(x)$, $U_2(x)$, ... of the first exit time T_{ab}.

Exercise 11.2 *Determine expressions for $U_1(x) := \mathbb{E}_x[T_{ab}]$:*

(a) *by using (11.24) for the particular case $h(x) \equiv 1$;*
(b) *by using (11.26) with $n = 1$, taking advantage of $U_0(x) \equiv 1$.*

Show that the two expressions are equivalent.
Suggestion: Exchange the order of integration in (b) and try to reach the result obtained in (a).

Exercise 11.3 *By appropriate choice of functions h and q in Property 11.3, we can obtain as a particular case of (11.13) the function $R(\lambda, x) = \mathbb{E}_x[\exp(-\lambda T_{a,b})]$, which is the* Laplace transform *of the pdf of $T_{a,b}$, also abbreviated as the Laplace transform of $T_{a,b}$. Then, as a particular case of (11.14), get an ODE for that Laplace transform. A more direct method can be seen in Cox and Miller (1965). Note that $\frac{\partial^n R(0,x)}{\partial \lambda^n} = (-1)^n \mathbb{E}_x[(T_{a,b})^n]$ ($n = 1, 2, ...$), so one can obtain the moments of $T_{a,b}$. The pdf of $T_{a,b}$ can be obtained by inverting the Laplace transform. Unfortunately, only in rare cases can we get an explicit expression for the pdf by this method, but one can still use numerical inversion of the Laplace transform to obtain an approximate expression for the pdf of $T_{a,b}$.*

In Exercise 11.1 we obtained a nice interpretation for the scale function. In order to obtain an interpretation for the speed measure, consider a process in natural scale (i.e. such that $S(x) = x$ is a scale function). Let $\varepsilon > 0$ and put

$a = x - \varepsilon$, $b = x + \varepsilon$. From (11.23), we get $u(x) = P_x[T_{x+\varepsilon} < T_{x-\varepsilon}] = 1/2$ and, from (11.24) with $h(x) \equiv 1$, we get

$$\mathbb{E}_x[T_{x-\varepsilon,x+\varepsilon}] = \int_x^{x+\varepsilon} (x + \varepsilon - \xi)m(\xi)d\xi + \int_{x-\varepsilon}^x (\xi - x + \varepsilon)m(\xi)d\xi,$$

so that

$$\lim_{\varepsilon \downarrow 0} \frac{1}{\varepsilon^2} \mathbb{E}_x[T_{x-\varepsilon,x+\varepsilon}]$$
$$= \lim_{\varepsilon \downarrow 0} \frac{1}{\varepsilon^2} \int_x^{x+\varepsilon} (x + \varepsilon - \xi)m(\xi)d\xi + \frac{1}{\varepsilon^2} \int_{x-\varepsilon}^x (\xi - x + \varepsilon)m(\xi)d\xi = 2m(x).$$

This shows that, for small ε, if the process is in natural scale and takes the value x, it will on average occupy the interval $(x - \varepsilon, x + \varepsilon)$ during a timespan of $2m(x)\varepsilon^2$ before exiting the interval. It would have been more natural to call M the 'occupation measure' instead of the official name of 'speed measure'.

11.3 Boundary behaviour of Itô diffusions, stationary densities, and first passage times

Here we will follow Karlin and Taylor (1981). Consider the framework of the beginning of Section 11.2. Assume that $X(0) = x \in]r_1, r_2[$ and $r_1 < a < x < b < r_2$.

For y in the state space, we have defined the first passage time $T_y = \inf\{t \geq 0 : X(t) = y\}$, which is by convention (on the definition of the infimum) $= +\infty$ for trajectories of $X(t)$ that never reach y. Let $T_{r_1^+} = \lim_{a\downarrow r_1} T_a$ and $T_{r_2^-} = \lim_{b\uparrow r_2} T_b$. If r_1 is in the state space, then $T_{r_1^+} = T_{r_1}$ due to the continuity of the trajectories and similarly, if r_2 is in the state space, then $T_{r_2^-} = T_{r_2}$.

Note that the scale measure $S(]c, d])$ of any interval $]c, d]$ (or, for that matter, also $[c, d]$, $[c, d[$ or $]c, d[$), with $r_1 < c < d < r_2$, is finite and positive since it is the integral on a finite interval (with positive length) of a function $s(x)$ which is positive and continuous on that interval. The same holds for the speed measure.

Boundary classification

To look at the boundary behaviour, we need to know if the scale measure of a small neighbourhood (contained in $]r_1, r_2[$) of the boundary is finite or $+\infty$. For that purpose, from the previous property it does not matter how large or small the neighbourhood is. In fact, if we have two neighbourhoods of r_1, say $]r_1, c]$ and $]r_1, d]$ with $r_1 < c < d < r_2$, $S(]r_1, c])$ and $S(]r_1, d])$ are both finite or both $= +\infty$ because the difference between the two is $S(]c, d])$, which is finite as we have seen. The same holds for neighbourhoods $]c, r_2[$ and $]d, r_2[$ of r_2.

So, choosing freely one neighbourhood $]r_1, c]$ of r_1, with fixed c such that $r_1 < c < r_2$, its scale measure $S(]r_1, c])$ is either finite or $= +\infty$. It is easy to see that $S(]r_1, c])$ is finite or $+\infty$ according to whether $S(r_1^+) := \lim_{c\downarrow r_1} S(c)$ is finite or $= -\infty$. In fact, assuming $\varepsilon < x_0$, we have $S(r_1^+) = \lim\limits_{\varepsilon \downarrow r_1} \int_{x_0}^{\varepsilon} s(\xi)d\xi = \lim\limits_{\varepsilon \downarrow r_1} S(\varepsilon) - S(x_0) = \lim\limits_{\varepsilon \downarrow r_1} -S(]\varepsilon, x_0]) = -S(]r_1, x_0]$.

Suppose $S(]r_1, c]) = +\infty$ for some c such that $r_1 < c < r_2$ (as we have seen, if true for some c, this is true for all), which is equivalent to $s(x)$ being non-integrable in a right-neighbourhood of r_1 and equivalent to $S(r_1^+) = -\infty$. Then

$$P_x[T_{r_1^+} < T_b] = 0 \quad \text{for all } r_1 < x < b < r_2. \tag{11.27}$$

To prove this, note that, from (11.23), we have $P_x[T_{r_1^+} < T_b] = \frac{S(b) - S(x)}{S(b) - S(r_1^+)} = \frac{S(]x, b])}{S(]r_1, b])} = 0$, since $S(]x, b])$ is positive finite and $S(]r_1, b] = +\infty$.

If the boundary r_1 satisfies (11.27), we say that it is an *non-attractive boundary*.

This is equivalent to $P_x[T_{r_1^+} \geq T_b] = 1$ and so, starting at any $X(0) = x$ (where x is an interior point of the state space), we have with probability one that $T_{r_1^+} \geq T_b$. This means that the process does not move first close to the boundary but rather moves first to any point b further away from the boundary.

Suppose now that $S(]r_1, c]) < +\infty$ for some c such that $r_1 < c < r_2$ (as we have seen, if true for some c, this is true for all), which is equivalent to $s(x)$ being integrable in a right-neighbourhood of r_1 and equivalent to $S(r_1^+)$ finite. Then

$$P_x[T_{r_1^+} \leq T_b] > 0 \quad \text{for all } r_1 < x < b < r_2. \tag{11.28}$$

To prove this, note that, from (11.23), we have $P_x[T_{r_1^+} \leq T_b] = \frac{S(]x, b])}{S(]r_1, b])} > 0$ since $S(]x, b])$ and $S(]r_1, b])$ are both positive finite.

If the boundary r_1 satisfies (11.28), we say that it is an *attractive boundary*. In this case, there is a positive probability that the process moves first close to the boundary.

We have seen that the necessary and sufficient condition for the boundary r_1 to be attractive is that $s(x)$ be integrable in a right-neighbourhood of r_1 (meaning that $S(]r_1, c])$ is finite for some $c \in]r_1, r_2[$), which is equivalent to having $S(r_1^+)$ finite. Otherwise, the r_1 boundary is non-attractive.

Analogously, we say that the boundary r_2 is non-attractive when

$$P_x[T_{r_2^-} < T_a] = 0 \quad \text{for all } a < x < r_2 \tag{11.29}$$

and is attractive when

$$P_x[T_{r_2^-} \leq T_a] > 0 \quad \text{for all } a < x < r_2. \tag{11.30}$$

A necessary and sufficient condition for the boundary r_2 to be attractive is that $s(x)$ be integrable in a left-neighbourhood of r_2 (meaning that $S(]c, r_2[)$ is finite for some $c \in]r_1, r_2[$), which is equivalent to having $S(r_2^-)$ finite. Otherwise, the r_2 boundary is non-attractive.

A boundary r_i ($i = 1, 2$) is *attainable* if it is attractive and the probability of reaching it in finite time is positive, i.e. $P_x[T_{r_i^+} < +\infty] > 0$ for all $r_1 < x < r_2$. A boundary that is not attainable is said to be *unattainable*.

A necessary and sufficient condition (see Khashminsky (1969)) for r_1 to be attainable is that

$$\Sigma(r_1) := \int_{r_1}^{c} S(]r_1, \xi]) dM(\xi) < +\infty \tag{11.31}$$

for some c such that $r_1 < c < r_2$.[5] Alternative equivalent expressions for $\Sigma(r_1)$ that may sometimes be more convenient are easily obtained (recurring to the exchange of the order of integration):

$$\Sigma(r_1) := \int_{r_1}^{c} S(]r_1, \xi]) dM(\xi) = \int_{r_1}^{c} \int_{r_1}^{\xi} s(\eta) d\eta m(\xi) d\xi$$

$$= \int_{r_1}^{c} \int_{\eta}^{c} m(\xi) d\xi s(\eta) d\eta = \int_{r_1}^{c} M(]\eta, c]) dS(\eta). \tag{11.32}$$

A necessary and sufficient condition for r_2 to be attainable is that

$$\int_{d}^{r_2} S(]\xi, r_2[) dM(\xi) < +\infty \tag{11.33}$$

for some d such that $r_1 < d < r_2$.

An attractive boundary that is unattainable cannot be reached in finite time a.s., but it might be reached in infinite time, i.e. we might have $P[\lim_{t \uparrow +\infty} X(t) = r_i] > 0$ for any $X(0) = x$ with $r_1 < x < r_2$. A non-attractive boundary cannot a.s. be reached in finite or in infinite time (see Karlin and Taylor (1981)).

There are other aspects involved in boundary classification that we do not deal with here. For example, if a boundary is attainable, we need to specify, besides the drift and diffusion coefficients, the behaviour at the boundary (absorbing, reflecting, sticky, etc.).

Stationary density

In the case in which both boundaries r_1 and r_2 are non-attractive, we can, loosely speaking, say that there is a tendency for the trajectories that get close to a boundary to be pushed away towards the interior of the state space. Since all these interior points communicate among themselves, the transient distribution (the distribution of $X(t)$) may have a pdf $p(t, y) = f_{X(t)}(y)$ $(r_1 < y < r_2)$. An important question is whether this transient distribution stabilizes in the sense of converging in distribution as $t \to +\infty$ to a limit distribution. If so, let X_∞ be a r.v. with such limit distribution with pdf $p(y) = f_{X_\infty}(y)$ $(r_1 < y < r_2)$, the so-called

---

5 The value of c is irrelevant to know whether $\Sigma(r_1)$ is finite or not, so the fixed value of c can be chosen freely.

stationary density. This is a real possibility since the diffusion process is homogeneous and therefore the rules of transition between states depend only on the duration of the transition, not on the initial and final moments of the transition. Of course, non-attractiveness of the boundaries is also important, for otherwise the existence (with positive probability) of trajectories that are attracted to one or both boundaries would lead to the progressive concentration of the probability distribution in the vicinity of the attracting boundary(ies), spoiling the stabilization of a probability density in the interior of the state space. However, the condition that the two boundaries are non-attractive is not, by itself, sufficient. Anyway, in the case where there is a stationary density, one intuitively expects it to be proportional to the occupation time of each state, i.e. proportional to $m(y)$ (we will check that later), and this is only possible if $\int_{r_1}^{r_2} m(\xi)d\xi$ is finite. Otherwise, anything proportional to $m(y)$ has also an infinite integral instead of an integral equal to one, a requirement for it to be a probability density.

For a more formal treatment, suppose r_1 and r_2 are non-attractive boundaries. If there is a stationary density $p(y)$, it needs to be a time invariant density, and so it must satisfy (5.20). We can write (5.20) in the form

$$\frac{d(a(y)p(y))}{dy} - \frac{1}{2}\frac{d^2(b(y)p(y))}{dy^2} = 0.$$

Integratig in order to y, we get $-2a(y)p(y) + \frac{d(b(y)p(y))}{dy} = C$ (C is the integration constant). Let us multiply both sides by the integrating factor $s(y)$ to get, after simple computations, $b(y)p(y)\frac{ds(y)}{dy} + \frac{d(b(y)p(y))}{dy}s(y) = Cs(y)$, so that $\frac{d(b(y)p(y)s(y))}{dy} = Cs(y)$. Integrating again, we get $b(y)p(y)s(y) = CS(y) + D$ (D is the integrating constant), so that

$$p(y) = m(y)(CS(y) + D).$$

Since the boundaries are non-attractive, $S(r_1^+) = -\infty$ and $S(r_2^-) = +\infty$. This implies in particular (see Exercise 11.1) that the process $Y(t) = S(X(t))$ has zero drift coefficient and that its state space is $] - \infty, +\infty[$. So, it is possible to keep $p(y)$ non-negative for all $y \in]r_1, r_2[$, as is required for probability densities, if and only if $C = 0$. Therefore, if there is a stationary density, it must have the form $p(y) = Dm(y)$, with D conveniently chosen to make $\int_{r_1}^{r_2} p(\xi)d\xi = 1$, another requirement of probability densities. But that is only possible if

$$M(r_1, r_2) := \int_{r_1}^{r_2} m(\xi)d\xi < +\infty,$$

i.e. if M is a finite measure. If that happens, we will have

$$p(y) = \frac{m(y)}{\int_{r_1}^{r_2} m(\xi)d\xi} \qquad (r_1 < y < r_2), \qquad (11.34)$$

proportional to the speed density $m(x)$, as suggested before. This is an invariant density. But the existence of an invariant density is not enough. We want it to also be a stationary or limiting density, so that the distribution of $X(t)$ converges to a distribution with such density. A proof of that can be seen in Gihman and Skorohod (1972), which also proves that $X(t)$ is ergodic. In conclusion:

Property 11.4   *Under the assumptions of this section, if both boundaries of the state space are non-attractive and the speed measure is finite (i.e. $\int_{r_1}^{r_2} m(\xi)d\xi < +\infty$), then there exists a stationary distribution towards which the distribution of $X(t)$ converges, having as pdf the stationary density (11.34). Furthermore, $X(t)$ is an* ergodic *process.*

Often in applications we only have a single realization of the stochastic process, so the ergodic property, when it holds, becomes very important. A process is ergodic when the time averages of certain functions along a single sample path converge to the mathematical expectation (which is an average over all sample-paths) of the same function under the stationary distribution. More specifically, if $h(x)$ is a bounded Borel-measurable function and $X(0) = x \in]r_1, r_2[$, then (note that the limit is a mean square limit)

$$\underset{T \to +\infty}{\text{l.i.m.}} \frac{1}{T} \int_0^T h(X(t))dt = \mathbb{E}_x[h(X_{+\infty})] = \int_{r_1}^{r_2} h(y)p(y)dy. \tag{11.35}$$

To obtain almost sure ergodicity instead of the m.s. ergodicity, i.e. to obtain

$$\lim_{T \to +\infty} \frac{1}{T} \int_0^T h(X(t))dt = \mathbb{E}_x[h(X_{+\infty})] = \int_{r_1}^{r_2} h(y)p(y)dy \quad \text{a.s.,} \tag{11.36}$$

it is not necessary that h be bounded, it is sufficient that

$$\mathbb{E}_x[|h(X_{+\infty})|] = \int_{r_1}^{r_2} |h(y)|p(y)dy < +\infty.$$

Ergodic one-dimensional processes have also the nice property that every state in the interior of the state space will be visited infinitely often.

Let us give two examples.

Black–Schloles model

An example of an autonomous SDE with no stationary density is the *Black–Scholes* model (8.1), with $X(0) = x > 0$. One recognizes that the state space is $]0, +\infty[$, so the boundaries are $r_1 = 0$ and $r_2 = +\infty$. In fact, from (8.5) and (8.6), we can see that $Z(t) = \ln \frac{X(t)}{x}$ has a normal distribution and so can take any real value, so that $X(t) = xe^{Z(t)}$ can take any real positive value but not the values 0 or $+\infty$ (so the boundaries are unattainable).

For the case $r := R - \sigma^2/2 > 0$, from (8.8) we have $X(t) \to +\infty$ and so the boundary $r_2 = +\infty$ is attracting and reached in infinite time, with an increasing mean and variance of $X(t)$; the other boundary $r_1 = 0$ is non-attracting.

For the case $r < 0$, we have $X(t) \to 0$ and the boundary $r_1 = 0$ is attracting and reached in infinite time, with the probability distribution progressively concentrated at values closer and closer to zero; the other boundary $r_2 = +\infty$ is non-attracting.

In neither case is there a stationary density.

We can reach the same conclusions without the need to use the explicit solution, by using the methods of this section. We have $a(x) = Rx$, $b(x) = \sigma^2 x^2$, $s(\xi) = \xi^{-2R/\sigma^2}$ (by choosing $\xi_0 = 1$), $m(\xi) = \frac{1}{\sigma^2} \xi^{2R/\sigma^2 - 2}$.

We see that, when $r > 0$, $s(\xi)$ is not integrable in a right-neighbourhood of the 0 boundary and is integrable in a left-neighbourhood of the $+\infty$ boundary, so we conclude likewise that 0 is non-attractive and $+\infty$ is attractive.

When $r < 0$ the opposite happens.

In the case $r = 0$, both boundaries are non-attractive, but still there is no stationary density since $m(x)$ is not integrable; in this case we know from (8.6) that $\ln(X(t)/x)$ has a normal distribution with mean $rt = 0$ and variance $\sigma^2 t \to +\infty$ when $t \to +\infty$.

The Vasicek model

An example of an autonomous SDE with a stationary density is the *Vasicek model* (11.3), for which we have already obtained the stationary density in Section 11.1 by solving the equation explicitly, determining the transient distribution at time t and letting $t \to +\infty$. The stationary distribution is a normal distribution with mean R and variance $\frac{\sigma^2}{2\alpha}$. Notice that the state space is $] - \infty, +\infty[$, with boundaries $r_1 = -\infty$ and $r_2 = +\infty$.

For this model we were fortunate and could obtain explicitly the solution, but there are cases in which we are unable to do this. However, using the methods of this section, we can directly obtain the stationary density without the need to explicitly solve the equation. Let us see how to do that using the Vasicek model as an example. We have $a(x) = -\alpha(x - R)$, $b(x) = \sigma^2$, $s(\xi) = \exp\left(\frac{\alpha}{\sigma^2}(\xi - R)^2\right)$ (choosing $\xi_0 = R$), and $m(\xi) = \frac{1}{\sigma^2}\exp\left(-\frac{(\xi - R)^2}{2(\sigma^2/(2\alpha))}\right)$. Since $s(\xi) \to +\infty$ as $\xi \to \pm\infty$, we see that it is not integrable in neighbourhoods of the boundaries, resulting in the boundaries being non-attractive.

It is clear that the speed measure is finite since $m(y)$ is integrable; in fact, it is basically, apart from a multiplicative constant, a Gaussian density with mean R and variance $\sigma^2/(2\alpha)$, so its integral is equal to one multiplied by the multiplicative constant. Of course, the stationary density, being proportional to $m(x)$ and having its integral equal to one, can be nothing else than the Gaussian pdf with mean R and variance $\sigma^2/(2\alpha)$. Obviously, this coincides with the result previously obtained in Section 11.1.

CIR model

Exercise 11.4   *For the CIR model (11.4) with $\alpha > 0$, $R > 0$, $\sigma > 0$, and $X(0) = x > 0$, determine a condition on the parameters guaranteeing the existence of a* *stationary density and determine the expression for the stationary density using the methods of this section.*

First passage times

Assume the assumptions of Property 11.4 are satisfied. Following the methods in Braumann *et al.* (2009) and Carlos *et al.* (2013), we can obtain recursive expressions for the moments $U_{n,a}(x) := \mathbb{E}_x[(T_a)^n]$ of the first passage time by the lower threshold a or moments $U_{n,b}(x) := \mathbb{E}[(T_b)^n]$ of the first passage time by the upper threshold b. In fact, since the process is ergodic, the mentioned first passage times are proper finite r.v. and, since the boundaries are non-attractive, T_a coincides with the first exit time of the interval $]a, r_2[$ and T_b with the first exit time of the interval $]r_1, b[$. So,

$$\begin{aligned}
U_{n,a}(x) &:= \mathbb{E}_x[(T_a)^n] = \lim_{b\uparrow r_2} U_n(x) \quad (a < x) \\
U_{n,b}(x) &:= \mathbb{E}_x[(T_b)^n] = \lim_{a\downarrow r_1} U_n(x) \quad (b > x),
\end{aligned} \tag{11.37}$$

where $U_n(x)$ is given by (11.26), with $u(x)$ given by (11.23).

Since the boundary r_1 is non-attractive, we have $S(r_1^+) = -\infty$ and so $\lim_{a\downarrow r_1} u(x) = \frac{S(x)-S(a)}{S(b)-S(a)} = 1$. Therefore, letting $a \downarrow r_1$, from (11.26) and (11.37), one gets

$$\begin{aligned}
U_{n,b}(x) := \mathbb{E}_x[(T_b)^n] = 2n \Big(&- \int_{r_1}^x \int_{r_1}^\xi U_{n-1,b}(\zeta)m(\zeta)d\zeta \, s(\xi)d\xi \\
&+ \int_{r_1}^b \int_{r_1}^\xi U_{n-1,b}(\zeta)m(\zeta)d\zeta \, s(\xi)d\xi \Big),
\end{aligned}$$

from which one obtains

$$U_{n,b}(x) := \mathbb{E}_x[(T_b)^n] = 2n \int_x^b \int_{r_1}^\xi U_{n-1,b}(\zeta)m(\zeta)d\zeta \, s(\xi)d\xi \quad (b > x). \tag{11.38}$$

Since $U_{0,b}(x) = 1$, we can use expression (11.38) to iteratively obtain the successive moments $U_{1,b}(x)$, $U_{2,b}(x)$, ... of the first passage time T_b.

Since the boundary r_2 is non-attractive, we have $S(r_2^-) = +\infty$. Note that the second line in (11.26) becomes $u(x) \int_a^b \int_a^\xi U_{n-1}(\zeta)m(\zeta)d\zeta \, s(\xi)d\xi = (S(x) - S(a)) \frac{\int_a^b \left(\int_a^\xi U_{n-1}(\zeta)m(\zeta)d\zeta \right) s(\xi)d\xi}{\int_a^b s(\xi)d\xi}$ and, when $b \uparrow r_2$, the fraction in the last expression has both the numerator and denominator going to $+\infty$. Using L'Hôpital's rule, one sees that

$$u(x) \int_a^b \int_a^\xi U_{n-1}(\zeta)m(\zeta)d\zeta \, s(\xi)d\xi \rightarrow (S(x) - S(a)) \int_a^{r_2} U_{n-1,a}(\zeta)m(\zeta)d\zeta.$$

From (11.26) and (11.37), one gets

$$
\begin{aligned}
U_{n,a}(x) := \mathbb{E}_x[(T_a)^n] &= 2n\left(-\int_a^x \int_a^\xi U_{n-1,a}(\zeta)m(\zeta)d\zeta\, s(\xi)d\xi\right.\\
&\quad\left. + (S(x)-S(a))\int_a^{r_2} U_{n-1,a}(\zeta)m(\zeta)d\zeta\right)\\
&= 2n\left(-\int_a^x \int_a^\xi U_{n-1,a}(\zeta)m(\zeta)d\zeta\, s(\xi)d\xi\right.\\
&\quad\left. + \int_a^x \int_a^{r_2} U_{n-1,a}(\zeta)m(\zeta)d\zeta\, s(\xi)d\xi\right),
\end{aligned}
$$

from which one obtains

$$
U_{n,a}(x) := \mathbb{E}_x[(T_a)^n] = 2n \int_a^x \int_\xi^{r_2} U_{n-1,a}(\zeta)m(\zeta)d\zeta\, s(\xi)d\xi \quad (a < x).
$$
(11.39)

Since $U_{0,a}(x) = 1$, we can use expression (11.39) to iteratively obtain the successive moments $U_{1,a}(x)$, $U_{2,a}(x)$, ... of the first passage time T_a.

In particular, assuming the assumptions of Property 11.4 are valid, from expressions (11.38) and (11.39) and using $U_{0,b}(x) = U_{0,a}(x) = 1$, we easily obtain expressions for the expected first passage time

$$
\begin{aligned}
U_{1,b} &:= \mathbb{E}_x[T_b] = 2\int_x^b s(\xi)(M(\xi) - M(r_1^+))d\xi \quad (b > x)\\
U_{1,a} &:= \mathbb{E}_x[T_a] = 2\int_a^x s(\xi)(M(r_2^-) - M(\xi))d\xi \quad (a < x).
\end{aligned}
$$
(11.40)

Starting from (11.40) and using (11.38) and (11.39), one can also obtain expressions for $U_{2,b}(x)$ and $U_{2,a}(x)$. Using such expressions and doing some simplifications, Carlos *et al.* (2013) obtained relatively simple expressions for the variances *VAR* $[T_b]$ and *VAR* $[T_a]$ of first passage times.

Exercise 11.5 *Apply (11.40) to the Vasicek model (11.3), with $\alpha > 0$, $\sigma > 0$, and $X(0) = x$, to obtain*

$$
\begin{aligned}
U_{1,b} &:= \mathbb{E}_x[T_b] = \frac{1}{\alpha}\int_{\frac{\sqrt{2\alpha}}{\sigma}(x-R)}^{\frac{\sqrt{2\alpha}}{\sigma}(b-R)} \frac{\Phi(z)}{\phi(z)}\, dz \quad (b > x)\\
U_{1,a} &:= \mathbb{E}_x[T_a] = \frac{1}{\alpha}\int_{-\frac{\sqrt{2\alpha}}{\sigma}(x-R)}^{-\frac{\sqrt{2\alpha}}{\sigma}(a-R)} \frac{\Phi(z)}{\phi(z)}\, dz \quad (a < x),
\end{aligned}
$$
(11.41)

where $\phi(u) = \frac{1}{\sqrt{2\pi}}e^{-u^2/2}$ and $\Phi(z) = \int_{-\infty}^z \phi(u)du$ are, respectively, the pdf and the d.f. of a standard Gaussian distribution.

Suggestion: Notice that the speed density $m(\xi)$ and the speed function $M(\xi)$ are, apart a multiplicative constant, the pdf and the d.f. of a non-standard Gaussian distribution, respectively, and then make a change of variable to standardize them; notice also that $s(\xi) = \frac{1}{\sigma^2}\frac{1}{m(\xi)}$.

Note: In Carlos *et al.* (2013), you can see nice simple expressions for the variances of T_b and T_a.

12

Some biological and financial applications

12.1   The Vasicek model and some applications

The Vasicek model (11.3) is studied both in terms of its applications in finance (mainly on exchange rates or on interest rates) and its application in growth of individual living organisms (using as state variable an appropriate function of the size of an individual organism). We have studied the transient distributions in Section 11.1 and obtained the stationary density as a limit when $t \to +\infty$. Since both boundaries $r_1 = -\infty$ and $r_2 = +\infty$ are non-attractive and the speed measure is finite, we have seen in Section 11.3 that the stationary density exists and is proportional to the speed density, so we can also obtain it directly using the speed density. Furthermore, the process is ergodic.

We have seen in Chapter 11 that, for the Vasicek model, the stationary distribution is Gaussian with mean equal to the reference rate R and variance equal to $\frac{\sigma^2}{2\alpha}$.

Exercise 12.1   *Suppose that $X(t)$, the exchange rate British pound (£) – euro (€), follows a Vasicek model (11.3) and that $X(0) = x = 1.0657 \, €/£$, $R = 1.10 \, €/£$, $\alpha = 0.2/year$, and $\sigma^2 = 0.04/year$.*

(a) *Determine the probabilities that the exchange rate will be below $1.00 \, €/£$ one year from now and 10 years from now.*

(b) *Assuming that half a year from now the exchange rate is $1.07 \, €/£$, what will be the conditional probability of that rate being below $1.00 \, €/£$ one year from now?*

(c) *Determine the probability in the long term (i.e. in the stationary regimen) of the exchange rate being below $1.00 \, €/£$.*

(d) *Determine the probability that the exchange rate will reach $a = 1.00 \, €/£$ before reaching $b = 1.10 \, €/£$. Suggestion: This is $1 - u(x)$, where $u(x)$ is given by (11.23). The resulting expression is the ratio of two integrals that you may compute by numerical methods.*

Introduction to Stochastic Differential Equations with Applications to Modelling in Biology and Finance,
First Edition. Carlos A. Braumann.
© 2019 John Wiley & Sons Ltd. Published 2019 by John Wiley & Sons Ltd.
Companion Website: www.wiley.com/go/braumann/stochastic-differential-equations

The Ornstein–Uhlenbeck model (11.2), used as an improved model for the Brownian motion of a particle suspended in a fluid, can be considered a particular case of the Vasicek model with $R = 0$ and describes the dynamical behaviour of the particle's speed along an axis.

ODE models have been used to study the growth of individual living beings, like farm animals, trees or fish. When the growth occurs in environments with random variations, SDE models have been proposed as, for example, in Garcia (1983) and Lv and Pitchford (2007). In Filipe (2011) and Filipe *et al.* (2010, 2015), several SDE models were used and methodologies developed to study the growth of individual farm animals, with an application to meat cows of the mertolengo breed raised in pasture (supplemented by silage when required) in the Alentejo region of Portugal. The models describe the evolution of an animal's weight $X(t)$ (measured in number of kilograms) with age t (measured in years). We will mention here, in particular, a stochastic Gompertz model, for which it is more convenient to work with $Y(t) = \ln X(t)$ because $Y(t)$ satisfies the Vasicek model $dY(t) = -\alpha(Y(t) - R)dt + \sigma dW(t)$, already studied above. Here $R = \ln A$, where A is an asymptotic weight around which, for a large age t, a cow's weight would fluctuate, and α regulates the speed of approach to the asymptotic size. For each cow there were weight measurements at several ages (ages that vary from cow to cow; data taken for 97 cows). Each cow was considered a different realization of the stochastic process, so different cows would correspond to different ω and independence among cows was assumed. The parameter values, estimated by maximum likelihood methods (see Section 12.2), have approximate 95% confidence intervals $A = 411.2 \pm 8.0$ (in kg), $\alpha = 1.676 \pm 0.057$/year, and $\sigma = 0.302 \pm 0.009/\sqrt{\text{year}}$.

Exercise 12.2   *For the model just mentioned for a cow's growth, with parameter values $A = 411.2$ (in kg), $\alpha = 1.676$/year, and $\sigma = 0.302/\sqrt{\text{year}}$, suppose a farmer has bought for €200 a cow with 7 months $= \frac{7}{12}$ years of age (approximate age of weaning) weighing 160 kg, to raise and later sell it. Determine:*

(a) *The expected weight and variance of that cow at the age of one year. Note that these are conditional moments given that $X(\frac{7}{12}\text{years}) = 160$, which is equivalent to $Y(\frac{7}{12}\text{years}) = \ln 160$.*
(b) *The probability that the cow will weigh over 250 kg at the age of one year.*
(c) *The probability that the cow will weigh over 420 kg at a large age (i.e. under the stationary distribution).*

Exercise 12.3   *Under the same conditions as in Exercise 12.2, assume the farmer wants to sell the cow when it reaches the age $t > 7$ months $= \frac{7}{12}$ years. That has fixed and variable costs and a revenue from selling the cow. Consider the situation used as an example in Filipe et al. (2015), in which* *the fixed costs are $C_1 = €227.45$ (the initial cost of €200, plus transportation, commercialization, and sanitation costs) and the variable costs are the feeding and care costs of $c_2 = €320.16$ per year. The revenue is $PDX(t)$, where $P = €3.50$ is the selling price per kilogram of the animal's weight and $D = 0.5$ is the dressing proportion (the usable part of the animal). The profit is the difference between the revenue and the costs, $\Pi(t) = PDX(t) - C_1 - c_2(t - \frac{7}{12})$.*

(a) *Determine the farmer's expected profit if the selling age is $t = 16$ months.*
(b) *Determine the probability of a positive profit if the selling age is $t = 16$ months.*
(c) *Determine the selling age that leads to the maximum expected profit.*

In Exercise 12.3, suppose that the farmer decides to sell the animal not when it reaches some fixed age t, but rather when it will reach a fixed weight $B > x$ for the first time. Let us count the time T_B, starting from the moment the farmer bought the cow (at age $\frac{7}{12}$ years with weight $X(\frac{7}{12}) = 160$), required for the cow to reach the weight $X = B$ for the first time; this is a random variable and the cow will first reach weight B at the age of $\frac{7}{12} + T_B$ years. The farmer's profit will be $\Pi^*(B) = PDB - C_1 - c_2 T_B$ and the expected profit will be $\mathbb{E}[\Pi^*(B)] = PDB - C_1 - c_2\mathbb{E}[T_B]$. Of course, due to the homogeneity of the process, T_B is the first passage time T_B^X of process $X(t)$ by B when the initial condition is $X(0) = x = 160$ (the weight of the animal at the moment we start counting), so that $\mathbb{E}[T_B] = \mathbb{E}_x[T_B^X]$. Since $Y(t) = \ln X(t)$, one sees that this is equal to $\mathbb{E}_y[T_b^Y]$, where T_b^Y is the first passage time of process $Y(t)$ by $b = \ln B$ when the initial condition is $Y(0) = y = \ln 160$.

Since $Y(t)$ follows a Vasicek model and $b > y$, from (11.41) we have

$$\mathbb{E}[T_B] = \mathbb{E}_y[T_b^Y] = \frac{1}{\alpha} \int_{\frac{\sqrt{2a}}{\sigma}(y-R)}^{\frac{\sqrt{2a}}{\sigma}(b-R)} \frac{\Phi(z)}{\phi(z)} \, dz.$$

Using this expression, we can numerically compute the integral. We can even determine, as was studied in Filipe *et al.* (2015), the value of B that maximizes the expected profit and also the standard deviation of the profit. Filipe *et al.* (2015) also shows that maximizing the expected profit with respect to a deterministic selling weight B gives a higher profit than maximizing with respect to a deterministic selling age (as in Exercise 12.3c). The gain, however, may not compensate the extra complication of frequently weighing the animal to check if it has already reached the optimal weight. For example, using the parameter values of Exercises 12.2 and 12.3, the expected profit for $B = 309$ is €105.95, and that is the optimal profit; the time to reach 309 kg is a r.v. with expected value $\mathbb{E}[T_B] = 0.65$ years, corresponding to an average selling age of the animal of $\frac{7}{12} + 0.65 = 1.23$ years.

The Gompertz model is also used to model the growth of some wildlife populations, where $X(t)$ now represents the population size. We will cover population growth applications in Section 12.3.

12.2 Monte Carlo simulation, estimation and prediction issues

Monte Carlo simulation of SDEs

In some situations we may not be able to obtain explicit expressions for the solution $X(t)$ of an SDE, or for the probabilities of certain events or for mathematical expectations, standard deviations, and quantiles of related random variables of interest to us. We can recur to *Monte Carlo simulations* and perform, on the computer, simulations of many trajectories of the solution of the SDE. These sample paths can then be used to approximate the distributions, probabilities, expected values, and quantiles by their sample equivalents.

Of course, one cannot simulate values for all times t in some interval, say $[0, T]$, and so usually one simulates the values $X_k = X(t_k)$ at a grid of points $0 = t_0 < t_1 < t_2 < \cdots < t_n = T$. Often, $X(0)$ has a known value x_0, in which case it does not need to be simulated. Although not required, it is usually convenient that the grid points are equidistant, i.e. $t_k = k\Delta$ ($k = 0, 1, 2, \ldots, n$) with $\Delta = T/n$.

If one has an explicit expression for the solution $X(t)$ of the SDE involving the Wiener process only at time t, one can use the technique used in Section 8.1 for the Black–Scholes model. One simulates trajectories of the Wiener process $W(t)$ by the method used in Section 4.1 and plugs the values in the expression for $X(t)$, thus obtaining the corresponding simulated trajectories of $X(t)$.

Sometimes the explicit expression for the SDE solution $X(t)$ involves $W(t)$ through a stochastic integral and the above plug-in technique cannot be used. For example, in the Vasicek model (11.3), the explicit expression of the solution is $X(t) = R + e^{-\alpha t}(x_0 - R) + \sigma e^{-\alpha t} \int_0^t e^{\alpha s} dW(s)$, which leads to

$$X_k = R + e^{-\alpha \Delta_k}(X_{k-1} - R) + \sigma e^{-\alpha \Delta_k} \int_{t_{k-1}}^{t_k} e^{\alpha s} dW(s).$$

Since the stochastic integral has a deterministic integrand and therefore a Gaussian distribution with mean 0 and variance $\int_{t_{k-1}}^{t_k} e^{2\alpha s} ds = \frac{1}{2\alpha}(e^{2\alpha t_k} - e^{2\alpha t_{k-1}})$, we can see that the transition distribution between times t_{k-1} and t_k is

$$\mathcal{N}\left(R + e^{-\alpha \Delta_k}(X_{k-1} - R), \frac{\sigma^2}{2\alpha}(1 - e^{-2\alpha \Delta_k})\right).$$

Its pdf is

$$p(t_k, X_k \mid t_{k-1}, X_{k-1}) =$$

$$\frac{1}{\sqrt{2\pi \frac{\sigma^2}{2\alpha}(1 - e^{-2\alpha\Delta_k})}} \exp\left(-\frac{(X_k - (R + e^{-\alpha\Delta_k}(X_{k-1} - R))^2}{2\frac{\sigma^2}{2\alpha}(1 - e^{-2\alpha\Delta_k})} \right).$$

The simulations take advantage of the Markov property. We start (step 0) by simulating X_0 using random number generation of the corresponding probability distribution. Quite often we know the exact value $X(0) = x_0$, in which case no simulation of it is required. Then we go by steps.

At step number k ($k = 1, 2, \ldots, n$), the simulated values of $X_0, X_1, \ldots, X_{k-1}$ are already available and we want to simulate X_k. Since X_{k-1} is known, the Markov property says that the probability distribution of X_k given the previous simulated values is just the transition probability distribution between times t_{k-1} and t_k. So we simulate a random number from that transition probability distribution (using the random number generators described in Section 2.2) and that becomes the simulated value of X_k. In the Vasicek model, we generate a random number from a Gaussian distribution with mean $R + e^{-\alpha\Delta_k}(X_{k-1} - R)$ and variance $\frac{\sigma^2}{2\alpha}(1 - e^{-2\alpha\Delta_k})$.

The technique works for any model as long as we have the transition probability distributions. Of course, we can repeat the above procedure to simulate as many trajectories as we need.

There are, however, many situations in which we cannot obtain the transition probability distributions. This is often the case when we cannot obtain an explicit expression for $X(t)$ or that expression is not very helpful (e.g. when it involves a stochastic integral with non-deterministic integrand).

In Section 7.1 we presented an approximate method to preform the simulations, the *Euler method*, also called *Euler–Maruyama* method.

Basically, what one does is the same procedure by the steps we have just described, the difference being that, since we do not know the expression of the transition probability distribution, we use a Gaussian approximation. Even if we just need to simulate a low number n of grid points, when using an approximate method we should work with a high value of n to reduce the approximation errors.

The steps are the same as before except that we replace the transition probability distribution by a Gaussian approximation. For an SDE $dX(t) = X_0 + f(t, X(t))dt + g(t, X(t))dW(t)$, as described in Section 7.1, the functions $f(t, X(t))$ and $g(t, X(t))$ are approximated in the interval $[t_{k-1}, t_k[$ by constants, namely by their values at the beginning of the interval (to insure the non-anticipative property of the Itô calculus). This leads to the first-order approximation scheme

$$X_k \simeq X_{k-1} + f(t_{k-1}, X_{k-1})(t_k - t_{k-1}) + g(t_{k-1}, X_{k-1})(W(t_k) - W(t_{k-1})),$$

where the transition probability distribution from time t_{k-1} to time t_k is Gaussian with mean $f(t_{k-1}, X_{k-1})(t_k - t_{k-1})$ and variance $g^2(t_{k-1}, X_{k-1})(t_k - t_{k-1})$.

So, at step k we can proceed as before, simulating a random number generated from this approximate Gaussian transition distribution to determine an approximate simulated value of X_k.

This is equivalent to simulating the Wiener process increment $\Delta W_k = W(t_k) - W(t_{k-1}) \frown \mathcal{N}(0, \Delta_k)$ and replacing the value in the expression $X_k \simeq X_{k-1} + f(t_{k-1}, X_{k-1})(t_k - t_{k-1}) + g(t_{k-1}, X_{k-1})\Delta W_k$ to get the approximate simulated value X_k.

In Section 7.1, together with some references, we have mentioned alternative higher-order methods, like the Milstein method, that converge faster (so require a not so high number n of grid points) than the Euler method.

Parameter estimation

Due to the Markov property, the likelihood function is just the product of the initial density and of the transition densities: $p(0, X_0) \prod_{k=1}^{n} p(t_k, X_k \mid t_{k-1}, X_{k-1})$. If the initial value X_0 is known, as is the case in the applications shown in this book, then the term $p(0, X_0)$ should be dropped and the *log-likelihood function* is

$$L(p_1, p_2, \dots, p_m) = \sum_{k=1}^{n} \ln p(t_k, X_k \mid t_{k-1}, X_{k-1}),$$

where p_1, p_2, $\dots$, p_m are the parameters of the model.

For the Vasicek model and for any other model for which we can obtain an explicit expression of the transition densities, we can obtain explicitly the log-likelihood function and use it to obtain maximum likelihood estimators of the parameters. For the Vasicek model, the parameters are R, α, and σ. For the Black–Scholes model we have used these estimation procedures in Section 8.1; the parameters are r and σ^2.[1]

The *maximum likelihood estimators* are the parameter values $\hat{p}_1, \dots, \hat{p}_m$ that maximize the log-likelihood. To obtain them, we can solve the system of equations $\frac{\partial L}{\partial p_i} = 0$ $(i = 1, \dots, m)$, being careful to check if the obtained solution is indeed a local maximum. A way to check is to use the Hessian matrix $\mathbf{H}$ and see if the matrix $-\mathbf{H}$ is positive definite; we are reminded that the Hessian matrix has entries $H_{ij} = \frac{\partial^2 L(\hat{p}_1, \dots, \hat{p}_m)}{\partial p_i \partial p_j}$ $(i, j = 1, \dots, m)$.

Often, the log-likelihood functions have a complicated expression, sometimes also with a high number m of parameters. In that case, one may use numerical software packages (there are several available in R), either to solve numerically the system of (usually non-linear) equations $\frac{\partial L}{\partial p_i} = 0$ $(i = 1, \dots, m)$ or to determine directly local maxima of the log-likelihood function L (or,

---

1 Since the maximum likelihood estimators have functional invariance, one can work alternatively with any equivalent set of parameters, like R and σ.

equivalently, local minima of the function $-L$). In the first case, one may get a local minima of L instead of a local maxima, so the solution should be checked (e.g. by slightly changing the values in the solution in different ways and seeing whether the corresponding values of L decrease). In both cases, the packages usually require an initial guess of the maximum to start the algorithm and one should be careful because a global maximum may not exist or the result might be a local but not a global maximum. Trying different initial guesses may lead to more local maxima (if there are several) and hopefully one of them will be the global maxima. The algorithm may not converge for some initial guesses, in which case one should try others.

The *Fisher information matrix* $\mathbf{F}$ is an $m \times m$ matrix with entries $F_{ij} = \mathbb{E}\left[-\frac{\partial^2 L(p_1,\dots,p_m)}{\partial p_i \partial p_j}\right]$ $(i,j = 1,\dots,m)$. Quite often, the expectations are difficult to obtain and one can use instead the empirical Fisher information matrix $\hat{\mathbf{F}} = -\mathbf{H}$ with entries $\hat{F}_{ij} = -\frac{\partial^2 L(\hat{p}_1,\dots,\hat{p}_m)}{\partial p_i \partial p_j}$ obtained without taking expectations and using the maximum likelihood parameter estimates as approximations of the true parameter values. If we use software packages to determine the maximum of L (or the mininum of $-L$), some packages give us at the end an approximation of $\mathbf{H}$ based on numerical approximations of the second-order derivatives.

The maximum likelihood estimators form a random vector $[\hat{p}_1,\dots,\hat{p}_m]^T$ that, under appropriate regularity conditions, has asymptotically (as $n \to +\infty$) a m-dimensional Gaussian distribution with mean vector $[p_1,\dots,p_m]^T$ (the true parameter values) and variance-covariance matrix equal to the inverse of the Fisher information matrix $\mathbf{C} = \mathbf{F}^{-1}$. So, these estimators are asymptotically unbiased and efficient. Using the asymptotic distribution as an approximation of the exact distribution, we can obtain approximate confidence intervals. Since we do not know the exact parameter values, we do not know the exact variance-covariance matrix $\mathbf{C}$, but we can approximate it by $\hat{\mathbf{C}} = \hat{\mathbf{F}}^{-1}$. For example, an approximate 95% confidence interval for the parameter $p_2 = \alpha$ of the Vasicek model is $\hat{\alpha} \pm 1.96\hat{c}_{22}$, where $\hat{c}_{22}$ is the second diagonal element of matrix $\hat{\mathbf{C}}$. That is how we have obtained the parameter estimates of the stochastic Gompertz cow growth model in Section 12.1 (we have worked there with the logarithms, which follow a Vasicek model).

In classical statistics the sample data are independent random variables coming from a common distribution. Here, however, the observations are dependent, so regularity conditions may need to be adjusted and the speed of convergence may be slower. Particularly for small n, maximum likelihood estimators may have relevant biases and variances may deviate appreciably from the asymptotic values. In that case, one may simulate trajectories to use in a bootstrap procedure in order to correct biases and variance discrepancies, thus obtaining more reliable confidence intervals, as did Filipe (2011) for the cow growth model.

If one cannot obtain the transition distributions explicitly, one can, like we did in the Euler method, replace them by approximations and use a *pseudo-likelihood function*. This has the advantage of using a Gaussian distribution. However, unlike simulations in which we can choose a large number n of grid points, in estimation the sample size n (and so the grid points available) depend on the data and may be relatively small, thus affecting the quality of the estimation. In such cases, particularly for small sample size n, there are alternative techniques, recurring simultaneously to estimation and to simulations, to improve the quality of the estimation.

So far, we have considered the estimation based on data of one sample path, which is the typical situation in most applications. In fact, typically we only have the trajectory ω 'chosen' by nature or the markets and cannot repeat the experiment with other 'natural' or 'market' conditions. In classical statistics, having no replications (i.e. a sample of one) is fatal, but here this is compensated by the fact that we have observations of the single sample path for different time instants; they come from different distributions but such distributions have relationships and this creates a kind of imperfect replication. If $X(t)$ is ergodic, as is the case for the Vasicek model and others with a stochastic equilibrium, we know that averaging continuously in time along a single trajectory converges as $t \to +\infty$ to the mathematical expectations that are usually estimated by averaging along many trajectories. But one cannot usually average continuously in time because there is only a finite number of observations at a discrete set of time instants. Fortunately, for ergodic and also non-ergodic models, other methods like maximum likelihood can be used.

We did not talk about it because we are working with a different type of applications, but there are situations, particularly in some engineering applications, in which a variable is measured continuously. In that case, if the process is ergodic, continuously averaging over time would give excellent results, particularly in estimating the function $g(x)$ (in theory, it could be estimated without error). In practice, averaging continuously may only be possible approximately; in reality, if computations are required, one can only use a discrete finite set of observations in such computations, although we may choose a very small time interval between the used observations.

There are, occasionally, other experimental situations in which we can replicate the experiment and have several sample paths (several ω values) available. That is the case for the mertolengo cows mentioned in Section 12.1, in which there were 97 different sample paths (one for each cow) with several observations each. In this case, each cow has its own log-likelihood function and, assuming the cows have independent growth processes with the same model and parameter values, the joint log-likelihood for the 97 cows is simply the sum of the 97 individual log-likelihoods. This joint log-likelihood function was used to obtain the parameter estimates and their confidence intervals displayed in Section 12.1. Of course, the estimates are much more precise than the ones based on just one cow.

There are many other situations one might consider. For example, the observations may not be exact due to the measurement error of the instrument or method used to obtain them. In this case, one can use some filtering techniques.

Another common situation is the one in which we know that an autonomous SDE $dX(t) = f(X(t))dt + g(X(t))dW(t)$ is an appropriate model and we know that the solution is an ergodic process, but we do not know the exact type of functions $f(x)$ and $g(x)$ that govern the dynamics. So, unlike parametric models, for example the stochastic Gompertz model, in which we know (or assume we know) the type of functions (and need only estimate the unknown parameters in those functions), now we do not have parameters to estimate. Rather, we need to estimate the whole functions $f(x)$ and $g(x)$ for all values of x in the state space. Since the process is ergodic, the process will assume all these state values x infinitely often. So, given enough time, the process $X(t)$ will assume any concrete value x several times and we could use the sample mean and variance of the quotient ratio of the immediately following changes as estimates of the infinitesimal mean $a(x) = f(x)$ and the infinitesimal variance $b(x) = g^2(x)$ at that state x. The problem again is that usually we only have a finite number of discrete-time observations. So, a particular value of x is likely not to be among those observed. However, we may observe values close to x which, assuming the functions are smooth, would provide approximations of $f(x)$ and $g(x)$. If we have a relatively large number of observations, we could, using appropriate smoothing techniques, estimate the whole functions $f(x)$ and $g(x)$. This is the object of *non-parametric estimation*.

The traditional methods for non-parametric estimation were developed for the typical situation of data coming from a single trajectory. See Florens-Zmirou (1993), Jiang and Knight (1997), Bandi and Phillips (2003), and Nicolau (2003). In Filipe *et al.* (2010) and Filipe (2011) these methods were extended to the case of several trajectories and applied to the mertolengo cow data, for which there were 97 trajectories. The shapes of the functions $f(x)$ and $g(x)$ obtained show some curves and counter-curves but do not differ much from the stochastic Gompertz model mentioned in Section 12.1. The non-parametric estimates of $f(x)$ and $g(x)$ can provide suggestions for the choice of appropriate parametric function models. The advantage of using parametric models is quite obvious, since they allow easy computation of many important probabilities, predictions, and other quantities of interest.

For more on estimation and other statistical inference issues on SDEs, we refer the reader to Basawa and Rao (1980), Kessler *et al.* (2012), Sørensen (2004), Sørensen (2009), Dacunha-Castelle and Florens-Zmirou (1986), Küchler and Sørensen (1997), and Aït-Sahalia (2002, 2008).

Prediction

We will give just the basic idea, following the example shown in Secton 8.1 for the Black–Scholes model on how to predict, at a time t_n, a future value $X(t_n + \tau)$ with $\tau > 0$, and also how to obtain a confidence prediction interval.

We may work directly with the variable of interest $X(t)$, as we assume here, or, for convenience, with some normalizing transformation of it (like we did in the Black–Scholes model using the logarithmic transformation to get Gaussian distributions). Here, we will, for convenience, assume Gaussian approximations, but one can use the exact distributions whenever available.

Basically, the point prediction could be the conditional expectation $\hat{X}(t_n + \tau) = \mathbb{E}[X(t_n + \tau) \mid X(t_n)]$. Due to the Markov property, there is no need to condition with respect to the past. Suppose we have an explicit expression of $X(t)$ in terms of $X(t_n)$, of the form $X(t) = X(t_n) + M$. Then $\hat{X}(t_n + \tau) = X(t_n) + \mathbb{E}[M \mid X(t_n)]$. This predictor has an error, which variance V_1 is the conditional variance of M. If we know the true parameter values, this provides, assuming M is approximately Gaussian, the approximate 95% confidence prediction interval $\hat{X}(t_n + \tau) \pm 1.96\sqrt{V_1}$.

If, as is usually the case, we do not know the true parameter values and replace them by their maximum likelihood estimators in the expression of $\mathbb{E}[M \mid X(t_n)]$, this introduces an extra prediction error, which is independent of the previous prediction error because the parameter estimators only used present and past information. Using the approximate Gaussian distribution with zero mean and variance-covariance matrix $\hat{\mathbf{C}} = \hat{\mathbf{F}}^{-1}$ of the estimators and using the expression of M, one can obtain approximately the mean B (bias, which might be 0) and variance V_2 of the extra prediction error. Correcting the point estimator for the bias if necessary, we may again use a Gaussian approximation, so that the total prediction error is approximately Gaussian with mean 0 and variance $V = V_1 + V_2$, leading to the approximate confidence interval: point predictor $\pm 1.96\sqrt{V}$. If V depends on some unknown parameters, we approximate them by their maximum likelihood estimators (introduces an extra error, usually negligible).

What happens if we do not have an explicit expression for $X(t_n + \tau)$ as a function of $X(t_n)$? We can use an approximation of the type used in the Euler method if τ is small. If τ is larger, then we can simulate many trajectories from time t_n to time $t_n + \tau$ (using also in each trajectory an independently simulated value of each parameter estimator according to the distribution of the parameter estimators). The simulations could use the Euler method. The sample mean of the trajectories at time $t_n + \tau$ can be used as the point prediction of $X(t_n + \tau)$. Choosing the empirical quantiles 2.5% and 97.5% of the trajectories at time $t_n + \tau$ gives an approximate 95% confidence prediction interval.

Model choice

One may test if some parameter p of a model is significant by using a likelihood ratio test, so that one can eliminate parameters that contribute little to explain the variability in the data. The null hypothesis $H_0 : p = p_0$, where p_0 is a concrete numerical value (often $p = 0$), will be tested against the alternative $H_1 : p \neq p_0$. Let us consider the log-likelihood L as a function of all parameters, including p. Then one determines the maximum likelihood estimators of the parameters in two ways. One way considers all parameters in the maximization of L, resulting in a maximum log-likelihood $\hat{L}_1$. The other fixes the value of p at p_0 in L and maximizes with respect to the other parameters, obtaining the maximum log-likelihood $\hat{L}_0$. Then, for n large, the test statistic $-2(\hat{L}_0 - \hat{L}_1)$ has asymptotically a chi-square distribution with one degree of freedom, which critical value for a 5% significance is 3.841. So, if the test statistic is above 3.841, one rejects H_0 at the 5% significance level and keeps the parameter p in the model. If the test statistic is below 3.841, one accepts H_0 and fixes the value of p at p_0 in the model, therefore reducing the number of parameters by one. Of course other significance levels can be used. One can test two or more parameters simultaneously, in which case the test statistic has two or more degrees of freedom (and different critical values).

Sometimes, one wants to choose between two different models, neither of which can be obtained from the other by fixing the value of a particular parameter. In that case, there is no test, but one can obtain the maximum log-likelihoods under the two models and compare the models by the commonly used statistical criteria (AIC, BIC, etc.). Of course, one can also compare more than two models. An alternative way of comparing models, which might be preferable, is to look at their performance in predicting data values not used in the parameter estimation procedure.

Mixed models

An example of mixed models can be seen in Filipe (2011). The cow weight $X(t)$ was modelled by a stochastic Gompertz model. So, $Y(t) = \ln X(t)$ satisfies the Vasicek model $dY(t) = -\alpha(Y(t) - R)dt + \sigma dW(t)$, with $R = \ln A$, where A is a kind of average asymptotic size of a cow. However, the mertolengo cows of the studied strand, although relatively similar, do have some genetic differences that affect their parameter R and so we may assume that different cows have a different R. Let the probability distribution of R values among these cows be Gaussian with mean value R^* and variance V. Then a randomly chosen cow will follow the SDE model $dY(t) = -\alpha(Y(t) - R)dt + \sigma dW(t)$, but with a random R value. That is an example of a mixed model.

The parameters of the mixed model are now α, σ, R^*, and V. Parameter estimation methods for this mixed model and comparisons with the classical stochastic Gompertz model (which is the particular case $V = 0$) were developed in Filipe (2011).

12.3 Some applications in population dynamics

In Chapter 3, for the growth of a population living in an environment with random fluctuations affecting the *per capita* growth rate (abbreviated to growth rate), we have proposed the *Malthusian growth model*. That assumes the resources are abundant no matter what is the size $X(t)$ of the population.

In a deterministic environment, the (instantaneous) growth rate would be a constant R, so that the growth rate in a small time interval $]t, t + \Delta t]$ $(\Delta t > 0)$ will be approximately $R\Delta t$ and the change in population size will be $\Delta X(t) :=$ $X(t + \Delta t) - X(t) \simeq (R\Delta t)X(t)$.

In a random environment (see Chapter 3 for details), the rationale is that the effect of environmental random fluctuations on the growth rate (noise) can be described approximately by a white noise, so that its accumulated effect up to time t can be approximated by a Wiener process $\sigma W(t)$. This leads to $\Delta X(t) \simeq (R\Delta t + \sigma\Delta W(t))X(t)$, with $\Delta W(t) = W(t + \Delta t) - W(t)$, and, letting $\Delta t \to 0$, one gets the stochastic Malthusian growth model

$$dX(t) = RX(t)\,dt + \sigma X(t)\,dW(t), \qquad X(0) = x_0 > 0.$$

R was loosely interpreted in the literature as an average growth rate.

We have used Itô calculus and obtained and studied the solution in Section 8.1. The solution is the *geometric Brownian motion* process

$$X(t) = x_0 \exp\left(\left(R - \frac{\sigma^2}{2}\right)t + \sigma W(t)\right).$$

In Section 9.1 we have seen that the loose interpretation of R as an unspecified average growth rate could be dangerous and was the reason behind the controversy regarding which calculus, Itô or Stratonovich, is more appropriate for applications. In fact, if one used the Statonovich SDE

$$(S) \quad dX(t) = R_S X(t)\,dt + \sigma X(t)\,dW(t), \qquad X(0) = x_0 > 0,$$

where R_S was also loosely interpreted as average growth rate, the solution would be apparently different:

$$X(t) = x_0 \exp(R_S t + \sigma W(t)).$$

Also, the asymptotic qualitative behaviour as $t \to +\infty$ would look different. Population extinction, i.e. $X(t) \to 0$, occurs under Stratonovich calculus when $R_S < 0$, while under Itô calculus it occurs when $R < \frac{\sigma^2}{2}$ and so occurs even for positive values of the so-called average growth rate R. When $R_S > 0$, under Stratonovich calculus we have $X(t) \to +\infty$, the same happening under Itô calculus when $R > \frac{\sigma^2}{2}$.

As we have seen in Section 9.1, in fact the so-called unspecified average growth rate turns out to be two different growth rates if one properly looks at the true meaning in terms of the population size trajectories. We have seen that, under Stratonovich calculus, R_S is the *geometric average growth rate* R_g, while under Itô calculus, R is the *arithmetic average growth rate* R_a (and $r = R - \frac{\sigma^2}{2}$ is the geometric average growth rate R_g). So, in terms of concrete well-defined average growth rates, the solutions under the two calculi are identical:

$$X(t) = x_0 \exp(R_g t + \sigma W(t)).$$

The qualitative behaviour is obviously also identical, having population extinction when the geometric average growth rate R_g is negative and having growth without bounds when R_g is positive.

However, wildlife populations have limitations in food, territory, and other resources, so the amount of resources available for the survival and reproduction of each individual decreases when the population size increases. Therefore, we should have *density-dependence*, i.e. the *per capita* growth rate (abbreviated to growth rate) should not be a constant but rather a decreasing function $R(X)$ of the population size. So, the Itô SDE model becomes

$$dX(t) = R(X(t))X(t)\,dt + \sigma X(t)\,dW(t), \qquad X(0) = x_0 > 0,$$

where, as seen in Chapter 9, $R(x)$ is the arithmetic average growth rate $R_a(x)$ when the population has size x. If one uses the Stratonovich SDE model (S) $dX(t) = R_S(X(t))X(t)\,dt + \sigma X(t)\,dW(t)$, we have seen that $R_S(x)$ is the geometric average growth rate $R_g(x)$ when the population has size x. If one uses the correct average for the corresponding calculus, the two calculi give exactly the same results and one can use one or the other indifferently. Results and properties can, of course, be indifferently expressed in terms of the arithmetic or the geometric average but, since population growth is a multiplicative phenomenon, analogies with the deterministic model are more striking if one expresses them in terms of the geometric average growth rate.

We will use here the Itô calculus and so $R(x)$ will mean the arithmetic average growth rate when the size of the population is x. The geometric average growth rate will be $R_g(x) = R(x) - \frac{\sigma^2}{2}$.

Throughout this section we will assume that there are indeed environmental random fluctuations so, unless otherwise stated, we will assume $\sigma > 0$. The exception is of course the case of the deterministic environment, corresponding to $\sigma = 0$ and to a deterministic model.

Among the many decreasing functions $R(x)$, the simplest one is a linear function $R(x) = r(1 - \frac{x}{K})$. The parameter $r > 0$ is called the *intrinsic growth rate*, which is the maximum growth rate and occurs when population size approaches 0, precisely when the resources available to each individual are the highest. The parameter $K > 0$ is called the *carrying capacity of the environment*, i.e. the population that the environment can carry in a sustainable way; in fact, K is a stable equilibrium for the deterministic model. An interpretation is that the total amount of resources is proportional to K and the amount of resources used by the population is proportional to the population size x, so $\frac{K-x}{K} = 1 - \frac{x}{K}$ is the proportion of resources still available for growth and so the growth rate $R(x)$ should be proportional to $1 - \frac{x}{K}$. The deterministic environment model

$\frac{dX}{dt} = rX(1 - \frac{X}{K})$ for $X = X(t)$ is the most popular population growth model, called the *logistic* or *Pearl–Verhulst* model, with a history dating back to 1838.

In a stochastic environment, we have the *stochastic logistic model*, first studied by Levins (1969) and Capocelli and Ricciardi (1974), which takes the form $dX = rX(1 - \frac{X}{K})dt + \sigma X dW$ or, making time dependence explicit:

$$dX(t) = rX(t)\left(1 - \frac{X(t)}{K}\right) dt + \sigma X(t)\, dW(t), \qquad X(0) = x_0 > 0.$$

$$(12.1)$$

As mentioned in Section 8.1, this is an *environmental stochasticity* model and is not appropriate to deal with *demographic stochasticity*.[2]

The stochastic logistic model is an autonomous SDE with coefficients of class C^1, and so, as mentioned in Section 7.3, even though the drift coefficient does not satisfy a Lipschitz condition, there is a unique solution up to a possible explosion time. However, since the boundaries of the state space are $r_1 = 0$ and $r_2 = +\infty$, we can show that there is no explosion by showing that the boundary $+\infty$ is unattainable; in fact, it is even non-attractive. Therefore, there are no explosions and the solution exists and is unique for all $t \geq 0$. We even know that the solution is a homogeneous diffusion process with drift and diffusion coefficients, respectively,

$$a(x) = rx\left(1 - \frac{x}{K}\right) \quad \text{and} \quad b(x) = \sigma^2 x^2.$$

To see that $r_2 = +\infty$ is non-attractive, note that the scale density is, from (11.16), $s(\xi) = \exp\left(-\int_{\xi_0}^{\xi} \frac{2r\eta(1-\eta/K)}{\sigma^2\eta^2} d\eta\right) = \exp\left(-\frac{2r}{\sigma^2}\int_{\xi_0}^{\xi}\left(\frac{1}{\eta} - \frac{1}{K}\right) d\eta\right) = \exp\left(-\frac{2r}{\sigma^2}\left(\ln\frac{\xi}{\xi_0} - \frac{1}{K}(\xi - \xi_0)\right)\right) = C\xi^{-2r/\sigma^2}\exp\left(\frac{2r}{\sigma^2 K}\xi\right)$, where $C > 0$ is a constant. Since $s(\xi) \to +\infty$ as $\xi \to +\infty$, for every constant $A > 0$ we have $s(\xi) \geq A$ for ξ larger than some x_A. So $S(]x_A, +\infty[) = \int_{x_A}^{+\infty} s(\xi)d\xi \geq \int_{x_A}^{+\infty} A d\xi = +\infty$. So, the boundary $+\infty$ is non-attractive.

Unlike the deterministic model, in (12.1) we do not expect population size to approach a stable equilibrium point when $t \to +\infty$, since the population size will continue changing due to the random environmental variations driven by $\sigma X dW$. Figure 12.1 gives an illustration, showing, for a particular case, two different $X(t)$ trajectories, where we have used the Euler method described in Section 7.1 to simulate the trajectories. What we may expect is the existence

---

2 Demographic stochasticity is the random sampling variations in the number of births and deaths that occur even when the environment is steady and therefore birth and death rates are not influenced by environmental conditions. For large population sizes, the *per capita* effect of demographic stochasticity, which is of the order of $X^{-1/2}$, is negligible compared to the constant order *per capita* effect of environmental stochasticity. However, to study extinction issues, when population size is low, we have to be more careful.

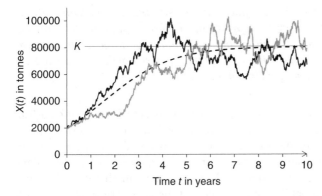

Figure 12.1 The solid irregular lines represent two simulated trajectories (two different values of ω) of population size $X(t)$ (in tonnes) in the time interval [0, 10] years, assuming a stochastic logistic model (12.1) with parameters of Exercise 12.4d and initial population size $X(0) = x_0 = 20000$ tonnes. Time was discretized in steps of $\Delta t = 0.01$ years and the values simulated by an Euler scheme were joined by straight lines. The dashed line is the solution of the deterministic logistic model, corresponding to $\sigma = 0$; the solid horizontal line shows the carrying capacity K.

of a *stochastic equilibrium* in the sense that the probability distribution of $X(t)$ will approach an equilibrium distribution with a stationary density $p(x)$ when $t \to +\infty$.

Exercise 12.4 *For the stochastic logistic model (12.1) with $r > 0$ and $K > 0$:*

(a) *Show that the extinction boundary $r_1 = 0$ is non-attractive if $r > \sigma^2/2$ and is attractive if $r < \sigma^2/2$.*
    Suggestion: Note that you do not need to actually compute $S(x)$ at the boundary, just check whether it is finite or infinite.

(b) *Using Property 11.4, show that, for $r > \sigma^2/2$, there is a stochastic equilibrium and the process is ergodic. Determine the stationary density $p(x)$ (defined in the state space, i.e. for $x > 0$).*
    Suggestion: Show that the speed density is $m(x) = Dx^{(2r/\sigma^2-1)-1} \exp\left(-\frac{2r}{\sigma^2}x\right)$, with $D > 0$ constant. Therefore, when $r > \sigma^2/2$, $m(x)$ is, apart a multiplicative constant, a gamma pdf with shape parameter $\frac{2r}{\sigma^2} - 1$, so its integral over the state space is finite. $p(x)$ is proportional to $m(x)$ and the proportionality constant can be obtained from the properties of the gamma distribution.

(c) *For $r > \sigma^2/2$, determine the mean and the mode of the stationary distribution, which are close to the deterministic equilibrium K when σ is small. Determine also the variance of the stationary distribution.*

(d) *For the particular case of a Pacific halibut (Hippoglossus hippoglossus) fishery, Hanson and Ryan (1998) estimated the parameters for the logistic growth model as $r = 0.71$/year and $K = 8.05 \times 10^4$ tonnes. Assume* *the same parameters for the stochastic logistic model and suppose that $\sigma = 0.2/\sqrt{year}$. If fishing is stopped and the fish population left growing towards the stochastic equilibrium, determine, for large t, an interval around the mean value where the population size will be with a 95% probability (coincidence prediction interval).*

For $r > \sigma^2/2$, the extinction boundary is non-attractive, so 'mathematical' extinction is excluded (i.e. the population size will not reach 0 nor converge to 0 as $t \rightarrow +\infty$) and, furthermore, the process is ergodic with a stochastic equilibrium having a stationary density. For $r < \sigma^2/2$, the extinction boundary is attractive and there is no stationary density; in fact, 'mathematical' extinction will occur with probability one. Notice that the geometric average growth rate $R_g(x) = r(1 - \frac{x}{K}) - \frac{\sigma^2}{2}$ converges as $x \downarrow 0$ to $R_g(0^+) = r - \frac{\sigma^2}{2}$, which can be interpreted as the *geometric average intrinsic growth rate*. It is the growth rate when population size approaches 0 and therefore the *per capita* availability of resources is the highest. In summary, we have 'mathematical' extinction when the geometric average intrinsic growth rate $R_g(0^+)$ is negative and we have a stochastic equilibrium with a stationary density and no 'mathematical' extinction when $R_g(0^+)$ is positive.

The *stochastic Gompertz model* corresponds to $R(x) = r \ln \frac{K}{x}$ with $r > 0$ and $K > 0$, leading to the Itô SDE

$$dX(t) = rX(t) \ln \frac{K}{X(t)} dt + \sigma X(t) \, dW(t), \qquad X(0) = x_0 > 0. \qquad (12.2)$$

The geometric average growth rate is $R_g(x) = r \ln \frac{K}{x} - \frac{\sigma^2}{2}$ when population size is x and the geometric average intrinsic growth rate is $R_g(0^+) = +\infty > 0$. The parameter K is still the carrying capacity, i.e. the stable equilibrium of the deterministic model. Of course, again, there is no stable equilibrium for the stochastic Gompertz model, but rather a stochastic equilibrium with a stationary density.

To facilitate the study of the Gompertz model, one can work with the transformed variable $Y(t) = \ln X(t)^3$ and apply Itô calculus rules to obtain the equivalent SDE

$$dY(t) = -r \left(Y - \left(\ln K - \frac{\sigma^2}{2r} \right) \right) dt + \sigma \, dW(t) \qquad Y(0) = y_0 = \ln x_0, \qquad (12.3)$$

---

3 If the population size $X(t)$ has physical units like 'tonnes' (used for fish populations) or 'individuals per cubic centimetre' (used for micro-organisms) or simply 'individuals', one cannot apply to it transcendental function transformations like the logarithm. In that case, one can use a transformation like $Z(t) = \ln \frac{X(t)}{x_0}$ or, more simply, use, as we did, $Z(t) = \ln X(t)$ and consider $X(t)$ with no physical units, interpreting it as 'number of tonnes of the population', 'number of individuals in a cubic centimetre' or 'number of individuals in the population'.

which is the already studied Vasicek model. The boundaries of $X(t)$ are again $r_1 = 0$ and $r_2 = +\infty$, and correspond to the boundaries $-\infty$ and $+\infty$ of the $Y(t)$ process, which we know from Section 11.3 are both non-attractive, the same happening, of course, to the $X(t)$ boundaries. We also know that the Y process has a stochastic equilibrium, so the same happens, of course, to the X process. The stationary density of the population size can be easily obtained from the already known $\mathcal{N}\left(\ln K - \frac{\sigma^2}{2r}, \frac{\sigma^2}{2r}\right)$ stationary density $p_Y(y)$ of the Y process. We simply have $p_X(x) = p_Y(y)\left|\frac{dy}{dx}\right| = \frac{1}{x}p_Y(y)$, and so, since $y = \ln x$, the stationary density of the X process is lognormal:

$$p_X(x) = \frac{1}{x\sqrt{2\pi\sigma^2/(2r)}}\exp\left(-\frac{(\ln x - (\ln K - \sigma^2/(2r)))^2}{2\sigma^2/(2r)}\right) \quad (x > 0).$$

Exercise 12.5 *For the stochastic Gompertz model (12.2) with $r > 0$ and $K > 0$:*

(a) *Determine the mean, the median, and the mode of the stationary distribution of the population size, which are close to the carrying capacity K of the deterministic model if σ is small. Determine also the variance of the stationary distribution.*

(b) *From Section 11.1, we know the explicit solution of the SDE (12.3) and also that the $Y(t)$ process has for finite $t > 0$ a Gaussian transient distribution with mean $\left(\ln K - \frac{\sigma^2}{2r}\right) + e^{-rt}\left(y_0 - \left(\ln K - \frac{\sigma^2}{2r}\right)\right)$ and variance $\frac{\sigma^2}{2r}(1 - e^{-2rt})$. Using such results, obtain the explicit solution in terms of $X(t)$ and the pdf of $X(t)$, as well as the mean and variance of $X(t)$.*

(c) *For the particular case of a Bangladesh shrimp (Penaeus monodon) fishery, Kar and Chakraborty (2011) estimated the parameters for the deterministic Gompertz growth model as $r = 1.331$/year and $K = 11.4 \times 10^3$ tonnes. Assume the same parameters for the stochastic Gompertz model and suppose that $\sigma^2 = 0.16$/year. Suppose that the current shrimp population size is $X(0) = x_0 = 5.0 \times 10^3$ (measured in tonnes). If fishing is stopped and the shrimp population left growing, determine the mean and standard deviation of $X(5$ years$)$ and the asymptotic mean and standard deviation as $t \to +\infty$, i.e. at the stationary equilibrium.*

(d) *If $X(1$ year$) = 8.7 \times 10^3$ tonnes, what is the conditional probability that $X(5$ years$)$ will exceed 10.8×10^3 tonnes?*

(e) *Simulate two trajectories of $X(t)$ for $t \in [0, 10]$ years using the parameter values in (c).*

Suggestion: Simulate the $Y(t)$ process, which follows a Vasicek model, using the method described in Section 12.2 and obtain $X(t) = \exp(Y(t))$.

Besides the logistic and Gompertz population growth models, there are many other growth rate functions $R(x)$ that have been proposed in the literature for populations with density-dependence in order to provide better adjustment to particular populations. However, the 'true' growth rate function $R(x)$ followed by the actual population is not exactly known and we would like to see whether the qualitative behaviour of the stochastic logistic and Gompertz models are just characteristic of these models or are valid for all other appropriate growth rate functions $R(x)$, including the 'true' unknown one followed by the population under study. For that, we need to show that such qualitative properties are valid for a general model in which, contrary to what we did before, we do not specify the form of $R(x)$.

Of course, we need to make some general assumptions on $R(x)$ to exclude functions that are not reasonable to describe the growth of wildlife populations under density-dependence. There are some technical assumptions, namely that $R(x)$ is defined for $x > 0$, is of class C^1 and the limit $R(0^+) := \lim_{x \downarrow 0} R(x)$ exists. The main biological assumption is the resource limitation assumption behind density-dependence, namely that the amount of resources available to each individual for survival and reproduction is smaller for larger sizes of the populations. So, we will assume that the (*per capita*) growth rate $R(x)$ is a strictly decreasing function of x. Notice that we say nothing concerning the shape this decrease will assume so that we cover all possibilities. As a result of resource limitation, we must also have $R(+\infty) := \lim_{x \to +\infty} R(x) < 0$, i.e. the environment cannot sustain an extremely large population, which will then have a death rate higher than the birth rate, resulting in a negative growth rate. We are assuming closed populations, with no immigration, so we may make the assumption that the total growth rate $xR(x)$, i.e. the growth rate of the whole population (not the *per capita* growth rate), converges to 0 when x converges to 0. Remember that the geometric average intrinsic growth rate is $R_g(0^+) = R(0^+) - \frac{\sigma^2}{2}$. We will see what happens when $R_g(0^+) > 0$ and when $R_g(0^+) < 0$. We will assume that $R_g(0^+) = 0$, being non-generic, cannot really happen in nature (the *a priori* probability of nature 'choosing' exactly the value zero for a given population is zero). Of course, one could undergo the useless study of that case, but one would need to know the specific form of $R(x)$ and little can be said in a general model setting.

Using only these very few assumptions, Braumann (1999b) (which also considers the case of harvested populations) shows that qualitative results very similar to those obtained above for specific models also hold for the *general stochastic population growth model* (see also Braumann (2007a)):

$$dX(t) = R(X(t))X(t)\,dt + \sigma X(t)\,dW(t), \qquad X(0) = x_0 > 0. \qquad (12.4)$$

Namely, the following results hold:

- In a deterministic environment (i.e. when $\sigma = 0$), the solution exists and is unique for all $t \geq 0$ and takes non-negative values. If the intrinsic growth rate $R_g(0^+) = R(0^+)$ is negative, the population will approach extinction size 0 as $t \to +\infty$. If $R_g(0^+)$ is positive, there is a stable equilibrium K (carrying capacity) towards which the population size will converge when $t \to +\infty$, where K is the only positive root of the equation $R(x) = 0$.
- In a randomly varying environment ($\sigma > 0$), the boundaries of the state space are $r_1 = 0$ and $r_2 = +\infty$ and, since the $+\infty$ boundary is always non-attractive, the solution of the SDE exists and is unique for all $t \geq 0$.
- In a randomly varying environment ($\sigma > 0$), if the geometric average intrinsic growth rate $R_g(0^+) = R(0^+) - \frac{\sigma^2}{2}$ is negative, the boundary 0 is attractive and the population will, with probability one, approach extinction size 0 as $t \to +\infty$.
- In a randomly varying environment ($\sigma > 0$), if $R_g(0^+) = R(0^+) - \frac{\sigma^2}{2}$ is positive, the boundary 0 is non-attractive and 'mathematical' extinction has a zero probability of occurring. There is no stable equilibrium, but rather a stochastic equilibrium with a stationary density and the process is ergodic. Furthermore, when σ is small, the mode of the stationary distribution approximately coincides with the carrying capacity of the deterministic model.

These results were even generalized (see Braumann (2002), which considers the case of harvested populations too) to models where the noise intensity may also depend on population size, i.e. we have $\sigma(x)$ instead of a constant σ. The assumptions are that $\sigma(x)$ is defined and positive for $x > 0$, is of class C^2, $x\sigma(x)$ converges to 0 when x converges to 0 (no immigration), and $\sigma(x)$ is bounded (or even unbounded but satisfying some other technical conditions).

So, for quite general SDE models of the growth of wildlife populations in randomly varying environments, we have seen that, when the geometric average intrinsic growth rate is positive, the probability of 'mathematical' extinction is zero and there is a stochastic equilibrium with a stationary density.

We shall remember that all models are approximations, hopefully good ones. So, if we are measuring population size in terms of number of individuals, $X(t)$ should be a discrete value taking non-negative integer values. However, we use as approximations ordinary differential equation (ODE) models if the environment is deterministic or SDE environmental stochasticity models if the environment is randomly varying, despite the fact these models use a continuous state variable $X(t)$ for the population size at time t. This is practically irrelevant if the population size is large, since we can approximate a non-integer model value by the nearest integer with just an insignificant relative error. Even though these models do not consider at all demographic stochasticity (sampling variations in births and deaths which need integer-valued variables to be treated properly), the effects of demographic stochasticity can be practically ignored for large population sizes, since the relative error committed in doing so is insignificant (of the order of $1/\sqrt{X(t)}$). However, the situation is quite different for small population sizes, and that holds both for ODE and SDE appropriate population models. Besides a large relative approximation error, we may have a model population value of, say, 0.7 individuals (or, if we measure population in biomass, we may have a value smaller that the smallest possible weight of an individual). This may happen if there is 'mathematical' extinction, and it is interesting to note that, in such a case, $\lim_{t \to +\infty} X(t) = 0$, but $X(t)$ never actually becomes 0, rather takes small positive values approaching zero. Such a low value may also happen, although with a very small probability, if 'mathematical' extinction does not occur. In any case, when the model gives a value like 0.7, should we say that the population is extinct or not? Furthermore, these models do not incorporate *Allee effects*[4] that affect some populations and, when they are of the strong type and the population has few individuals, lead almost automatically to extinction. A proper treatment of extinction in randomly varying environments should take also into account demographic stochasticity and possible Allee effects and would require much more sophisticated models, but we can take a shortcut approximation that keeps the most relevant qualitative features.

The idea is to replace the concept of 'mathematical' extinction, which is harder to achieve than real biological extinction, by what we call *'realistic' extinction*. We set up an adequate low *extinction threshold* value $a > 0$ below which we consider the population to be extinct. There are some criteria for the choice of an appropriate a used in wildlife preservation of endangered species. Otherwise, one might choose a value for which demographic stochasticity effects will lead to extinction with high probability or it could simply be $a = 1$ individual (we cannot really have populations smaller than that) or $a = 2$ individuals (in sexual populations, if the size is smaller than 2, extinction is unavoidable). If there are strong Allee effects not considered in the model growth rate $R(x)$, it could be the threshold a below which the real $R(x)$ would become negative (since we know that populations below that threshold will become extinct). We may assume that the initial population $X(0) = x_0 > a$ is not yet extinct. The first passage time T_a will be called the *extinction time*.

---

4 *Allee effects* (see Allee *et al.* (1949)) are exhibited by some populations when the density $X(t)$ is low due, for instance, to the difficulty of finding mating partners or the inefficiency of group defence against predators. For small populations sizes x, the growth rate $R(x)$ is now the result of two contradictory effects, the Allee effects (that fade away as the population gets larger, and would lead to $R(x)$ being an increasing function) and the resource limitation effects (that would lead to $R(x)$ being a decreasing function of x). When the result of the two effects leads to a $R(x)$ decreasing for low values of x and decreasing for higher values of x, the model is declared an Allee effects model. There are two types of Allee effects, strong and weak Allee effects, depending on whether $R(0^+)$ is negative or positive.

Going back to our general SDE density-dependence model for the growth of wildlife populations living in a randomly varying environment, satisfying the very general assumptions we have mentioned above, one can see in Braumann (1999b, 2002) and it is quite trivial to show that *'realistic' extinction will occur with probability one.* So, the issue is no longer whether extinction will occur or not (in 'realistic' terms, it will always occur), but how long it will take.

In fact, if the geometric average intrinsic growth rate $R_g(0^+)$ is negative, 'mathematical' extinction occurs with probability one, and so $X(t)$, a continuous function, on its approach to 0, will cross the extinction threshold a, and 'realistic' extinction will occur with probability one. If the geometric average intrinsic growth rate $R_g(0^+)$ is positive, 'mathematical' extinction has zero probability of occurring, but, since the process is ergodic, it will visit every point in the state space sooner or later; in particular, it will sooner or later cross the extinction threshold a and so 'realistic' extinction also occurs with probability one.

To study the *extinction time T_a* we can use the techniques of Section 11.3.

Exercise 12.6 *For the stochastic Gompertz model, obtain an expression for the mean extinction time.*

Suggestion: Take advantage of (11.41) and the fact that $Y(t) = \ln X(t)$ follows a Vasicek model.

Figure 12.2 shows, for a particular choice of parameters, an example of the behaviour of the mean and standard deviation of the extinction time for the stochastic Gompertz model as a function of $z = x_0/a$. Of course, for $z = 1$, we start with an extinct population and the expected extinction time $\mathbb{E}_{x_0}[T_a] = \mathbb{E}_a[T_a] = 0$.

General results and graphs showing the behaviour of the mean and standard deviation of the extinction time for several parameter values of the stochastic logistic and Gompertz models can be seen in Braumann (2008), Carlos and Braumann (2005, 2006, 2014), and Carlos (2013).

There are some general features an interested reader can observe there. As expected, the extinction time is larger for larger values of the geometric average intrinsic growth rate, for larger values of K (carrying capacity of the deterministic model and the value for which $R(K) = 0$), and for lower values of the noise intensity σ.

There are, however, two features that might be unexpected. One, which you can observe in Figure 12.2, is that the standard deviation of T_a is of the same order of magnitude as the mean extinction time. This means that the extinction time T_a is a r.v. with quite high relative variability, so that different populations with similar parameter values are likely to have quite different extinction times. The second feature, which can also be observed in Figure 12.2, is the behaviour of the mean (and similar behaviour holds for the standard deviation)

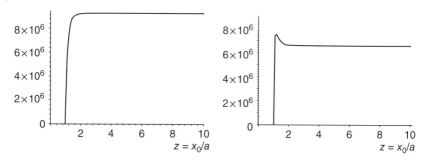

Figure 12.2 Mean (left) and standard deviation (right) of the extinction time T_a for the stochastic Gompertz model (12.2) as a function of $z = x_0/a$ (values of z on the horizontal axis) for $K = 100a$ and $r = \sigma^2 = 1$ per time unit. Time units are not specified here to allow one to still be able to use the figures when $r \neq 1$ in the original time unit: one just needs to change the time unit so that $r = 1$ per (changed) time unit, remembering that the extinction time values (vertical axes) refer to the changed time unit. Courtesy of Clara Carlos.

of the extinction time as a function of the ratio $z = x_0/a$. When the population size x_0 is very close to the extinction threshold (i.e. when z is small), the mean extinction time is strongly influenced by z, but otherwise it is an almost constant function of z (so it does not really matter whether the population size x_0 is large or small). One would like to have an idea of what is the value of this (almost) constant for different parameter values of the model. Of course, this expected extinction time can be of the order of a few generations, for highly varying (large σ) environments with low values of the intrinsic growth rate and carrying capacity. But it can also take an astronomical value, far exceeding the current estimated age of the universe, for the opposite situation. Typical situations for wildlife populations are usually somewhere in between these two extremes.

Of course, one can compute the expected value and standard deviation of T_a even when a is not an extinction threshold, but some other value $a < x_0$ of interest. For example, in fisheries or wildlife preservation, one may be worried about keeping the population size above some value a that is not the extinction threshold but rather an economical viable value for fishing or a minimal fruition level for nature lovers.

One can also study, using the techniques of Section 11.3, the first passage time T_b by a high threshold $b > x_0$. That may be interesting for setting up food product validity dates by considering the time T_b some spoiling bacterial population needs to reach a safety risk threshold b. In agriculture, if an insect pest population is damaging to a crop when it reaches some size level b, we will also be interested in the value of T_b in order to take appropriate timely pest control measures.

We have not directly considered the case of wildlife populations subjected to Allee effects and living in randomly varying environments. First accounts can be seen in Dennis (1989, 2002). In Carlos (2013) and Carlos and Braumann (2017), you can find a thorough qualitative study of general SDE population growth models with Allee effects in what concerns extinction and existence of a stochastic equilibrium with a stationary density. You can also find the particular case of the stochastic logistic-like model with Allee effects, as well as its comparison with the classical (no Allee effects) stochastic logistic model (12.1). For convenience, the study uses Stratonovich calculus, but it can be easily adapted to the use of Itô calculus.

A final word about population growth modelling. For the qualitative properties concerning the extinction and existence of a stochastic equilibrium with a stationary density, we have obtained general results that are valid whatever form the growth rate $R(x)$ assumes, as long as it follows the very general assumptions that characterize density-dependent growth. However, one often needs to make numerical predictions concerning the mean and standard deviation of the population size or the probability of a given event at a future date, or the mean and standard deviation of the extinction time, or how long it will take for the population to reach a given size. For such purposes, we need to use specific models like the stochastic logistic or the stochastic Gompertz model (or some other specific model) for which, based on the parameter values (either given or estimated), we compute the numerical values required for such predictions. We should be, however, aware that the model used in the computations is really an approximation of the 'true' unknown model governing the population growth and so our numerical predictions are also approximations. We would like to have bounds for the approximation error, so we know how accurate our predictions are. You can find out how to do this, as well as some examples, in Carlos (2013) and Carlos and Braumann (2014).

In nature, usually a population lives in a community of populations that interact in different ways, like predator–prey or parasite–host interactions, competition or mutualism. Interactions between two populations or among the populations of a whole community have been extensively studied in deterministic environments. The models used are systems of two or more ODEs, one for each of the interacting populations, describing how its growth rate is influenced by its own population size and by the sizes of the other populations. In a randomly varying environment, we have a system of two or more SDEs, i.e. a multidimensional SDE (see the end of Section 7.3), but the available studies are limited and more complex and will not be dealt with in this book. However, for the interested reader, we suggest as examples Rudnicki and Pichór (2007), Pinheiro (2016), Liu *et al.* (2018), and Hening and H. Nguyen (2018).

12.4   Some applications in fisheries

We will now study the growth of a population living in a randomly varying environment and being harvested. We will think of a fish population under fishing, but the same type of models can be applied in forestry to a tree population or in hunting to wildlife populations. Deterministic models of growth and fishing go way back and are commonly used in the management of fisheries. A good account of this vast field in bioeconomics, which includes also forestry, the treatment of uncertainty, structured populations, control theory, and other issues, can be seen in Clark (2010). Specific stochastic models of fisheries started with the pioneer work of Beddington and May (1977). In Braumann (1985) several specific models and the issue of yield optimization at the steady-state are covered.

Typically, a *stochastic fishing model* is an SDE population growth model with an extra term to account for the fish harvesting mortality. The *fishing effort* $E(t) \geq 0$ at time t measures the amount of effort invested in the fishing activity, such as hours trawled per day, number of hooks set per day or number of hauls of a beach seine per day; of course, there are different technologies and so it is common to use standardized fishing units (SFU) (see Hanson and Ryan (1998)). The fraction of the fish population size harvested per unit of effort and per unit of time is called the *catchability coefficient* or simply *catchability* and is denoted by $q > 0$. Therefore, the fraction of the fish population size harvested per unit of time is $qE(t)$ and, in terms of the fish population, corresponds to an additional mortality rate. What matters now to the population dynamics is not the natural growth rate but rather the *net growth rate* $R(x) - qE(t)$, which is the difference between the fish natural growth rate $R(x)$ and the additional mortality rate $qE(t)$ due to harvesting. The *yield* or harvesting rate is the amount of fish harvested per unit time:

$$H(t) = qE(t)X(t). \tag{12.5}$$

Although other models (like, for example, the Gompertz or the generalized logistic) have been used, it is very common to assume that the natural growth rate is of a logistic type $R(x) = r\left(1 - \frac{x}{K}\right)$. If the fishing effort is constant, $E(t) \equiv E$, this leads to the well-know model

$$dX(t) = rX(t)\left(1 - \frac{X(t)}{K}\right)\,dt \,-\, qEX(t)dt \,+\, \sigma X(t)\,dW(t), \tag{12.6}$$

$$X(0) = x_0 > 0,$$

which we can name *stochastic Gordon–Schaefer model*, following the name of the corresponding deterministic model. You can see an illustration on the left of Figure 12.3, which shows two different simulated trajectories (the two irregular lines, corresponding to two different states of nature ω) and also an approximation (the thick solid line, based on averaging over a 1000 simulated

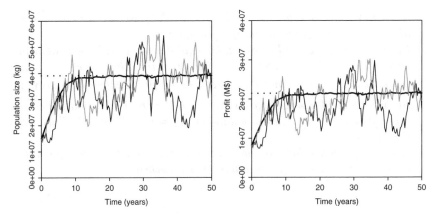

Figure 12.3 Population size $X(t)$ in kilograms (left) and profit per unit time $\Pi(t)$ in millions of dollars (right) for the stochastic logistic model with constant effort fishing (12.6). Population parameters are the same as in Exercise 12.4d, namely $r = 0.71$ year^{-1}, $K = 8.05 \times 10^7$ kg, and $\sigma = 0.2$ year$^{-1/2}$, and the initial population used is $X(0) = x_0 = 1.5 \times 10^7$ kg. The effort E used is the optimal sustainable effort $E^* = 104540$ SFU. The right-hand figure shows the corresponding profit $\Pi^*(t)$, i.e. the profit given by (12.8) when $E = E^*$; we are assuming, as estimated in Hanson and Ryan (1998), $q = 3.30 \times 10^{-6}$ SFU^{-1} year^{-1}, $p_1 = \$1.59$ kg^{-1}, $p_2 = \$0$ kg^{-2} year, $c_1 = \$96 \times 10^{-6}$ SFU^{-1} year^{-1}, and $c_2 = \$0.10 \times 10^{-6}$ SFU^{-2} year^{-1}. In each figure, the two thin irregular lines are two randomly chosen simulated trajectories (two different ω values) and the thick solid line is an estimate of the mean value based on averaging over a 1000 simulated trajectories; the dotted horizontal line shows the asymptotic (when $t \to +\infty$) mean value. Courtesy of Nuno M. Brites.

trajectories) of the expected value $\mathbb{E}[X(t)]$. Of course, what fishermen 'see' is not the expected value but one trajectory (the one actually 'chosen' by nature, similar to the two we have simulated). For the Monte Carlo simulations, we have used the Euler method (see Section 7.1).

But let us start with the deterministic Gordon–Schaeffer model

$$\frac{dX(t)}{dt} = rX(t)\left(1 - \frac{X(t)}{K}\right) - qEX(t),$$

obtained by putting $\sigma = 0$ in (12.6). Exercise 12.7 invites the reader to study this deterministic model.

Exercise 12.7 *For the deterministic Gordon–Schaefer model ((12.6) with $\sigma = 0$), qualitative behaviour is determined by the sign of the net intrinsic growth rate $R(0^+) - qE = R_g(0^+) - qE = r - qE$.*

(a) *If $r - qE < 0$, i.e. if the fishing effort $E > \frac{r}{q}$, show that 'mathematical' extinction will occur and we may say that there is overfishing.*

(b) *If there is no overfishing, i.e. when $r - qE > 0$, which is equivalent to $E < \frac{r}{q}$, show that 'mathematical' extinction will not occur and, as $t \to +\infty$, the population size $X(t)$ will converge to the stable deterministic equilibrium*

$X_\infty = K(1 - \frac{qE}{r})$. *Of course, when there is no fishing, we have $E = 0$ and population size converges to the carrying capacity K.*

(c) *For $r - qE > 0$, the yield at the stable equilibrium, called the* sustainable yield, *is $H_\infty = qEX_\infty = qEK(1 - \frac{qE}{r})$. Show that the fishing effort that maximizes it is $E_{MSY} = \frac{r}{2q}$, that the corresponding yield, called the* maximum sustainable yield, *abbreviated to MSY, is $\hat{H}_\infty = rK/4$, and that the corresponding stable equilibrium population is $\hat{X}_\infty = K/2$. Figure 12.4 illustrates these facts.*

Note: E_{MSY} usually does not maximize the yield $H(t)$ at a time t for which the stable population equilibrium is not yet reached.

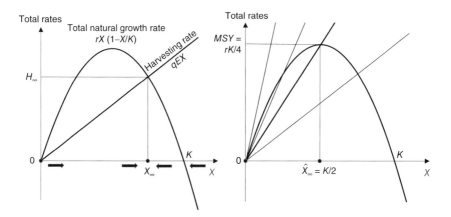

Figure 12.4 Illustration of the deterministic Gordon–Schaeffer model $\frac{dX(t)}{dt} = rX(t)\left(1 - \frac{X(t)}{K}\right) - qEX(t)$. On the left, the total rates (rates of change of the whole population, not the *per capita* rates) are shown as functions of population size X. The parabolic curve represents the natural total growth rate $rX\left(1 - \frac{X}{K}\right)$ and the straight line the harvesting rate qEX for a constant fishing effort $E < \frac{r}{q}$; the difference between the two is the net total growth rate of the population, which is positive or negative according to whether the curve is above or below the straight line. The equilibria occur when the curve meets the straight line (when the natural total growth rate equals the harvesting rate) and that happens when population size is 0 or is $X_\infty = K(1 - \frac{qE}{r})$. The arrows show the direction of the population size changes (population increases or decreases according to whether the net total growth rate is positive or negative) and one can see that, in this case, the equilibrium 0 is unstable and the equilibrium X_∞ is stable. At the stable equilibrium X_∞, the sustainable yield is the corresponding harvesting rate $H_\infty = qEK(1 - \frac{qE}{r})$. On the right, we show the same situation for several values of the effort E. One can see that, under overfishing $E > \frac{r}{q}$ (the steepest straight line), the only equilibrium population is 0 and is stable (even globally, i.e. the curve is always below the line and the population will move towards extinction, making the yield converge to zero). One can also see that the MSY is reached when the effort is such that the straight line meets the curve at the highest point (which is illustrated by the thick solid line).

(d) *For the data in Exercise 12.4d, namely $r = 0.71\,year^{-1}$ and $K = 8.05 \times 10^7\,kg$, and for $q = 3.30 \times 10^{-6}\,SFU^{-1}\,year^{-1}$ (SFU = Standardized Fishing Unit), obtain the MSY and the corresponding effort E_{MSY} at the stable equilibrium.*

(e) *Determine the explicit solution $X(t)$.*

Suggestion: Use the change of variable $Z(t) = \dfrac{e^{(r-qE)t}}{X(t)}$ and work with the ODE for $Z(t)$, which is easy to solve. Unfortunately, this trick does not help to get a useful explicit expression for the solution of the stochastic Gordon–Schaefer model.

Sometimes, however, the purpose is not to maximize the yield, but rather the profit, which is the difference between the revenue from selling the fish caught and the costs of fishing. The simplest profit structure assumes the following, per unit time:

- The revenues are proportional to the yield, $R(t) = pH(t)$, p being the sale price per unit of fish, e.g. if $X(t)$ is in kilograms and the sales are in dollars, p would be in $ kg^{-1} units.
- The costs are proportional to the fishing effort, $C(t) = cE(t)$, where c is the cost per unit effort and per unit of time, e.g. if the effort is in SFU and time is in years, c will be in $ $\mathrm{SFU}^{-1}\,\mathrm{year}^{-1}$.
- The profit is $\Pi(t) = R(t) - C(t)$.

We allow here a more general profit structure:

- We allow p to vary with the yield, since, when the yield is higher, fish sale prices tend to be lower. We assume for simplicity that $p = p(H) = p_1 H - p_2 H^2$ with $p_1 > 0$ and $p_2 \geq 0$, but obviously other functions $p(H)$ can be considered.
- We allow c to vary with the effort since, when the effort is high, back-up vessels that are not as efficient might have to be used and fishermen may have to work overtime with higher salary costs. We assume for simplicity that $c = c(E) = c_1 E + c_2 E^2$ with $c_1 > 0$ and $c_2 \geq 0$, but obviously other functions $c(E)$ can be considered.
- The profit is, therefore, attending to $H(t) = qE(t)X(t)$,

$$\Pi(t) = (p_1 - p_2 H(t))H(t) - (c_1 + c_2 E(t))E(t),\text{ with } H(t) = qE(t)X(t).$$
$$(12.7)$$

In the particular case of constant effort E, the profit is

$$\Pi(t) = (p_1 - p_2 H(t))H(t) - (c_1 + c_2 E)E,\text{ with } H(t) = qEX(t). \quad (12.8)$$

Remember, however, that in the above expressions $X(t)$ also depends on the effort.

- Optimization of the yield $H(t)$ is a particular case of profit $\Pi(t)$ optimization when $p_1 = 1$ and $p_2 = c_1 = c_2 = 0$.

Exercise 12.8 *For the deterministic Gordon–Schaefer model ((12.6) with $\sigma = 0$), assuming that the constant effort satisfies $E < \frac{r}{q}$, there is a stable equilibrium, as seen above. Considering the cost structure (12.8), the purpose now is to maximize not the yield but the profit at the stable equilibrium*

$$\Pi_\infty = (p_1 - p_2 H_\infty) H_\infty - (c_1 + c_2 E) E, \text{ with } H_\infty = qEX_\infty. \tag{12.9}$$

Remember that, as seen in Exercise 12.7b, $X_\infty = K(1 - \frac{qE}{r})$, so that Π_∞ is a fourth-degree polynomial function of the effort E.

(a) *Assuming additionally that $p_2 = 0$ and $E^* > 0$, show that:*

$$E^* = \frac{r}{2q} \frac{1 - \frac{c_1}{p_1 qK}}{1 + \frac{c_2 r}{p_1 q^2 K}}, \qquad \Pi_\infty^* = \frac{rKp_1}{4} \frac{\left(1 - \frac{c_1}{p_1 qK}\right)^2}{1 + \frac{c_2 r}{p_1 q^2 K}},$$

$$X_\infty^* = \frac{K}{2} \frac{\left(1 + \frac{2c_2 r}{p_1 q^2 K}\right) + \frac{c_1}{p_1 qk}}{1 + \frac{c_2 r}{p_1 q^2 K}}, \tag{12.10}$$

where Π_∞^* and X_∞^* are the values of Π_∞ and X_∞, respectively, when $E = E^*$. Suggestion: To determine E^*, one just needs to take the derivative with respect to E in the expression of Π_∞ and equate it to zero, checking whether the solution $E^* > 0$ is indeed a maximum of Π_∞. Then simple replacement in the expressions of Π_∞ and X_∞ and some algebra lead to the above result, which is a particular case of the result obtained by Brites (2017) in the stochastic case. If $p_2 > 0$, the derivative of Π_∞ with respect to E is a third-degree polynomial and it is probably easier to obtain E^* by numerical methods.

(b) *Assume the population parameters are the same as in Exercise 12.4d, namely $r = 0.71 \ year^{-1}$ and $K = 8.05 \times 10^7 \ kg$. Assume also, as estimated in Hanson and Ryan (1998), $q = 3.30 \times 10^{-6} \ SFU^{-1} \ year^{-1}$, $p_1 = \$1.59 \ kg^{-1}$, $p_2 = \$0 \ kg^{-2} \ year$, $c_1 = \$96 \times 10^{-6} \ SFU^{-1} \ year^{-1}$, and $c_2 = \$0.10 \times 10^{-6} \ SFU^{-2} \ year^{-1}$. Using the results of (a), determine E^*, Π_∞^*, and X_∞^*.*

For the stochastic Gordon–Schaefer model (12.6), the use of the techniques of Section 11.3, leads to the following conclusions:

(a) The boundary $+\infty$ is non-attractive and the solution exists and is unique for all $t \geq 0$, being a homogeneous diffusion process with drift coefficient $a(x) = rx \left(1 - \frac{x}{K}\right) - qEx$ and diffusion coefficient $b(x) = \sigma^2 x^2$.

(b) The *geometric average intrinsic net growth rate* is $R_g(0^+) - qE = R(0^+) - \frac{\sigma^2}{2} - qE = r - \frac{\sigma^2}{2} - qE$. It is negative or positive according to whether the constant effort E is, respectively, larger (overfishing) or smaller than $\frac{r - \sigma^2/2}{q}$.

(c) If $E > \frac{r - \sigma^2/2}{q}$ (overfishing), the 0 boundary is attractive and 'mathematical' extinction will occur with probability one.

(d) If $E < \frac{r - \sigma^2/2}{q}$, 'mathematical' extinction will not occur and there is an equilibrium distribution with a stationary gamma density (see, for example, Beddington and May (1977) or Braumann (1985))

$$p(x; E) = \frac{1}{\Gamma(\rho(E))} \theta^{\rho(E)} x^{\rho(E)-1} e^{-\theta x} \quad (x > 0),$$
$$\text{with } \rho(E) = \frac{2(r - qE)}{\sigma^2} - 1, \; \theta = \frac{2r}{\sigma^2 K}. \tag{12.11}$$

So, as $t \to +\infty$, the population size converges in distribution to a r.v. X_∞ with pdf $p(x; E)$; the stationary mean (which coincides with the stable equilibrium of the deterministic model) and second moment are

$$\mathbb{E}[X_\infty] = \frac{\rho(E)}{\theta} = K\left(1 - \frac{qE}{r}\right), \quad \mathbb{E}[X_\infty^2] = \frac{\rho(E)(1 + \rho(E))}{\theta^2}.$$

(e) If $E < \frac{r - \sigma^2/2}{q}$, the profit per unit time $\Pi(t)$ converges in distribution to the r.v. Π_∞, the *sustainable profit* given by (note that $H_\infty = qEX_\infty$)

$$\Pi_\infty = (p_1 - p_2 H_\infty)H_\infty - (c_1 + c_2 E)E = p_1 qEX_\infty - p_2 q^2 E^2 X_\infty^2 - c_1 E - c_2 E^2. \tag{12.12}$$

The expected value of the sustainable profit is then

$$\mathbb{E}[\Pi_\infty] = p_1 qE\mathbb{E}[X_\infty] - p_2 q^2 E^2 \mathbb{E}[X_\infty^2] - c_1 E - c_2 E^2. \tag{12.13}$$

Note that $\mathbb{E}[X_\infty]$ and $\mathbb{E}[X_\infty^2]$ also depend on the effort E.

(f) One cannot optimize the profit, since it is randomly varying, but one can optimize the expected profit. Brites (2017) and Brites and Braumann (2017), for $E < \frac{r - \sigma^2/2}{q}$, determine the *optimal sustainable constant effort* E^* that maximizes the expected value $\mathbb{E}[\Pi_\infty]$ of the sustainable profit. Let

$$\Pi^*(t) = p_1 qE^* X^*(t) - p_2 q^2 E^{*2} X^{*2}(t) - c_1 E^* - c_2 E^{*2} \tag{12.14}$$

and

$$\Pi_\infty^* = p_1 qE^* X_\infty^* - p_2 q^2 E^{*2} X_\infty^{*2} - c_1 E^* - c_2 E^{*2}, \tag{12.15}$$

where $X^*(t)$ is the solution of the SDE model (12.6) with $E = E^*$ and where X_∞^* has pdf $p(x; E^*)$.

Note that $\Pi^*(t)$ converges in distribution to Π_∞^*. Note also that $\Pi^*(t)$ is, at time t, a r.v. corresponding to the profit per unit time of the fishery when the fishing policy uses the optimal sustainable constant effort E^*, but its expected value is not optimal at time t; in fact, our optimization was not done at time t but at the stochastic equilibrium (i.e. when $t \to +\infty$). The right-hand side of Figure 12.3 illustrates, for a particular case, the behaviour of $\Pi^*(t)$ for two different simulated trajectories (two different possible states of nature ω). As you can see, the profit of the fishery per unit time may vary quite a bit. The dotted horizontal line represents the *optimal expected sustainable profit* $\mathbb{E}[\Pi_\infty^*]$.

Exercise 12.9 *For the stochastic Gordon–Schaefer model (12.6), assuming that the constant effort satisfies* $E < \frac{r - \sigma^2/2}{q}$:

(a) *Derive expression (12.11) of the stationary density* $p(x; E)$.
(b) *Assuming additionally that* $p_2 = 0$, *show, as in Brites (2017) and Brites and Braumann (2017), the analogue of (12.10) for the stochastic environment:*

$$E^* = \frac{r}{2q} \frac{1 - \frac{\sigma^2}{2r} - \frac{c_1}{p_1 qK}}{1 + \frac{c_2 r}{p_1 q^2 K}}, \qquad \mathbb{E}[\Pi_\infty^*] = \frac{rKp_1}{4} \frac{\left(1 - \frac{\sigma^2}{2r} - \frac{c_1}{p_1 qK}\right)^2}{1 + \frac{c_2 r}{p_1 q^2 K}},$$

(12.16)

$$\mathbb{E}[X_\infty^*] = \frac{K}{2} \frac{\left(1 - \frac{\sigma^2}{2r}\right)\left(1 + \frac{2c_2 r}{p_1 q^2 K}\right) + \frac{c_1}{p_1 qk}}{1 + \frac{c_2 r}{p_1 q^2 K}}.$$

(c) *Assume the same parameters and cost structure as in Exercise 12.8b, but now under a randomly varying environment with* $\sigma = 0.2/\sqrt{year}$. *Determine* E^*, $\mathbb{E}[\Pi_\infty^*]$, *and* $\mathbb{E}[X_\infty^*]$. *Compare with the results of Exercise 12.8b.*

An alternative model is the *Fox model*, which is based on the Gompertz model for the natural growth of the population. The *stochastic Fox model*

$$dX(t) = r \ln \frac{K}{X(t)} \, dt - qEX(t)dt + \sigma X(t) \, dW(t), \qquad (12.17)$$
$$X(0) = x_0 > 0,$$

has also been studied for yield optimization (see, for instance, Braumann (1985)). Results on profit optimization can be seen in Brites (2017), where one can also find results on stochastic fisheries logistic-like models with Allee effects.

Going back to the stochastic logistic model with constant effort, i.e. the Gordon–Schaefer model (12.6), we have determined the optimal sustainable constant effort E^* that maximizes the expected profit $\mathbb{E}[\Pi_\infty]$ at the stochastic equilibrium. One can, however, use a more general variable effort $E(t)$ and work with the model

$$dX(t) = rX(t)\left(1 - \frac{X(t)}{K}\right) dt - qE(t)X(t)dt + \sigma X(t) \, dW(t), \quad (12.18)$$
$$X(0) = x_0 > 0.$$

Then one may try to use the effort $E(t)$ as a control variable and maximize the expected accumulated profit over some time interval $[0, T]$. One should use a discount rate $\delta > 0$, since there are depreciation, loss of opportunity, and other social costs that make a given amount of money now more valuable than the same amount in the future.[5] So what one tries to optimize is the *present value* of the expected profits accumulated during $[0, T]$:

$$V := \mathbb{E}\left[\int_0^T e^{-\delta t}\Pi(t)dt\right]. \tag{12.19}$$

One wants to determine a non-anticipative function $E^+(t)$, corresponding to the *variable fishing effort* that should be applied at each moment in order to maximize V. This is beyond the scope of this book, but the interested reader can consult the pioneering work of Lungu and Øksendal (1997) and Alvarez and Shepp (1998) for yield optimization. For more general profit structures, like (12.7) with $c_2 > 0$, stochastic optimal control theory could be used (see, for example, Fleming and Rishel (1975), Hanson (2007), Øksendal and Sulem (2007), and Clark (2010)). Their use in fishing optimization can be seen, for example, in Hanson and Ryan (1998), Suri (2008), and Clark (2010). The fishing policies with optimal variable effort $E^+(t)$ obtained using optimal control theory are quite simple and work quite well in deterministic environments (when $\sigma = 0$). Unfortunately, in randomly varying environments ($\sigma > 0$), the optimal effort $E^+(t)$ is very irregular and keeps changing all the time, accompanying the random variations of population size; the uninterrupted changes frequently require fishing to stop ($E^+(t) = 0$, which poses social problems like unemployment) or fishing at maximum capacity. These policies are operationally not applicable. There is an added complication even if they were applicable, which is the requirement of knowing the population size $X(t)$ at any time t in order to determine the value of the effort $E^+(t)$ to apply at that moment. This is not possible, since estimating the population size of fish is a complicated, inaccurate, and costly process. Furthermore, the profit per unit time is wildly variable (and is zero at the periods where $E^+(t) = 0$).

In Brites (2017) and Brites and Braumann (2017), comparisons were made between the optimal variable effort $E^+(t)$ fishing policy based on optimal control theory and the optimal *sustainable constant effort E^* policy* we have presented here and which poses no implementation problems. Contrary to the variable effort policy, the effort is always the same, so it is very easy to implement and without social problems, and one does not need to know what the population size is to determine the fishing effort. The comparisons were extensive and used real population and profit parameters, and were accompanied by a sensitivity study on such parameters. For logistic natural

---

5 If nothing else, a certain amount now could be invested in a riskless asset and become a larger amount at a future date.

growth models ((12.6) and (12.18)) you can see the comparisons in Brites (2017) and Brites and Braumann (2017); the real population data came from Hanson and Ryan (1998) on a Pacific halibut (*Hippoglossus hippoglossus*) fishery and is described in Figure 12.3. For Gompertz natural growth models ((12.17) and similar equations with variable $E(t)$) you can see the comparisons in Brites (2017); the real population data came from Kar and Chakraborty (2011) on a Bangladesh shrimp (*Penaeus monodon*) fishery. In Brites (2017), comparisons were also made for models with Allee effects.

One might think that, because we are restricting the effort to be constant, the optimal sustainable constant effort E^* policy would lead to a much lower profit V than the optimal variable effort policy $E^+(t)$ based on optimal control theory. This is not the case, the differences in the value of V being less than 5% for the logistic growth and less than 2% for the Gompertz growth. This slight disadvantage in profit is largely compensated by the fact that optimal constant effort policies can be applied (are even extremely easy to apply) and pose no social problems, contrary to what happens with the optimal variable effort policies. Even the slight disadvantage in profit is likely to be much smaller than the hidden cost (costs not considered in the model nor in the comparisons) of the variable effort policies, namely unemployment compensation costs during periods where $E^+(t)$ is low and the high costs of estimating the population size frequently.[6]

The natural growth rate $R(x)$ of the harvested population is not known and we would like to see whether the qualitative results we have obtained above for the Gordon–Schaefer model (12.6) hold for *general stochastic fishing models*

$$dX(t) = R(x)X(t)\,dt \ - \ qEX(t)dt \ + \ \sigma X(t)\,dW(t), \qquad (12.20)$$
$$X(0) = x_0 > 0,$$

where $R(X)$ satisfies the general density-dependent assumptions (based on resource limitation and no immigration) considered in Section 12.3. As can be seen for (a) to (d) in Braumann (1999b), the conclusions are similar:

(a) The boundary $+\infty$ is non-attractive and the solution exists and is unique for all $t \geq 0$, being a homogeneous diffusion process with drift coefficient $a(x) = R(x)x - qEx$ and diffusion coefficient $b(x) = \sigma^2 x^2$.

(b) The *geometric average intrinsic net growth rate* is $R_g(0^+) - qE = R(0^+) - \frac{\sigma^2}{2} - qE$. It is negative or positive according to whether the constant effort E is, respectively, larger (overfishing) or smaller than $\frac{R(0^+) - \sigma^2/2}{q}$.

---

6 There is another extra cost due to the fact that real field estimates are approximations and therefore one is not using the true optimal effort $E^+(t)$ but rather an approximation. This extra cost is not considered in the comparisons made. One can also reduce the estimation costs of population size by doing fewer field estimates, but that only worsens the approximation errors.

(c) If $E > \frac{R(0^+)-\sigma^2/2}{q}$ (overfishing), the 0 boundary is attractive and 'mathematical' extinction will occur with probability one.

(d) If $E < \frac{R(0^+)-\sigma^2/2}{q}$, 'mathematical' extinction will not occur and there is an equilibrium distribution with a stationary density $p(x; E)$ proportional to the speed density. So, as $t \to +\infty$, the population size converges in distribution to a r. v. X_∞ with pdf $p(x; E)$ proportional to the speed density; the mode of the stationary distribution is, for small σ, close to the stable equilibrium of the deterministic model.

(e) If $E < \frac{R(0^+)-\sigma^2/2}{q}$, the profit per unit time $\Pi(t)$ converges in distribution to the r.v. Π_∞, the *sustainable profit* given by (note that $H_\infty = qEX_\infty$)

$$\Pi_\infty = (p_1 - p_2 H_\infty)H_\infty - (c_1 + c_2 E)E = p_1 qEX_\infty - p_2 q^2 E^2 X_\infty^2 - c_1 E - c_2 E^2.$$

The expected value of the sustainable profit is then

$$\mathbb{E}[\Pi_\infty] = p_1 qE\mathbb{E}[X_\infty] - p_2 q^2 E^2 \mathbb{E}[X_\infty^2] - c_1 E - c_2 E^2.$$

Note that $\mathbb{E}[X_\infty]$ and $\mathbb{E}[X_\infty^2]$ also depend on the effort E.

(f) One cannot optimize the profit, since it is randomly varying, but, for $E < \frac{R(0^+)-\sigma^2/2}{q}$, one can optimize the expected profit. One can determine the *optimal sustainable constant effort* E^* that maximizes the expected value $\mathbb{E}[\Pi_\infty]$ of the sustainable profit. Let

$$\Pi^*(t) = p_1 qE^* X^*(t) - p_2 q^2 E^{*2} X^{*2}(t) - c_1 E^* - c_2 E^{*2}$$

$$\Pi_\infty^* = p_1 qE^* X_\infty^* - p_2 q^2 E^{*2} X_\infty^{*2} - c_1 E^* - c_2 E^{*2},$$

where $X^*(t)$ is the solution of the SDE model (12.20) with $E = E^*$ and X_∞^* has pdf $p(x; E^*)$. Note that $\Pi^*(t)$ converges in distribution to Π_∞^*.

These results were even generalized in Braumann (2002) to density-dependent noise intensities (i.e. with constant σ replaced by a variable $\sigma(X)$) satisfying some technical conditions (the same as mentioned in Section 12.3) and to density-dependent variable effort fishing policies (i.e. with constant effort E replaced by a variable $E(X)$) satisfying some technical assumptions. The technical assumptions for $E(X)$, a function defined for $X > 0$, consist of being non-negative of class C^1 and such that the limit $E(0^+) = \lim_{X \downarrow 0} E(X)$ exists and $\lim_{X \downarrow 0} XE(X) = 0$.

12.5 An application in human mortality rates

Human *mortality rates* are very important for prediction purposes in social security systems, life insurance, public and private pension plans, etc. The mortality rates of all age and sex groups have a tendency to decline over time,

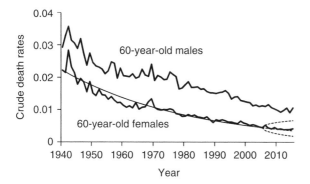

Figure 12.5 Crude death rates of 60-year-old females and males of the Portuguese population in the period 1940–2015 (data taken from the Human Mortality Database, University of California, Berkeley USA and Max Planck Institute for Demographic Research, Germany). For females, we have adjusted the geometric Brownian motion model (12.21), using only the data of the period 1940–2005 for parameter estimation and reserving the period 2006–2015 for prediction. The maximum likelihood parameter estimates are $\hat{r} = -0.02463$/year and $\hat{\sigma}^2 = 0.009239$/year. The adjusted thin solid line is based on the deterministic environment ($\sigma = 0$). The dashed lines show the predictions for 2006–2015; the middle dashed line (almost indistinguishable from the solid line of the true death rates) refers to the point predictions and the upper and lower dashed lines are the bounds of the 95% prediction band.

but they clearly show random fluctuations. This is illustrated in Figure 12.5, which shows the crude death rates (proportion of deaths during the calendar year) of 60-year-old females and males of the Portuguese population for each year of the period 1940–2015 (data taken from the Human Mortality Database, University of California, Berkeley USA and Max Planck Institute for Demographic Research, Germany). For convenience, we start counting time in 1940, so $t = 0$ corresponds to 1940. The tendency of death rates to decrease comes from the progress in medicine and in social conditions, and seems to follow an exponential-like decay. The random fluctuations are too big to be explained just by demographic stochasticity, i.e. by the sampling variations in births and deaths; in fact, the population of each age-sex group is always quite big, so the sample proportion of deaths, which is the crude death rate, would have a quite small standard deviation of approximately $\sqrt{d(1-d)/N}$, where d is the crude death rate and N the population size of the group. The random fluctuations are mostly explained by *environmental stochasticity* and so, representing by $X(t)$ the death rate at time t of a given age-sex group, one could use an SDE model (our familiar Black–Scholes or stochastic Malthusian model)

$$dX(t) = RX(t)dt + \sigma X(t)dW(t), \quad X(0) = x_0 > 0, \tag{12.21}$$

the solution of which is *geometric Brownian motion*

$$X(t) = x_0 \exp(rt + \sigma W(t)) \quad \text{with } r = R - \frac{\sigma^2}{2}. \tag{12.22}$$

Now, however, r, which is the geometric average of the rate of decay of $X(t)$, would be negative since the tendency is for $X(t)$ to decrease exponentially. σ measures the intensity of the effect of environmental fluctuations (weather conditions, epidemic diseases, social conditions, etc.) on the rate of change.

If one believes there is a minimum average death rate K that cannot be improved no matter what health and social conditions will progress, then a natural alternative model could be used, namely the stochastic Gompertz model (12.2) with $X(0) > K > 0$. In this case, death rates would have a tendency to asymptotically decrease towards K, although, of course, with some random fluctuations. That was the model used in Lagarto and Braumann (2014).

Lagarto (2014) has used both models and they both performed quite well in terms of predictive ability (using as criterion the mean square error of the predictions), with mixed results on which wins, although, for the vast majority of age-sex groups, the winner is the geometric Brownian motion model (12.21).

In Figure 12.5, we show the results of using model (12.21) for the crude death rates of 60-year-old Portuguese females. Notice the decline from $X(0) = 0.022298$/year in 1940 to $X(65) = 0.004480$/year in 2005 and $X(75) = 0.004147$/year in 2015. For parameter estimation, we only use the data of the period 1940–2005, reserving the period 2006–2015 for prediction purposes, putting us in the position of someone that only has the observed data up to 2005 and wants to predict the crude death rates of the following years.

We can apply the parameter estimation technique used in Section 8.1 for the Renault stock. In this case of mortality rates, we have time equidistant observations with $\Delta = 1$ year and $t_k = k\Delta$ years ($k = 0, 1, \ldots, n = 65$). Denoting by $X_k = X(t_k)$ and by $L_k = \ln \frac{X_k}{X_{k-1}}$ the log-returns, one gets for model (12.21) the *maximum likelihood* estimates of the parameters:

$$\hat{r} = \frac{1}{n\Delta} \sum_{k=1}^{n} L_k = \frac{1}{n\Delta} \ln \frac{X_n}{X(0)} = -0.02463/\text{year}$$

$$\hat{\sigma}^2 = \frac{1}{n\Delta} \sum_{k=1}^{n} (L_k - \hat{r}\Delta)^2 = 0.009239/\text{year}. \tag{12.23}$$

The 95% confidence intervals of the parameters are $\hat{r} \pm 1.96\sqrt{\sigma^2/(n\Delta)} = (-0.02463 \pm 0.02337)$/year and $\hat{\sigma}^2 \pm \sqrt{2\sigma^4/n} = (0.009239 \pm 0.003176)$/year.

Using in (12.22) the parameter value $\hat{r}$ obtained in (12.23) and putting $\sigma = 0$, as if we have a deterministic environment, we obtain an adjusted curve (the thin solid line in Figure 12.5) that gives an idea of the trend.

If one has the data up to 2005 (corresponding to $n = 65$ and $t_n = 2005 - 1940 = 65$ years) and wants to make *predictions* for the future period 2006–2015, one uses model (12.21) with the estimated parameter values (based only on data up to 2005), starting with the observed death rate of 2005, $X(t_n) = X(65) = 0.00448$/year. It is easier to work with $Y(t) = \ln X(t)$, since it is Gaussian. For $\tau > 0$ years, using the same technique as in Section 8.1, one gets the *point predictor*

$$\hat{Y}(t_n + \tau) = \ln 0.00448 + \hat{r}\tau = \ln 0.00448 - 0.02463\tau$$
$$\hat{X}(t_n + \tau) = \exp(\hat{Y}(t_n + \tau)) \tag{12.24}$$

and an approximate 95% *confidence prediction interval* for $Y(t_n + \tau)$

$$\hat{Y}(t_n + \tau) \pm 1.96 \sqrt{\hat{\sigma}^2 \tau (1 + \tau/t_n)},$$

from which extremes, by taking exponentials, one gets an approximate confidence prediction interval for $X(t_n + \tau)$. The three dashed lines in Figure 12.5 show the extremes of such a prediction interval (outer dashed lines) and the point prediction (middle dashed line). As you can see, the point prediction is quite good, almost undistinguishable from the true death rates observed in 2006–2015. Of course, the further into the future, the larger the uncertainty of the prediction.

As mentioned in Section 8.1, one can also be less ambitious and make a *step-by-step prediction*, not shown in the figure. At each time $n\Delta$, one *only* predicts the next unknown death rate at time $(n + 1)\Delta$ (the coming year in this case) and keeps updating the available information at each step. The prediction errors are smaller since we have always $\tau = \Delta$.

In the literature (see, for instance, Bravo (2007) and references therein), it is more common to model the mortality rates of cohorts. A cohort is a group of individuals born in the same year that will therefore all have the same age in the following years. The study consists of determining the evolution over time of the mortality rates of the cohort. However, mortality rates depend on age strongly and in a very complicated way, so the cohort models are also very complicated and, furthermore, the connections among models of different cohorts required for most applications are again complicated.

On the contrary, the transversal behaviour used here, i.e. the time evolution of the mortality rate of a given age-sex group, has the disadvantage of dealing with different individuals in different years, but has the advantage of requiring very simple models, like the geometric Brownian motion model. However, for applications, one needs to make connections between the models of the different age-sex groups.

One cannot simply consider independent models for different age-sex groups, since their yearly mortality rates are correlated, sometimes very strongly. For example, in Figure 12.5, one can recognize positive correlations between the crude death rates of 60-year-old females and 60-year-males. It seems that there is a tendency (with exceptions) for death rates of the two sexes to move towards the trend or away from the trend in the same years; we may say that there are bad years and good years. These positive correlations can also be observed (see Lagarto (2014)), with variable strength, when comparing the two sexes at different ages or even when comparing two different age groups (whether of the same sex or not).

In Lagarto and Braumann (2014) and Lagarto (2014) it is proposed that different age-sex groups follow their own model, but with correlated effects of environmental fluctuations. This is achieved by assuming that the Wiener processes driving the SDEs of the different age-sex groups are correlated instead of being independent. So we have a system of SDEs or multidimensional SDE models for the joint evolution of the death rates of the age-sex groups involved.

Lagarto and Braumann (2014) considered precisely the joint evolution of the crude death rates of two age-sex groups, although using the stochastic Gompertz model. Here, following the methods in Lagarto (2014), we will use for illustration purposes the geometric Brownian motion model and examine the case of the joint evolution of the two age-sex groups of females and males aged 60 in the Portuguese population for the period 1940–2005, the same period used above to estimate the parameters for the females. Denoting by $X_1(t)$ and $X_2(t)$ the crude death rates at time t of females aged 60 and of males aged 60, respectively, and the initial values at the year $t = 0$ (corresponding to 1940) by $X_1(0) = x_{01} = 0.022208/\text{year}$ and $X_2(0) = x_{02} = 0.029270/\text{year}$, the *bivariate geometric Brownian motion model* will be

$$dX_1(t) = R_1 X_1(t)dt + \sigma_1 X_1(t)dW_1^*(t)$$
$$dX_2(t) = R_2 X_2(t)dt + \sigma_2 X_2(t)dW_2^*(t),$$

(12.25)

where $W_1^*(t)$ and $W_2^*(t)$ have covariance ρt (with $-1 < \rho < 1$) and are standard Wiener processes. Therefore, since both have variance t, they have correlation ρ. This correlation is convenient since now the log-returns of the two groups, $L_{i,t} = \ln \frac{X_i(t)}{X_i(t-1)} = r_i + \sigma_i(W_i^*(t) - W_i^*(t-1))$ $(i = 1, 2)$, with $r_i = R_i - \sigma_i^2/2$, are $\frown \mathcal{N}(r_i, \sigma_i^2)$ (remember that here $\Delta = 1$ year). Although the log-returns of the same age-sex group at different time intervals are still independent, the log-returns of the two groups with respect to the same time interval are correlated, precisely with a correlation $CORR[L_{1,t}, L_{2,t}] = \rho$.

Sometimes it is more convenient to work with independent standard Wiener processes $W_1(t)$ and $W_2(t)$. The reader can confirm the result in Lagarto (2014) that

$$W_1^*(t) = \alpha W_1(t) + \beta W_2(t) \quad \text{and} \quad W_2^*(t) = \beta W_1(t) + \alpha W_2(t),$$

with

$$\alpha = \left(\frac{1 + (1 - \rho^2)^{1/2}}{2} \right)^{1/2} \quad \text{and} \quad \beta = \text{sign}(\rho)(1 - \alpha^2)^{1/2}$$

are indeed standard Wiener processes with the required correlation ρ. So model (12.25) can be written in the form

$$dX_1(t) = R_1 X_1(t)dt + \sigma_1 X_1(t)(\alpha dW_1(t) + \beta dW_2(t))$$
$$dX_2(t) = R_2 X_2(t)dt + \sigma_2 X_2(t)(\beta dW_1(t) + \alpha dW_2(t)).$$

(12.26)

The parameters r_1, r_2, σ_1, σ_2, and ρ can be estimated by maximum likelihood, using precisely the log-returns of the two groups. In the particular case of the Black–Scholes model with time equidistant observations (in this case even with time intervals of $\Delta = 1$), one may notice this can be treated by classical basic statistics methods. In fact, in this case the observed log-returns $(L_{1,t}, L_{2,t})$ $(t = 1, 2 \ldots, n)$ form a paired sample of bivariate independent Gaussian random variables having a common mean vector (r_1, r_2) and a common variance-covariance matrix (the variances being σ_1^2 and σ_2^2 and the covariance being $\rho\sigma_1\sigma_2$, where ρ is the correlation coefficient). Therefore, the sample means and variances of the log-returns of the two groups are estimates of the means and variances we need, and the sample correlation coefficient is an estimate of ρ.[7] For the example above where the two groups are the 60-year-old females and the 60-year-old males of the Portuguese population for the period 1940–2005, we obtain $\hat{r}_1 = -0.02463$/year, $\hat{r}_2 = -0.01473$/year, $\hat{\sigma}_1^2 = 0.009239$/year, $\hat{\sigma}_2^2 = 0.006977$/year, and $\hat{\rho} = 0.4087$. Note that, for consistency with the single-group treatment, we had used the maximum likelihood estimators $\hat{\sigma}_i^2$ and not the unbiased sample variance estimators $\tilde{\sigma}_i^2 = \frac{n}{n-1}\hat{\sigma}_i^2$ commonly used in paired samples statistical analysis.

Having the bivariate model does not change the point predictions of future death rates, but does change the probabilities of certain events concerning these rates.

One can perform a significance test for the null hypothesis $H_0 : \rho = 0$ that the true correlation coefficient ρ is zero using classical statistical methods for paired Gaussian samples. In our illustrative example, the test rejects the null hypothesis even at the 1% significance level. This test is of course specific of this simple situation. In general (and this technique also works here), if maximum likelihood methods can be used for estimation of the parameters under a given model, we can (assuming the number of data observations is high) perform a likelihood ratio test concerning a parameter of the model (see Section 12.2 for details).

However, in applications one needs to consider many sex-age groups simultaneously, and so one has to deal not with a bivariate but rather with a high-dimension multivariate SDE model, where the Wiener processes are

---

7 This treatment by classical basic statistical methods could already have been used when we treated above the single group of 60-year-old females, since the log-returns of that group formed a sample of independent Gaussian random variables with common mean r and common variance σ^2. So the sample mean and the sample variance of the log-returns would be estimates of r and σ^2.

correlated. For m sex-age groups, one has $m(m-1)/2$ correlations. If one considers all the yearly age groups (say for ages below 100), there are 200 groups (100 for each sex) and so there are $200 \times 199/2 = 19900$ correlations, plus 200 means r_i and 200 variances σ_i^2. The model will be hugely over-parametrized considering that the available data will consist of $n \times 200$ data values, where n is the number of calendar years with data. Estimation would be impossible, so, to allow reliable parameter estimation, one needs a low number of parameters compared to the available data. For that reason, Lagarto (2014) works with 5-year sex-age groups (females aged 0–4 years, males aged 0–4, females aged 5–9, males aged 5–9, females aged 10–14, males ages 10–14, etc.). The number of groups reduces to 40, but still would be 860 parameters, too many. One therefore needs to assume that the 780 correlations among groups have some structure (e.g. the correlation coefficient of two groups may decrease in an approximately predictable way with the age difference between the two groups) that allows one to express all of the correlations as a function of a much smaller number of parameters. Then one determines the log-likelihood of the corresponding model and estimates the parameters. One may also test if some parameters are significant by using a likelihood ratio test, so that one can eliminate parameters that contribute little to explain the variability in the data. Using the likelihood, models with different structures can also be compared in terms of performance (by some of the usual criteria AIC, BIC, etc.) or in terms of predictive ability. Lagarto (2014) has performed an extensive study of alternative structures and has determined, among those, which ones have a good performance. These can then be used in predictions and in applications.

In this way, it will be possible to evaluate, in insurance, pension systems and plans, and other situations, the risk due to the random variations in death rates produced by environmental stochasticity. This risk has been neglected in the past and it cannot be attenuated by the traditional methods used to control the risk of demographic stochasticity. These methods rely on the fact that the demographic stochasticity relative risk has a standard deviation of the order $N^{-1/2}$, where N is the number of people involved in the insurance or pension system; this is negligible if N is large. The risk can thus be diluted by pooling different systems into a larger system (that will therefore involve more people), using reinsurance or co-insurance schemes. The environmental stochasticity relative risk, on the contrary, is of order N^0 (constant) and cannot be diluted in that way, so neglecting it and failing to make provisions to deal with it can have very serious consequences.

13

Girsanov's theorem

13.1   Introduction through an example

Consider a complete probability space $(\Omega, \mathcal{F}, P)$ and a Wiener process defined on it. Consider an autonomous SDE

$$dX(t) = f(X(t))dt + g(X(t))dW(t), \quad X(0) = x \quad t \in [0, T] \tag{13.1}$$

or, in the equivalent integral form,

$$X(t) = x + \int_0^t f(X(s))ds + \int_0^t g(X(s))dW(s), \quad t \in [0, T], \tag{13.2}$$

with f and g satisfying a Lipschitz condition. Since the SDE is autonomous, this implies the existence and uniqueness of the solution.

The solution $X(t)$ is a homogeneous diffusion process with drift coefficient $a(y) = f(y)$ and diffusion coefficient $b(y) = g^2(y)$ and the solution is in $H^2[0, T]$. Since we are considering a deterministic initial condition, automatically independent of the Wiener process, we can work with the natural filtration of this process $\mathcal{M}_s = \sigma(W(u) : 0 \leq u \leq s)$, as we shall assume in this chapter. Since we will be exclusively working in the time interval $[0, T]$ and $\mathcal{M}_s \subset \mathcal{M}_T$ for $s \in [0, T]$, we can restrict the Wiener process to the interval $[0, T]$ and work with the probability space $(\Omega, \mathcal{M}_T, P)$, where we use the same letter P for the restriction of the original probability to $\mathcal{M}_T$. We will do that throughout this chapter.

If the drift coefficient $f(x)$ was zero, we would have $X(t) = x + \int_0^t g(X(s))dW(s)$ and, from Property 6.2, $X(t)$ would be a martingale (with respect to the filtration $\mathcal{M}_s$). Martingales have very nice properties and so working with martingales in theory and in applications has obvious advantages.

In Section 11.2 we have seen a method to change the drift coefficient to zero through a change of variable $Y(t) = S(X(t))$, where S is a scale function. The process $Y(t)$ has zero drift coefficient and is a martingale, but the initial process $X(t)$ keeps its original drift coefficient. We will look now at another method

Introduction to Stochastic Differential Equations with Applications to Modelling in Biology and Finance, First Edition. Carlos A. Braumann.
© 2019 John Wiley & Sons Ltd. Published 2019 by John Wiley & Sons Ltd.
Companion Website: www.wiley.com/go/braumann/stochastic-differential-equations

in which we reach a zero drift coefficient (or, for that matter, any other drift coefficient we desire) without having to change the original process $X(t)$. This is particularly useful in financial applications.

Of course, to change the drift coefficient of $X(t)$ without changing $X(t)$ is not possible unless we change something else. In the SDE, therefore, besides changing the drift coefficient $f(x)$ to the new drift coefficient we desire, say $f^*(x)$, we also need to change the Wiener process $W(t)$ to some new process $W^*(t)$ in such a way that the solution of the new SDE is the same solution $X(t)$ of the original SDE (13.1). Remember that $W(t)$ is a Wiener process with respect to the probability P but we should not expect the same to be true for the new process $W^*(t)$. Fortunately, it is possible to change the original probability P to some modified probability P^* such that $W^*(t)$ is a Wiener process with respect to this new probability.

The new probability P^* is even *equivalent* to P. The term equivalent means that P^* is *absolutely continuous* with respect to P, which we denote by $P^* << P$, and means that $P(N) = 0$ implies $P^*(N) = 0$, and also that $P << P^*$.[1] By saying that P and P^* are equivalent, we just mean that the two have the same sets of zero probability; however, the sets with positive probability will usually have different probabilities under P and P^*.

The *Radon–Nikodym theorem* shows that $P^* << P$ if and only if P^* has a P-density $h(\omega)$, which is also called *Radon–Nikodym derivative* and is represented by $h = \frac{dP^*}{dP}$. This means that h is non-negative and that $P^*(A) = \int_A h \, dP$ for all measurable sets A. This derivative is P-a.s. uniquely defined. Furthermore, for arbitrary measurable functions $Z(\omega)$, we have the substitution rule $\int_\Omega Z \, dP^* = \int_\Omega Z \frac{dP^*}{dP} \, dP$, as long as one of the integrals is defined.[2] In this case, since also $P << P^*$, the Radon–Nikodym derivative $\frac{dP}{dP^*}$ also exists.

Basically, we have compensated the change of the drift coefficient from $f(x)$ to $f^*(x)$ by changing the probability from P to P^*. In the particular case where $f(x)$ is non-null and $f^*(x) \equiv 0$, the solution $X(t)$ of the SDE, which was not a martingale for the original probability P, becomes a martingale for the new probability P^*.

The property of being martingale is quite powerful, particularly in financial applications. In fact, if some stochastic process $Y(t)$ ($t \in [0, T]$) is a martingale for a probability P^* we have $Y(t) = \mathbb{E}^{P^*}[Y(T)|\mathcal{M}_t]$, where $\mathbb{E}^{P^*}$ represents the mathematical expectation with respect to the probability P^*, and so we can obtain $Y(t)$ ($t \in [0, T]$) from $Y(T)$. This comes in handy in determining

---

1 These concepts apply also to measures μ and μ^* defined on the same measurable space $(\Omega, \mathcal{F})$.
2 The Radon–Nikodym theorem and derivative also apply to σ-finite measures μ and μ^* instead of probabilities. Note that probabilities, being finite measures, are just particular cases. In a σ-finite measure we do not require that all measurable sets have finite measure, just that any measurable set is a countably union of measurable sets with finite measure. The pdf of an absolutely continuous r.v. X is just the Radon–Nikodym derivative of its probability distribution P_X with respect to the Lebesgue measure.

the value $Y(t)$ of some financial assets, for example European call options (see Chapter 14), for which we know the expression at time T.

It is important to keep in mind that, although P can be interpreted as giving the real probabilities $P(A)$ of the different measurable sets A of market scenarios ω, the new probability P^* will give fictional probabilities $P^*(A)$, i.e. will distort the real market probabilities in such a way that the drift coefficient of $X(t)$ becomes $f^*(x)$.

In the particular case that $f^*(x) \equiv 0$, $X(t)$ will be a martingale for P^* and P^* is called *equivalent martingale probability measure* or *equivalent martingale measure*.

To better understand the method, we are going to show its use for a very simple particular case of *Brownian motion with drift*, in which the drift coefficient and diffusion coefficients are constant. For the sake of simplicity, and since our focus is on the drift coefficient, we will use a diffusion coefficient equal to one. For this example, we have $f(x) \equiv \alpha$ and $g(x) \equiv 1$ and our SDE is

$$dX(t) = \alpha dt + 1 dW(t), \quad X(0) = x, \quad 0 \le t \le T,$$

or, written in integral form;

$$X(t) = x + \int_0^t \alpha ds + \int_0^t 1 dW(s), \quad 0 \le t \le T.$$

The solution is a stochastic process $X(t)$ ($t \in [0, T]$) in the probability space $(\Omega, \mathcal{M}_T, P)$ where we are working, as already mentioned. It is easy to verify that

$$X(t) = x + \alpha t + W(t) \frown \mathcal{N}(x + \alpha t, t).$$

Remember that $W(t)$ is a martingale for the probability P. Let $W^*(t) = \alpha t + W(t)$. Of course, $W^*(t) \frown \mathcal{N}(\alpha t, t)$ is no longer a Wiener process nor a martingale for the probability P. However, it has the same natural filtration as $W(t)$ (notice that its expression is a non-random function of $W(t)$).

Since $dW^*(s) = \alpha ds + dW(s)$, we easily recognize that $X(t)$ satisfies $X(t) = x + \int_0^t 0 ds + \int_0^t dW^*(s)$, i.e. in differential form,

$$dX(t) = 0 dt + 1 dW^*(t), \quad X(0) = x, \quad 0 \le t \le T,$$

which corresponds to $f^*(x) \equiv 0$.

Exercise 13.1 *Show that $M(t) = \exp\left(-\alpha W(t) - \frac{1}{2}\alpha^2 t\right)$ ($t \in [0, T]$) is a martingale for the probability P with respect to the filtration $\mathcal{M}_t$ ($t \in [0, T]$).*

Let us choose the new probability defined in the measurable space $(\Omega, \mathcal{M}_T)$ (the same used for P). We choose P^* as being the probability defined by the *Radon–Nikodym derivative* $dP^*/dP = M(T)$. This means that, given any $A \in \mathcal{M}_T$, we define

$$P^*(A) = \int_{\omega \in A} dP^*(\omega) = \int_{\omega \in A} \frac{dP^*}{dP} dP(\omega) = \int_{\omega \in A} M(T) dP(\omega).$$

Notice that, by the Radon–Nikodym theorem, P^* is absolutely continuous with respect to P. Note also that $M(T) > 0$ and so we can also obtain the Radon–Nikodym derivative $dP/dP^* = 1/M(T)$. This shows that P and P^* are equivalent (i.e. they are absolutely continuous with respect to each other) and, consequently, have the same sets of probability zero.

We now show that $W^*(t)$ is a Wiener process for the probability P^*, which implies that it is also a martingale for the probability P^*. In fact:

- $W^*(0) = 0$ P^*-a.s.
  This comes from $W^*(0) = W(0) = 0$ P-a.s. and the fact that P and P^* have the same sets of zero probability.
- W^* has independent increments.
  Since, for $s < t$, an increment $W^*(t) - W^*(s) = \alpha(t - s) + W(t) - W(s)$ of W^* is a non-random function of the corresponding increment $W(t) - W(s)$ of W and since $W(t)$ has independent increments, the same happens to $W^*(t)$.
- $W^*(t) - W^*(s) \overset{P^*}{\frown} \mathcal{N}(0, t - s)$ for $s < t$, where $\overset{P^*}{\frown}$ means the P^*-distribution. Since we are using two different probabilities P and P^*, we need to indicate which probability we are using. So, for example, we will use $F_Z^{P^*}(z) = P^*[Z \leq z]$, $f_Z^{P^*}(z)$ and $\mathbb{E}^{P^*}[Z]$, respectively, for the d.f., the pdf, and the expectation of a r.v. Z under the probability P^*. Let $A_x = \{\omega \in \Omega : W^*(t) - W^*(s) \leq x\} = \{\omega \in \Omega : W(t) - W(s) + \alpha(t - s) \leq x\} = \{\omega \in \Omega : W(t) - W(s) \leq x - \alpha(t - s)\}$, which is $\mathcal{M}_t$-measurable. The P^*-d.f. of $W^*(t) - W^*(s)$ is

$$F_{W^*(t)-W^*(s)}^{P^*}(x) = P^*(A_x) = \int_{A_x} dP^*(\omega) = \int_{A_x} \frac{dP^*(\omega)}{dP(\omega)} dP(\omega)$$

$$= \int_\Omega I_{A_x}(\omega) \frac{dP^*(\omega)}{dP(\omega)} dP(\omega) = \mathbb{E}^P \left[I_{A_x}(\omega) \exp\left(-\alpha W(T) - \frac{1}{2}\alpha^2 T\right) \right]$$

$$= \mathbb{E}^P \left[I_{A_x}(\omega) \mathbb{E}^P \left[\exp\left(-\alpha W(T) - \frac{1}{2}\alpha^2 T\right) | \mathcal{M}_t \right] \right],$$

where the last line results from the properties of conditional expectations. Since $M(t)$ is a martingale for P, we have $\mathbb{E}^P \left[\exp\left(-\alpha W(T) - \frac{1}{2}\alpha^2 T\right) | \mathcal{M}_t \right] = \exp\left(-\alpha W(t) - \frac{1}{2}\alpha^2 t\right)$. So, due to the independence of the increments of W,

$$F_{W^*(t)-W^*(s)}^{P^*}(x) = e^{-\frac{1}{2}\alpha^2 t} \mathbb{E}^P [I_{A_x}(\omega) e^{-\alpha W(t)}]$$

$$= e^{-\frac{1}{2}\alpha^2 t} \mathbb{E}^P [I_{A_x}(\omega) e^{-\alpha(W(t)-W(s))} e^{-\alpha W(s)}]$$

$$= e^{-\frac{1}{2}\alpha^2 t} \mathbb{E}^P [I_{A_x}(\omega) e^{-\alpha(W(t)-W(s))}] \mathbb{E}^P [e^{-\alpha W(s)}]$$

$$= e^{-\frac{1}{2}\alpha^2 t} \left(\int_{A_x} e^{-\alpha(W(t)-W(s))} dP(\omega) \right) e^{\frac{1}{2}\alpha^2 s}$$

$$= e^{-\frac{1}{2}\alpha^2(t-s)} \int_{-\infty}^{x-\alpha(t-s)} e^{-\alpha z} f^P_{W(t)-W(s)}(z) dz$$

$$= e^{-\frac{1}{2}\alpha^2(t-s)} \int_{-\infty}^{x-\alpha(t-s)} e^{-\alpha z} \frac{1}{\sqrt{2\pi(t-s)}} \exp\left(-\frac{z^2}{2(t-s)}\right) dz.$$

With a change of variable $y = z + \alpha(t-s)$ in the last integral, we get finally

$$F^{P^*}_{W^*(t)-W^*(s)}(x) = \int_{-\infty}^{x} \frac{1}{\sqrt{2\pi(t-s)}} \exp\left(-\frac{y^2}{2(t-s)}\right) dy,$$

which is precisely the d.f. of a r.v. with distribution $\mathcal{N}(0, t-s)$.

This concludes the proof that $W^*(t)$ is a Wiener process for the new probability P^*. Therefore, the solution of the SDE $dX(t) = 0dt + 1dW^*(t)$ is $X(t) = x + \int_0^t dW^*(s) = x + W^*(t)$ and is a martingale for the probability P^* (with respect to the filtration $\mathcal{M}_t$).

In the example, we chose to change the drift $f(x)$ to a new zero drift $f^*(x) \equiv 0$ so that the solution $X(t)$ would be a martingale with respect to the new probability P^*, but we can choose any other new drift we desire. This example of an autonomous SDE in which f and g are constant functions is, of course, too simple and the extension to the general case requires Girsanov's theorem, which is studied in the next section.

13.2 Girsanov's theorem

Let $0 < T < \infty$. Consider a complete probability space $(\Omega, \mathcal{F}, P)$ and a Wiener process $W(t)$ ($t \in [0, T]$) defined on it. Let $\mathcal{M}_t = \sigma(W(u) : 0 \leq u \leq t)$ be the natural filtration of the Wiener process. We are going to restrict P to $\mathcal{M}_T$, i.e. we are going to work with the probability space $(\Omega, \mathcal{M}_T, P)$. We have the:

Theorem 13.1 (Girsanov's theorem) *Let $X(t)$ be the solution of the stochastic differential equation*

$$dX(t) = f(X(t))dt + g(X(t))dW(t), \quad X(0) = x \quad (t \in [0, T]), \qquad (13.3)$$

with f and g satisfying a Lipschitz condition (which insures the existence and uniqueness of the solution).

Suppose that $f^*(x)$ also satisfies a Lipschitz condition. Suppose there is a function $\theta(x)$ such that[3]

$$g(x)\theta(x) = f(x) - f^*(x) \qquad (13.4)$$

---

3 In case $g(x) \neq 0$, we have $\theta(x) = \frac{f(x) - f^*(x)}{g(x)}$.

and let

$$M(t) = \exp\left(-\int_0^t \theta(X(s))dW(s) - \frac{1}{2}\int_0^t \theta^2(X(s))ds\right). \tag{13.5}$$

Suppose that $\theta(X(t)) \in M^2[0, T]^4$ and that $M(t)$ ($t \in [0, T]$) is a martingale for the probability P (with respect to the filtration $\mathcal{M}_t$). A sufficient condition for this to hold is that the Novikov *condition*

$$\mathbb{E}^P\left[\exp\left(\frac{1}{2}\int_0^T \theta^2(X(s))ds\right)\right] < +\infty \tag{13.6}$$

is satisfied.

Let P^* be a new probability measure in the measurable space $(\Omega, \mathcal{M}_T)$ defined by the Radon–Nikodym derivative

$$dP^*/dP = M(T)$$

and let

$$W^*(t) = W(t) + \int_0^t \theta(X(s))ds. \tag{13.7}$$

Then P^* is equivalent to P and $W^*(t)$ is a Wiener process with respect to the probability P^*.[5]

Furthermore, the process $X(t)$ satisfies the SDE

$$dX(t) = f^*(X(t))dt + g(X(t))dW^*(t), \quad X(0) = x \quad (t \in [0, T]). \tag{13.8}$$

Therefore, the finite-dimensional distributions of $X(t)$ with respect to P^* coincide with the finite-dimensional distributions with respect to P of the solution $Y(t)$ of the SDE[6]

$$dY(t) = f^*(Y(t))dt + g(Y(t))dW(t), \quad Y(0) = x \quad t \in [0, T]. \tag{13.9}$$

Observation 13.1 *In the important particular case of $f^* \equiv 0$, we get $X(t) = x + \int_0^t g(X(s)dW^*(s)$. Since g satisfies a Lipschitz condition, we know that $G(s, \omega) := g(X(s, \omega)) \in H^{2P^*}[0, T]$, where $H^{2P^*}[0, T]$ is the class of jointly measurable functions adapted to the filtration $\mathcal{M}_s$ such that $\mathbb{E}^{P^*}\left[\int_0^T g^2(s)ds\right] < +\infty$. So, due to Property 6.2, $X(t)$ will in this particular case be a martingale for P^* (with respect to the filtration $\mathcal{M}_t$). Therefore, we got a process of transforming*

---

4 Which means that (see Chapter 6) $\theta(t, \omega) := \theta(X(t, \omega))$ is a jointly measurable function adapted to the filtration $\mathcal{M}_t$ such that $\int_0^T \theta^2(t, \omega)dt < +\infty$ P-a.s., i.e. with P probability one.
5 It is not, in general, a Wiener process for the probability P.
6 Since only the finite-dimensional distributions are relevant to compute probabilities concerning the values of the process, it will be indifferent to work with the distributions of $Y(t)$ with respect to P or with the distributions of $X(t)$ with respect to P^*, which may be quite useful in determining weak solutions of stochastic differential equations.

the solution of an autonomous SDE, which in general is not a martingale for the original probability P, into a martingale for a new probability P*. The example in Section 13.1 is just the particular case of $f(x) \equiv \alpha$ and $g(x) \equiv 1$, for which we have $\theta(x) = \frac{f(x)-0}{g(x)} = \alpha$.

Observation 13.2 *It is not difficult to generalize Girsanov's theorem to multi-dimensional SDE in the terms mentioned in Section 7.3. Naturally, $\boldsymbol{\theta}$ is now a column vector with the same dimension m of the vector Wiener process and $M(t) = \exp\left(-\int_0^t \boldsymbol{\theta}^T(\mathbf{X}(s))d\mathbf{W}(s) - \frac{1}{2}\int_0^t \boldsymbol{\theta}^T(\mathbf{X}(s))\boldsymbol{\theta}(\mathbf{X}(s))ds\right)$. In the Novikov condition, one should replace θ^2 by $\boldsymbol{\theta}^T\boldsymbol{\theta}$. In the particular case of $n = m$ and of $\mathbf{g}$ being an invertible matrix, one can choose $\boldsymbol{\theta} = \mathbf{g}^{-1}(\mathbf{f} - \mathbf{f}^*)$.*

Observation 13.3 *Another relatively trivial generalization consists of allowing the replacement of $f(X(s))$ by a more general $F(s, \omega) \in M^2[0, T]$ (so that the initial SDE may not be autonomous), as long as there is a $\theta(s, \omega)$ such that $g(X(s, \omega))\theta(s, \omega) = F(s, \omega) - f^*(x)$. Now, it is required that $\theta(s, \omega) \in M^2[0, T]$ and that $M(t) = \exp\left(-\int_0^t \theta(s, \omega)dW(s) - \frac{1}{2}\int_0^t \theta^2(s, \omega)ds\right)$ is a martingale for P; a sufficient condition for that to happen is the Novikov condition $\mathbb{E}\left[\exp\left(\frac{1}{2}\int_0^T \theta^2(s, \omega)ds\right)\right] < +\infty$. Now we have $W^*(t) = W(t) + \int_0^t \theta(s, \omega)ds$.*

Assuming P gives the true probabilities of the events (e.g. sets of market scenarios or states of nature), we should be aware that P^* will distort the probabilities of the events to make it seem that the infinitesimal mean is $f^*(x)$ instead of the true infinitesimal mean $f(x)$. Usually, this is done for reasons of expediency and one chooses a $f^*(x)$ for which it is easier to obtain results, e.g. if we choose $f^*(t) \equiv 0$, we get a martingale with respect to P^* and martingales have nice properties to work with. However, we should be aware that, in so doing, the results we get, like probabilities of events, expected values or distributions of r.v., will be in terms of the distorted probability P^*. So, at the end, we should convert these results into results in terms of the true probability P to get the true probabilities, expected values or distributions of the r.v.

We now proceed to the proof of Girsanov's theorem, which not so interested readers may skip. We start by proving two useful Lemmas.

Lemma 13.1 *Let P and P^* be probabilities on the measurable space $(\Omega, \mathcal{G})$ such that $\frac{dP^*}{dP} = h(\omega)$. Let Y be a $\mathcal{G}$-measurable r.v. such that $\mathbb{E}^{P^*}[|Y|] = \int_\Omega |Y(\omega)|h(\omega)dP(\omega) < +\infty$ and let $\mathcal{H}$ be a sub-σ-algebra of $\mathcal{G}$. Then*

$$\mathbb{E}^P[hY|\mathcal{H}] = \mathbb{E}^{P^*}[Y|\mathcal{H}]\mathbb{E}^P[h|\mathcal{H}] \quad P\text{-a.s.}$$

Proof of Lemma 13.1: We follow the proof in Øksendal (2003). Let $H \in \mathcal{H}$. Using the definition of conditional probability and the properties of conditional expectations, we get

$$\int_H \mathbb{E}^{P^*}[Y|\mathcal{H}]h dP = \int_H \mathbb{E}^{P^*}[Y|\mathcal{H}] dP^* = \int_H Y dP^*$$

$$= \int_H Yh dP = \int_H \mathbb{E}^P[Yh|\mathcal{H}] dP.$$

On the other hand,

$$\int_H \mathbb{E}^{P^*}[Y|\mathcal{H}]h dP = \int_\Omega \mathbb{E}^{P^*}[Y|\mathcal{H}]h I_H dP = \mathbb{E}^P[\mathbb{E}^{P^*}[Y|\mathcal{H}]h I_H]$$

$$= \mathbb{E}^P[\mathbb{E}^P[\mathbb{E}^{P^*}[Y|\mathcal{H}]h I_H|\mathcal{H}]] = \mathbb{E}^P[I_H \mathbb{E}^{P^*}[Y|\mathcal{H}]\mathbb{E}^P[h|\mathcal{H}]]$$

$$= \int_H \mathbb{E}^{P^*}[Y|\mathcal{H}]\mathbb{E}^P[h|\mathcal{H}] dP.$$

Therefore,

$$\int_H \mathbb{E}^{P^*}[Y|\mathcal{H}]\mathbb{E}^P[h|\mathcal{H}] dP = \int_H \mathbb{E}^P[Yh|\mathcal{H}] dP$$

for any $H \in \mathcal{H}$, which proves the lemma. ∎

The fact of having in Girsanov's theorem $\frac{dP^*}{dP} = M(T)$ in $\mathcal{M}_T$ means that, given any $A \in \mathcal{M}_T$, we have

$$P^*(A) = \int_A dP^*(\omega) = \int_A \frac{dP^*}{dP} dP(\omega) = \int_A M(T) dP(\omega).$$

Is it possible to get a stronger result in the case that $A \in \mathcal{M}_t$ with $t \in [0, T]$? The answer is in:

Lemma 13.2 *Under the conditions of Girsanov's theorem 13.1 and for $t \in [0, T]$, we get $\frac{dP^*}{dP} = M(t)$ in $\mathcal{M}_t$, which means that, for any $A \in \mathcal{M}_t$, we have*

$$P^*(A) = \int_A dP^*(\omega) = \int_A M(t) dP(\omega).$$

Proof of Lemma 13.2: We follow again Øksendal (2003). Since $M(t)$ is a martingale for P (with respect to the filtration $\mathcal{M}_t$), we have

$$P^*(A) = \int_A dP^* = \int_A M(T) dP = \int_\Omega I_A M(T) dP = \mathbb{E}^P[I_A M(T)]$$

$$= \mathbb{E}^P[\mathbb{E}^P[I_A M(T)|\mathcal{M}_t]] = \mathbb{E}^P[I_A \mathbb{E}^P[M(T)|\mathcal{M}_t]]$$

$$= \mathbb{E}^P[I_A M(t)] = \int_A M(t) dP. \quad ∎$$

Proof of Girsanov's theorem 13.1 (sketch):

Novikov condition

Let

$$Z(t) = -\int_0^t \theta(X(s)dW(s) - \frac{1}{2}\int_0^t \theta^2(X(s))ds.$$

Applying Itô Theorem 6.1 to $M(t) = \exp(Z(t))$ and attending to $\theta(X(t)) \in M^2[0, t]$, one gets

$$dM(t) = e^{Z(t)}dZ(t) + \frac{1}{2}e^{Z(t)}(dZ(t))^2$$

$$= M(t)\left(-\theta dW(t) - \frac{1}{2}\theta^2 dt\right) + \frac{1}{2}M(t)\theta^2 dt$$

$$= -\theta(X(t))M(t)dW(t),$$

from which, since $M(0) = 1$,

$$M(t) = 1 - \int_0^t \theta(X(s))M(s)dW(s). \tag{13.10}$$

If the integrand function $\theta(X(s))M(s)$ is in $H^2[0, T]$, the properties of the stochastic integral as a function of the upper limit of integration imply that $M(t)$ is a martingale for P (with respect to the filtration $\mathcal{M}_t$). But nothing guarantees that indeed $\theta(X(s))M(s) \in H^2[0, T]$ is true. That is the reason why in Girsanov's theorem we have required that $M(t)$ be a martingale for P.

A sufficient condition for that to happen is the Novikov condition. It implies that

$$\exp\left(\mathbb{E}^P\left[\frac{1}{2}\int_0^T \theta^2(X(s))ds\right]\right) \leq \mathbb{E}^P\left[\exp\left(\frac{1}{2}\int_0^T \theta^2(X(s))ds\right)\right] < +\infty,$$

and, therefore, implies $\theta \in H^2[0, T] \subset M^2[0, T]$. The Novikov condition also implies that $M(t)$ is a martingale for P since it allows (13.10) to be approximated by a sophisticated truncation technique that ensures the martingale property. The interested reader can see the proof in Karatzas and Shreve (1991).

Equivalence

As we did in Section 13.1, it is enough to attend to the fact that $M(T) > 0$.

$X(t)$ satisfies (13.8)

Noting that $dW^*(t) = dW(t) + \theta(X(t))dt$ and replacing $dW(t)$ by $dW^*(t) - \theta(X(t))dt$ in (13.3), one obtains (13.8).

$W^*(t)$ is a Wiener process for P^*

We again follow Øksendal (2003). Let

$$R(t) = W^{*2}(t) - t.$$

Using *Lévy's characterization* (see Section 4.2) of the Wiener process, one only needs to show that $W^*(t)$ and $R(t)$ are martingales for P^* (with respect to the filtration $\mathcal{M}_t$).

Let

$$K(t) = M(t)W^*(t)$$

and

$$L(t) = M(t)R(t).$$

We have $K(0) = 0$ and

$$
\begin{aligned}
dK(t) &= M(t)dW^*(t) + W^*(t)dM(t) + dM(t)dW^*(t) \\
&= M(t)(\theta(X(t))dt + dW(t)) + W^*(t)(-\theta(X(t))M(t)dW(t)) \\
&\quad - \theta(X(t))M(t)dt \\
&= M(t)(1 - \theta(X(t))W^*(t))dW(t),
\end{aligned}
$$

from which

$$K(t) = \int_0^t M(s)(1 - \theta(X(s))W^*(s))dW(s)$$

is a martingale for P.

For $0 \leq s < t \leq T$ and attending to $\mathcal{M}_s \subset \mathcal{M}_t$, one gets, using the martingale property and Lemmas 13.1 and 13.2,

$$
\begin{aligned}
M(s)W^*(s) = K(s) &= \mathbb{E}^P[K(t)|\mathcal{M}_s] = \mathbb{E}^P[M(t)W^*(t)|\mathcal{M}_s] \\
&= \mathbb{E}^P\left[\frac{dP^*}{dP}W^*(t)|\mathcal{M}_s\right] = \mathbb{E}^{P^*}[W^*(t)|\mathcal{M}_s]\mathbb{E}^P\left[\frac{dP^*}{dP}|\mathcal{M}_s\right] \\
&= \mathbb{E}^{P^*}[W^*(t)|\mathcal{M}_s]\mathbb{E}^P[M(t)|\mathcal{M}_s] = \mathbb{E}^{P^*}[W^*(t)|\mathcal{M}_s]M(s).
\end{aligned}
$$

Since $M(s) \neq 0$, we get

$$W^*(s) = \mathbb{E}^{P^*}[W^*(t)|\mathcal{M}_s],$$

which proves that $W^*(t)$ is a martingale for P^*.

The proof that $R(t) = W^{*2}(t) - t$ is a martingale for P^* is similar, working now with $L(t)$ instead of $K(t)$. ∎

14

Options and the Black–Scholes formula

14.1 Introduction

A financial *option* is a contract that gives its holder the right, but not the obligation, to perform some financial transaction. The holder will or will not exercise that right according to what is more favourable. Let us look at a couple of examples:

- A company in the European Union has imported some goods at the cost of one million US dollars, to be paid in 60 days. Suppose that the current exchange rate for buying dollars is €0.85 but the importing company needs or wishes to buy the required dollars for payment at the payment expiration date. The company is, however, afraid that, at that date, the exchange rate might be at an unacceptable level, higher than €0.90. To solve this problem, the company can acquire a call option of one million dollars with expiration date in 60 days at the exercise price of €0.90 per dollar. That option will give the company the right to buy, at the expiration date, one million dollars at the price of €0.90 per dollar. Should the market exchange rate at the expiration date be higher than €0.90 per dollar, the company will exercise the option and will buy the dollars at the agreed price of €0.90 per dollar, since this is more advantageous. If, however, the market exchange rate at the expiration date is lower than €0.90 per dollar, the company is not obliged and will not exercise the option, but rather will buy the one million dollars at the market at the more favourable market exchange rate.
- A company needs to pay monthly the interests of a loan, the interest rate being the EURIBOR 6-month rate plus a spread of 1.5%. To avoid unpleasant negative deviations from its budget predictions, it would like to be sure that the EURIBOR rate, currently at (suppose) −0.250%, will not go above +0.500% or, if it does, the difference would not be supported by the company. The company could buy a basket of options with monthly expiration dates that, if exercised at a certain month, would make the company pay that

Introduction to Stochastic Differential Equations with Applications to Modelling in Biology and Finance,
First Edition. Carlos A. Braumann.
© 2019 John Wiley & Sons Ltd. Published 2019 by John Wiley & Sons Ltd.
Companion Website: www.wiley.com/go/braumann/stochastic-differential-equations

month's interest at the the rate of 0.500% plus the spread. Of course, if at a given month, the EURIBOR rate is above +0.500%, the company will exercise that month's option. If, however, the EURIBOR rate at a given month is below +0.500%, the company will not exercise that month's option and will pay that month's interest at the EURIBOR rate plus the spread.

These are just two examples of a great variety of options. There are options on stocks, on bonds, on commodities (coffee, soya, pork bellies, gold, zinc, etc.), exchange rates, stock indices, financial funds, catastrophes, to mention a few. We can have *call options*, which give the holder the right to buy some financial asset or good, or *put options*, which give the holder the right to sell. The options can be *European* if they can be exercised only at the agreed expiration date, or *American* if they can be exercised at any time up to the expiration date. There are also other derivative[1] products (warrants, futures, exotic options of several kinds, etc.)

These derivative products can be acquired to serve as a kind of insurance, as in the examples given at the beginning, but they are often acquired for speculative purposes by market players. Call options on a company's stock are also used by some companies as a premium offered to employees, giving them the right to buy stock at a future date at an attractive price, hopefully quite lower than the future market price one expects if the employees do help the company in achieving its goals. Another common situation where an investor buys a call option on a stock is when the investor thinks the stock is a good investment (expects the stock price to go considerably up) but does not have the financial means to buy the desired number of stock units now, so the investor buys a call option (which has a relatively small price) to buy that stock at a good price later (when the investor expects to have the needed funds). Put options are usually preferred when the investor expects the stock price to go down.

Here, we will concentrate on the classical European options, either call or put options, on stocks. These options are abundantly traded and their treatment by Black, Scholes and Merton (see Black and Scholes (1973) and Merton (1973)) was the stepping-stone of the modern discipline of financial mathematics, which has since progressed at an extraordinary pace. Merton and Scholes won the 1997 Nobel Prize in Economics for their contributions (Black was already deceased).

We will first deal with European call options on a stock. Consider two financial assets, one (the risky asset) being a stock traded in the stock market and the other being a riskless asset (that will serve as a reference and refuge), for example a bank account (numeraire) with fixed interest rate or a (guaranteed) bond with fixed return rate.

---

1 They are call derivative products because they are contracts on rights over the primary or underlying assets (stocks, bonds, commodities, etc.)

Let r be the instantaneous interest/return rate (we will simply call it the return rate) of the riskless asset. It is assumed that the interest/return is compounded continuously in time, so that the capital will be continuously updated to include the interest/return accumulated till then. Therefore, if $B(0)$ is the initial value of one unit of the riskless asset (one bond or one unit of the numeraire, say one euro or one dollar), then its value $B(t)$ at time $t \geq 0$ satisfies the ordinary differential equation $\frac{dB(t)}{dt} = rB(t)$. We can equivalently write

$$dB(t) = rB(t)dt, \tag{14.1}$$

which is quite suggestive in showing that, in a small time interval $]t, t + \Delta t]$, the corresponding return $\Delta B(t) = B(t + \Delta t) - B(t)$ will be approximately proportional to the value at the beginning of the interval and to the duration Δt of the interval, the constant of proportionality being the return rate r. Solving the equation, we get

$$B(t) = B(0)\exp(rt). \tag{14.2}$$

Note that, in a real bank account or a real bond with fixed rates, the interest/return may not be compounded continuously and you may be given not the instantaneous rate r but an yearly rate $\tilde{r}$ that is only realized at the end of regular periods of time p. To give an example, suppose $p = 1/2$ years (interest/returns are paid and compounded only at the end of each semester) and $\tilde{r} = 5\%$ per year. In this case, after one semester, the value of each bond (or of an initial unit of numeraire) is $B(0) + \frac{0.05}{2} \times B(0) = 1.025B(0)$ and that is the value $B(0.5 \text{ years})$ at which you start the second semester, so at the end of the year the value is $B(0.5) + \frac{0.05}{2} \times B(0.5) = 1.050625B(0)$. Suppose instead that, as we are assuming here, the return is compounded continuously with an instantaneous rate of return of $r = 5\%$ per year. In that case, the value at the end of the year will be higher, $B(1) = e^{0.05 \times 1}B(0) = 1.051271B(0)$. Furthermore, under semester compounding, the value one day before the end of the year would be equal to the value at the end of the first semester $1.025B(0)$, while under continuous compounding it would be $B(364/365) = e^{0.05 \times 364/365}B(0) = 1.051127B(0)$. Of course, these disadvantages of non-continuous compounding are relatively insignificant when the period p of compounding, instead of being a semester, is shorter, like a month or a day, in which case you get almost the same values.

Although it is possible to handle more complex situations, like non-continuous compounding, bank commissions, variable return rates, etc., for simplicity we will present the theory of European call options as is traditionally done, based on some simplifying assumptions concerning the riskless asset:

- *There is a riskless asset with constant continuously compounded return rate r during the period of the contract.*
- *The riskless asset can be freely bought and sold at any time, with the same price for buying and selling, and (frictionless markets) with no taxes or bank commissions in such transactions.*

If the asset is a fixed rate bond, that means that you can buy and sell bonds freely at any time at the same price $B(t) = B(0)e^{rt}$ for buying and selling. If the asset is a bank account, buying means making a deposit and selling means making a withdrawal and obviously there is no difference between buying and selling prices.

- *These transactions are so free that one may possess at any time t a positive, zero or negative number of units b(t) of the riskless asset and that number can even be fractional.*

One may have a *long position* (a positive number of bonds or a positive balance on the bank account) or a zero position, but one may also have a *short position*. For bank accounts this means a negative balance, equivalent to having a loan, which is assumed will have the same instantaneous interest rate r in favour of the bank, i.e. loans can be freely taken without obstacle and pay the same interest rate as the interest rate of deposits. For fixed rate bonds, a short position means that one can short-sell bonds to others, which means selling bonds one does not have; of course, the short-seller takes full responsibility for complying with the rights of the buyers on such bonds.

From now on, for convenience, we will speak of fixed rate (riskless) bonds for the riskless asset, but the reader should keep in mind that it can also be another riskless asset, for example a bank account.

Let $S(t)$ be the price of one stock at time t and assume it has a constant (arithmetic) average return rate μ and a constant volatility $\sigma > 0$, following a *Black–Scholes model*. See Chapters 3 and 8, and, for the interpretation of the average return rate, Chapter 9; notice, however, that we now have a different notation and are using the letter μ instead of R.

We will assume that $S(0)$ is a known positive real number (so it is no longer random). $S(t)$ satisfies the Itô SDE

$$dS(t) = \mu S(t)dt + \sigma S(t)dW(t), \tag{14.3}$$

for which the solution is, as we know,

$$S(t) = S(0)\exp((\mu - \sigma^2/2)t + \sigma W(t)). \tag{14.4}$$

Although it is possible to handle more complex situations (like modifications in the Black–Scholes model), we will present the theory of European call options as is traditionally done, based on some simplifying assumptions concerning the stock:

- *The price S(t) of the stock follows the Black–Scholes model (14.3).*
- *During the period of the option contract, the stock pays no dividends.*
- *The stock can be freely bought and sold at any time, with the same price for buying and selling, and (frictionless markets) with no taxes or bank commissions in such transactions.*

- *These transactions are so free that one may possess at any time t a positive, zero or negative number of units s(t) of the stock and that number can even be fractional.*

  Again, one may have a *long position* (a positive number of stocks), a zero position or a *short position* (a negative number of stocks). A short position means that one can short-sell stocks (stocks it does not have) to others, taking full responsibility for complying with the rights of the buyers on such stocks.

Besides the assumptions made on the riskless asset and on the stock (some of which can be relaxed later with appropriate model modifications), we will also assume, and this is a critical assumption, the principle of *no arbitrage*. In layman's terms, this simply means that *it is not possible to make a riskless profit with a null investment.*

This can be mathematically translated to the following definition of *arbitrage*: arbitrage is a portfolio value process $V(t)$ that has zero value now ($V(0) = 0$) and, at some future time T, verifies $P[V(T) \leq 0] = 0$ and $P[V(T) > 0] > 0$, i.e. with an initial null investment, using that portfolio one can a.s. never loose and have a positive probability of winning. The assumption of no arbitrage says that such a portfolio does not exist. By *portfolios* we mean a group of financial assets, for example stocks, bonds, and cash.

Another equivalent definition of no arbitrage is: *if two investment portfolios have, at a certain future time, a.s. the same value for all market scenarios ω, they must a.s. have the same value now (and also, using the same reasoning, they must have the same value for all times in between).*

To show the equivalence, imagine that the two portfolios have a different value now. The holder of the more valuable portfolio could exchange it with the less valued portfolio (by buying and selling the appropriate assets), retaining in their pocket the difference in their values. Since, later on, the two portfolios will be equally valued whatever the market scenario, that person would have at that time a sure profit (precisely the initial difference that he/she had put in their pocket) without any financial risk and without having made any investment.

The assumption of no arbitrage is based on the idea that the markets are efficient and their operators are equally informed and behave smartly. If so, any opportunity of arbitrage (such as in the example of the two portfolios with equal future value but different values now) would rapidly disappear because of a higher demand (and resulting price increase) of undervalued assets and lower demand (and resulting price decrease) of overvalued assets.

Of course, if there is inside trading (people using privileged information that is not public to make a sure profit without financial risk), this idea does not work and that is the reason why inside trading is a criminal offence in most countries.

We will now define the concept of the *trading strategy* $(b(t), s(t))$ $(t \geq 0)$ of an investor. We consider here the particular case of the two assets mentioned above, the fixed rate bond (or bank account) and the stock, but of course we can have many assets involved, including several types of stocks and bonds, options, etc. Here $b(t)$ represents the number of units of the bond and $s(t)$ the number of units of the stock the investor has at time t. Note that $b(t)$ and $s(t)$ can take real positive values (long positions) or real negative values (short positions), or even be zero. Knowing the trading strategy $(b(t), s(t))$ is equivalent to knowing the buying and selling trades to be performed at any time t. For example, if $s(t)$ decreases by a certain amount at time t, it means that at time t the investor sells as many units of the stock as the referred amount of the decrease. If it is an increase instead of a decrease, the investor buys stock in the amount of that increase. Of course, $(b(0), s(0))$ is the initial portfolio and its value is

$$b(0)B(0) + s(0)S(0).$$

The trading strategy is said to be *self-financing* if there is no additional input or output of funds. The investor may freely buy assets for the portfolio as long as this is financed by selling other assets in the portfolio. Likewise, the investor may sell assets from the portfolio as long as he/she buys other assets with the same value. Therefore, the value of the portfolio at any time t is the initial value plus the cumulative income (which can be positive or negative) up to time t generated by the stock in the portfolio, $\int_0^t s(u)dS(u)$, and generated by the bonds (or bank account) in the portfolio, $\int_0^t b(u)dB(u)$.[2] In mathematical terms, the strategy is self-financing if, for any $t \geq 0$, we have

$$b(t)B(t) + s(t)S(t) = b(0)B(0) + s(0)S(0) + \int_0^t b(u)dB(u) + \int_0^t s(u)dS(u).$$

$$(14.5)$$

We will now define a *European call option* on the stock (which is the underlying asset) having *expiration date* $T > 0$ (where T is a deterministic fixed time, also called *maturity*) and *strike price* $K > 0$. It is a contract in which there is a seller and a buyer, the holder of the option. That contract gives to its holder the right, but not the obligation, to buy from the contract seller one unit of the

---

2 At time u, the trading portfolio will have $s(u)$ units of stock, each one with value $S(u)$. In a small time interval $]u, u + \Delta u]$ (with $\Delta u > 0$), the value of a stock unit changes by $\Delta S(u) = S(u + \Delta u) - S(u)$, so the income generated by the stock units existing in the portfolio on that small interval is approximately $s(u)\Delta S(u)$. For the whole interval $]0, t]$ the cumulative income generated by the stock is approximately the sum of the incomes of these small time intervals that fill the whole interval. As $\Delta u \to 0$, this sum gives progressively better approximations and so its limit, which is the Riemann–Stieltjes integral $\int_0^t s(u)dS(u)$, is the desired cumulative income generated by stock up to time t. Similarly, $\int_0^t b(u)dB(u)$ is the cumulative income generated by the bond.

stock at time T by paying at that time the agreed price K.[3] It is traditional to say that the call option is *in the money, at the money* or *out of the money*, according to whether the current value of the stock (also known as its *spot price*) is, respectively, larger than, equal to or smaller than the strike price K, meaning that, if the expiration date was right now instead of at time T, the holder of the call option, should (s)he decide to exercise it, would make money, would be even, or would loose money, respectively.

But what matters is the price $S(T)$ of the stock at the expiration date T. If $S(T)$ is higher than K, the holder of the option will uphold his/her right and exercise the option, i.e. will buy at time T one unit of stock at the strike price K, immediately gaining a benefit of $S(T) - K > 0$. Note that this is a real benefit since the holder can, if so wishes, sell that unit of stock at the market at the price $S(T)$. If, however, $S(T)$ is lower than K, then the holder of the option will have a zero benefit because, being a rational person, will not exercise the option. Remember that there is no obligation to do this and exercising the option would result in a real loss because it would mean paying K for one unit of stock when the lower price $S(T)$ could be paid by buying the stock at the market. If $S(T) = K$, it is indifferent to exercise or not to exercise the option, since the benefit is zero either way.

In conclusion, whatever the market scenario, the holder of the call option will have a benefit given by

$$(S(T) - K)^+ := \max(S(T) - K, 0).$$

Looking from the point of view of the seller of the option, at best it will gain nothing and, at worst, should the buyer have a positive benefit, the seller would have a loss equal to the buyer's benefit. There should therefore be some compensation for the seller's risk and so the option should have a price to be paid by the buyer to the seller at the time $t = 0$ of the option's purchase. The question Black, Merton, and Scholes posed was what should be the fair price of the option. Since the holder of the option may, at a certain time $t \leq T$, wish to sell it to someone else, it would also be nice to know what should be the fair price of the option at that time. Let us denote the price of the option at time $t \in [0, T]$ by $c(t)$.[4]

---

3 For simplicity, we consider a contract for just one unit of stock and determine, following Merton, Scholes, and Black, the fair price of such an option; of course, the contract can be made for any number of units of the stock, the fair price in that case being obtained by just multiplying the fair price we are going to determine by that number of units.

4 Other notations are also used in the literature. It is quite common to write B_t, b_t, S_t, s_t, c_t instead of $B(t)$, $b(t)$, $S(t)$, $s(t)$, $c(t)$. Sometimes the values of K, T, and the price S of the stock at time t are made explicit in the notation of the option price, for example using the notation $c(t; S, K, T)$, or, since, as we shall see, the option price depends on times t and T only through the difference $\tau = T - t$, the notation $c(\tau; S, K)$.

14.2   The Black–Scholes formula and hedging strategy

To answer the question of the fair price $c(t)$ $(t \in [0, T])$ of a European call option on a stock, it is immediately obvious that the fair price at time T should be equal to the benefit of the holder at that time, so:

$$c(T) = (S(T) - K)^+.$$

Following the steps of Black, Scholes, and Merton, assume that there is a self-financing trading strategy $(b(t), s(t))$ $(t \in [0, T])$ which value at time T is precisely $c(T)$, i.e.

$$b(T)B(T) + s(T)S(T) = c(T).$$

Then the assumption of no arbitrage implies that

$$b(0)B(0) + s(0)S(0) = c(0).$$

In fact, one can consider two portfolios, one being $(b(t), s(t))$ with $b(t)$ units of the bond and $s(t)$ units of the stock, the other consisting of a single European call option. Since they have the same value at time T, they should have the same value at time 0, i.e. $b(0)B(0) + s(0)S(0) = c(0)$.[5] The same reasoning can be applied to any other $t \in [0, T]$, leading to

$$b(t)B(t) + s(t)S(t) = c(t) \quad (0 \le t \le T). \tag{14.6}$$

Notice that the value $B(t)$ of the bond is deterministic and can be obtained in advance for all t. So, $c(t) = c(t, \omega)$, as a function of the market scenario ω, can only depend on the only random asset, which is the stock price. But $c(t)$ must be non-anticipating and cannot depend on the future unknown values of $S(t)$, for otherwise we could not establish its value at time t as we need to for trading purposes.

However, we can reach a much stronger conclusion using the principle of non-arbitrage and the Markov property of $S(t)$. Let, at a given time $t \in]0, T[$, the stock price be $S(t) = x$. Given the present state x of $S(t)$, the value of $S(T)$ does not depend on past values $S(u)$ with $u < t$, due to the Markov property. Therefore, given the present state x of $S(t)$, $c(T) = (S(T) - K)^+$ does not depend

---

5 Let us look at this in more detail. Suppose that $b(0)B(0) + s(0)S(0) < c(0)$. A person or institution A having the second portfolio (consisting of one option) at time 0 could sell it and obtain the first portfolio of $b(0)$ units of the bond and $s(0)$ units of the stock, pocketing the price difference $c(0) - (b(0)B(0) + s(0)S(0)) > 0$ between the two portfolios. Then A would follow the self-financing strategy $(b(t), s(t))$ $(t \in [0, T])$, which requires no additional funding. At time T, A can liquidate its assets, with a value of $b(T)B(T) + s(T)S(T) = c(T)$, precisely the same value A would have at time T if A had not switch the portfolios at the beginning. So, by having switched, A obtained a sure gain (the amount A has pocketed at the beginning) with a null investment and no risk and so there is arbitrage. The same conclusion can be reached if $b(0)B(0) + s(0)S(0) > c(0)$ by considering a portfolio switch by the person or institution holding the first portfolio. Therefore, the only way to have no arbitrage is to have $b(0)B(0) + s(0)S(0) = c(0)$.

on such past values either, and, by the principle of no arbitrage, the same must happen to $c(t)$.

We conclude that $c(t)$ can only depend on the value $S(t)$ of the stock at the present time t and not on future or past values of the stock.

Therefore, there is a function $C(t, x)$ such that

$$c(t) = C(t, S(t)), \quad t \in [0, T]$$

and, since $c(T) = (S(T) - K)^+$, we must have $C(T, x) = (x - K)^+$.

Apply the Itô formula (see the Itô Theorem 6.1) either directly or, as we shall do for convenience of exposition, taking advantage of the mnemonic in Observation 6.3. We get, taking into account (14.2) and (14.3),

$$
\begin{aligned}
dc(t) &= \frac{\partial C(t, S(t))}{\partial t} dt + \frac{\partial C(t, S(t))}{\partial x} dS(t) + \frac{1}{2} \frac{\partial^2 C(t, S(t))}{\partial x^2} (dS(t))^2 \\
&= \left(\frac{\partial C}{\partial t} + \frac{\partial C}{\partial x} \mu S(t) + \frac{1}{2} \frac{\partial^2 C}{\partial x^2} \sigma^2 S^2(t) \right) dt + \frac{\partial C}{\partial x} \sigma S(t) dW(t).
\end{aligned}
\tag{14.7}
$$

Attending to (14.5) and (14.6), we have

$$
\begin{aligned}
c(t) &= b(t)B(t) + s(t)S(t) \\
&= b(0)B(0) + s(0)S(0) + \int_0^t b(u)dB(u) + \int_0^t s(u)dS(u),
\end{aligned}
\tag{14.8}
$$

and therefore

$$
\begin{aligned}
dc(t) &= b(t)dB(t) + s(t)dS(t) \\
&= b(t)rB(t)dt + s(t)(\mu S(t)dt + \sigma S(t)dW(t)) \\
&= (b(t)rB(t) + s(t)\mu S(t))dt + s(t)\sigma S(t)dW(t).
\end{aligned}
\tag{14.9}
$$

The drift and diffusion coefficients of $c(t)$ are uniquely defined, so we must have the same coefficients in (14.7) and (14.9). Therefore

$$b(t)rB(t) + s(t)\mu S(t) = \frac{\partial C}{\partial t} + \frac{\partial C}{\partial x} \mu S(t) + \frac{1}{2} \frac{\partial^2 C}{\partial x^2} \sigma^2 S^2(t) \tag{14.10}$$

$$s(t)\sigma S(t) = \frac{\partial C}{\partial x} \sigma S(t). \tag{14.11}$$

Since $S(t) > 0$, from (14.11) we have

$$s(t) = \frac{\partial C(t, S(t))}{\partial x}, \tag{14.12}$$

and so $C(t, S(t)) = c(t) = b(t)B(t) + s(t)S(t) = b(t)B(t) + S(t)\frac{\partial C(t, S(t))}{\partial x}$, from which we get

$$b(t) = \frac{1}{B(t)} \left(C(t, S(t)) - \frac{\partial C(t, S(t))}{\partial x} S(t) \right). \tag{14.13}$$

Replacing in (14.10) $b(t)$ and $s(t)$ by the expressions (14.12) and (14.13), we obtain, using x in lieu of $S(t)$ to simplify the notation, the partial differential equation valid for $t \in [0, T[$ and $x > 0$:

$$-\frac{\partial C(t,x)}{\partial t} = \frac{\partial C(t,x)}{\partial x}rx + \frac{1}{2}\frac{\partial^2 C(t,x)}{\partial x^2}\sigma^2 x^2 - rC(t,x), \qquad (14.14)$$

with boundary condition

$$C(T,x) = (x - K)^+. \qquad (14.15)$$

This is known as the *Black–Scholes equation*.

There are several techniques to solve this equation, but the use of the Feynman–Kac formula (see Property 10.3) seems to be the best. It is more convenient to work with the variable $\tau = T - t$ (time to expiration) than with the present time t. Let $D(\tau, x) = C(T - \tau, x)$, from which we have $c(t) = D(T - t, S(t))$. We have $\frac{\partial D}{\partial \tau} = \frac{\partial D}{\partial x}rx + \frac{1}{2}\frac{\partial^2 D}{\partial x^2}\sigma^2 x^2 - rD$ with boundary condition $D(0, x) = (x - K)^+$. One immediately recognizes this is a particular case of the Feynman–Kac formula (10.11) with $h(x) = (x - K)^+$ (which is not of compact support but can be approximated by functions of that type), $q(x) \equiv r$, a drift coefficient rx and a diffusion coefficient $\sigma^2 x^2$. So, attending to (10.10) and Observation 10.3, the solution $D(\tau, x)$ will be of the form

$$D(\tau, x) = \mathbb{E}_x\left[e^{-\int_0^\tau r\,ds}(X(\tau) - K)^+\right] = \mathbb{E}_x[e^{-r\tau}(X(\tau) - K)^+], \qquad (14.16)$$

where $X(t)$ is a diffusion process with initial value $X(0) = x$ and the already mentioned drift and diffusion coefficients. So, $dX(t) = rX(t)dt + \sigma X(t)dW(t)$, and consequently $X(\tau) = x\exp((r - \sigma^2/2)\tau + \sigma W(\tau))$. Notice that τ is deterministic and $Z(\tau) = \ln X(\tau) \frown \mathcal{N}(\ln x + (r - \sigma^2/2)\tau, \sigma^2\tau)$. Let us put $\alpha = r - \sigma^2/2$ and $\beta = r + \sigma^2/2$. Then

$$D(\tau, x) = e^{-r\tau}\mathbb{E}_x[(e^{Z(\tau)} - K)^+]$$

$$= e^{-r\tau}\int_{-\infty}^{+\infty}(e^z - K)^+\frac{1}{\sigma\sqrt{2\pi\tau}}\exp\left(-\frac{(z - \ln x - \alpha\tau)^2}{2\sigma^2\tau}\right)dz$$

$$= e^{-r\tau}\int_{\ln K}^{+\infty}(e^z - K)\frac{1}{\sigma\sqrt{2\pi\tau}}\exp\left(-\frac{(z - \ln x - \alpha\tau)^2}{2\sigma^2\tau}\right)dz$$

$$= \int_{\ln K}^{+\infty}e^{-r\tau}e^z\frac{1}{\sigma\sqrt{2\pi\tau}}\exp\left(-\frac{(z - \ln x - \alpha\tau)^2}{2\sigma^2\tau}\right)dz$$

$$- \int_{\ln K}^{+\infty}e^{-r\tau}K\frac{1}{\sigma\sqrt{2\pi\tau}}\exp\left(-\frac{(z - \ln x - \alpha\tau)^2}{2\sigma^2\tau}\right)dz$$

$$= \int_{\ln K}^{+\infty}x\frac{1}{\sigma\sqrt{2\pi\tau}}\exp\left(-\frac{(z - \ln x - \beta\tau)^2}{2\sigma^2\tau}\right)dz$$

$$- \int_{\ln K}^{+\infty}e^{-r\tau}K\frac{1}{\sigma\sqrt{2\pi\tau}}\exp\left(-\frac{(z - \ln x - \alpha\tau)^2}{2\sigma^2\tau}\right)dz$$

$$= x \int_{\frac{\ln K - \ln x - \beta\tau}{\sigma\sqrt{\tau}}}^{+\infty} \frac{1}{\sqrt{2\pi}} e^{-u^2/2} du - Ke^{-r\tau} \int_{\frac{\ln K - \ln x - \alpha\tau}{\sigma\sqrt{\tau}}}^{+\infty} \frac{1}{\sqrt{2\pi}} e^{-v^2/2} dv$$

$$= x\left(1 - \Phi\left(\frac{\ln K - \ln x - \beta\tau}{\sigma\sqrt{\tau}}\right)\right) - Ke^{-r\tau}\left(1 - \Phi\left(\frac{\ln K - \ln x - \alpha\tau}{\sigma\sqrt{\tau}}\right)\right),$$

where $\Phi(u) = \int_{-\infty}^{u} \frac{1}{\sqrt{2\pi}} e^{-v^2/2} dv$ is the distribution function of a standard Gaussian r.v.

After easy calculations this gives

$$C(t,x) = x\Phi(d_1(t,x)) - Ke^{-r(T-t)}\Phi(d_2(t,x)),$$
$$\text{with} \quad d_1(t,x) = \frac{\ln(x/K) + (r + \sigma^2/2)(T-t)}{\sigma\sqrt{T-t}} \tag{14.17}$$
$$d_2(t,x) = \frac{\ln(x/K) + (r - \sigma^2/2)(T-t)}{\sigma\sqrt{T-t}} = d_1(t,x) - \sigma\sqrt{T-t}.$$

This is the celebrated *Black–Scholes formula*.

To obtain the European call option price at time t, one just needs to know the price $x = S(t)$ of the stock at that time:

$$c(t) = C(t, S(t)) = S(t)\Phi(d_1(t)) - Ke^{-r(T-t)}\Phi(d_2(t)),$$
$$\text{with} \quad d_1(t) = \frac{\ln\left(\frac{S(t)}{K}\right) + (r + \frac{\sigma^2}{2})(T-t)}{\sigma\sqrt{T-t}} \tag{14.18}$$
$$d_2(t) = \frac{\ln\left(\frac{S(t)}{K}\right) + (r - \frac{\sigma^2}{2})(T-t)}{\sigma\sqrt{T-t}} = d_1(t) - \sigma\sqrt{T-t}.$$

The initial value of the option is obtained by putting $t = 0$ in this expression.

Exercise 14.1 *Check that C given by (14.17) satisfies the boundary condition (14.15) and so, from (14.18), $c(T) = (S(T) - K)^+$.*
Suggestion: For this, it may help to consider separately the possibilities $x < K$, $x = K$, and $x > K$. Note that, at time T and when $x \neq K$, $d_1(T, x)$ and $d_2(T, x)$ are both infinite, either $+\infty$ or $-\infty$ according to the sign of $\ln(x/K)$.

It is important to notice that the Black–Scholes formula does not depend on the average return rate μ of stock, which is a very difficult parameter to estimate with reasonable precision, but it rather depends on the return rate r of the riskless asset (fixed rate bond or bank account with fixed interest rate).

The above deduction of the Black–Scholes formula was based on the assumption that there exists a *self-financing trading strategy* $(b(t), s(t))$ $(t \in [0, T])$ such that $c(T) = b(T)B(T) + s(T)S(T)$. Does it exist? The answer is yes, as we shall see, and of course that strategy is given by expressions (14.12) and (14.13), from which, using (14.18) and performing some computations, we get

$$s(t) = \Phi(d_1(t)) \quad (t \in [0, T]) \tag{14.19}$$

and

$$b(t) = -\frac{1}{B(t)} Ke^{-r(T-t)} \Phi(d_2(t)) \quad (t \in [0, T]). \tag{14.20}$$

Exercise 14.2 *Show that, in fact and as claimed, the trading strategy given by (14.19) and (14.20) is indeed self-financing, i.e. it satisfies (14.5).*
Suggestion: Remember that $B(t)$ satisfies (14.1) and (14.2), $S(t)$ satisfies (14.3) and (14.4), and that you can use the Itô formula.

It is interesting to note that $s(t) > 0$ and $b(t) < 0$, which means that the self-financing strategy for the European call option has always a long position in stock and a short position in the riskless asset. This strategy at time t has therefore a capital invested in stock equal to $s(t)S(t) = S(t)\Phi(d_1(t))$ and a (negative) capital invested in bonds (or in a bank account, corresponding to a loan) $b(t)B(t) = -Ke^{-r(T-t)}\Phi(d_2(t))$, which are exactly the first and second terms of (14.18) for $c(t)$. Remember that, from (14.6), $c(t)$ is, at time t, simultaneously the price of the European call option and the value of the self-financing strategy.

Therefore, *the self-financing strategy consists in having, at time t, a capital of* $S(t)\Phi(d_1(t))$ *invested in stock and a capital of* $-Ke^{-r(T-t)}\Phi(d_2(t))$ *invested in the riskless asset.* This corresponds to having $s(t) = \Phi(d_1(t))$ units of the stock and, if the asset is a fixed rate bond, to having $b(t) = -\frac{1}{B(t)}Ke^{-r(T-t)}\Phi(d_2(t))$ units of the bond. *The total capital invested is exactly equal to the price c(t) of the option.*

That means that the seller of the option, having received at time 0 its price $c(0)$, can immediately invest that amount in bonds (or in a bank account) and stocks according to $(b(0), s(0))$ and keep trading bonds and stocks according to the self-financing strategy $(b(t), s(t))$. At time T, the value of the self-financing strategy the seller holds is $c(T)$, which is precisely the benefit of the option buyer that the seller is committed to uphold. *So, the seller of the option, following this self-financing strategy, is completely protected from the financial risks of the market and neither loses nor gains. This makes c(t), given by the Black–Scholes formula (14.18), the fair price for the European call option at time t and the one that respects the no arbitrage principle.*

For these reasons, the self-financing strategy given by (14.19) and (14.20) is called the *hedging strategy*, in analogy to a hedge (or fence made of bushes or shrubs) one can put around a house for protection.

One may think that the buyer, instead of buying the European call option, could invest the price $c(0)$ it would cost on the self-financing trading strategy, so that at time T it would have exactly the same benefit $(S(T) - K)^+$. That is indeed the case. Remember, however, that a self-financing strategy requires continuously trading the riskless asset for stock or vice versa according to the fluctuations of the stock price (fluctuations that keep happening), while an option requires only the initial buying and the final liquidation work. The continuous trading required by the self-financing strategy is only viable because we have assumed no transaction costs and adjustments are necessary if such costs exist.

14.3 A numerical example and the Greeks

Consider the example of the Renault stock at Euronext Paris between 2 January 2012 and 31 December 2015 (four years of data) that is shown in Figure 2.3. From the observed trajectory, parameter estimates for the Black–Scholes model (14.3) were obtained by the maximum likelihood method, as mentioned in Section 8.1, with the following estimates: $\hat{\sigma} = 0.387/\sqrt{\text{year}}$, with an approximate 95% confidence interval of 0.387 ± 0.017, and $\hat{m} = 0.303/\text{year}$[6] (where $m = \mu - \sigma^2/2$ is the geometric average return rate), with an approximate 95% confidence interval of $0.303 \pm 0.379/\text{year}$. Notice the low degree of precision in estimating the return rate. The arithmetic average return rate estimate is $\hat{\mu} = \hat{m} + \hat{\sigma}^2/2 = 0.378/\text{year}$. Just for curiosity, since it has no direct relevance on the computation of option prices, based on these numbers, at 31 December 2015 the point prediction for the stock price at 31 March 2016 is €99.92 and a 95% prediction interval is [€67.66, €147.57] (see Section 8.1). The actual stock price at 31 March 2016 turned out to be €87,32.

Let us assume that the return rate of a riskless asset was $r = 0.5\%$ per year. Suppose that, on 31 December 2015, which we will take as time 0, one buys a European call option on the Renault stock with expiration date 31 March 2016 (three months later), so $T = 0.25$ years approximately, and with strike price $K = \text{€}95$. Since the price of the stock on 31 December 2015 is $S(0) = \text{€}92.63$, this option is out of the money. Let us for simplicity assume that the riskless asset is a riskless bond and that its initial value is $B(0) = 1$ (if it is a bank account with fixed interest rate $r = 0.5\%$, instead of worrying about the value $B(t)$ of each unit and the number of units $b(t)$, it is more convenient to look at the capital in euros $b(t)B(t)$ at the bank account). Whoever bought this option could not know that on 31 March 2016 the actual stock price was going to be $S(T) = \text{€}87.32$, lower than the strike price, and so its benefit from the option at time $T = 0.25$ years was going to be zero. If, however, the price of the stock at $T = 0.25$ years had been, for example, €110, the holder of the option would have had a €15 benefit.

According to the Black–Scholes formula (14.18), and assuming the volatility σ has the value $= 0.387/\sqrt{\text{year}}$ (value that we have estimated from the four-year historical data prior to the issuing date of the bond), the price of this option at time 0 (31 December 2015) is $c(0) = \text{€}6.156$. The hedging strategy

---

6 In Section 8.1 we have denoted the stock price by $X(t)$, its arithmetic average return rate by R, and its geometric average return rate by r. Here, following traditional financial notation, we are using $S(t)$ to denote the stock price, μ to denote the arithmetic average return rate, and $m = \mu - \sigma^2/2$ to denote the geometric average return rate. Remember that here r has a very different meaning, namely the deterministic fixed rate of the riskless asset.

at time 0 would consist in having €45.304 in stock (which corresponds to $s(0) = \Phi(d_1(0)) = 0.489$ units of stock) and $-€39.148$ in the riskless asset (either in a bank account, meaning a money loan, or, if its is a fixed rate bond, assuming $B(0) = €1$, that means having $b(0) = -\frac{1}{B(0)}Ke^{-r(T-0)}\Phi(d_2(0)) = -39.148$ units of the bond).

Let us present a possible R code. The first line contains the values of 'tau', meaning $T - t$, and 'Scurrent' and 'Bcurrent', meaning the current values of the stock and of the riskless asset. We wrote the values at time $t = 0$ ('tau' equal to 0.25), but you can replace by the values at some other time; we have assumed that at time $t = 0$, $B(0) = 1$ (so we put 'Bcurrent' equal to 1) but, if this is not the case, just change the value. 'sS' and 'bB' are the values of the hedging strategy in the stock and in the riskless asset, and 'c' is the value of the call option. Remember that, if you choose a time t different from 0, 'Bcurrent' is given by $B(0)e^{rt}$. The R code could then be:

```
tau<-0.25; Scurrent<-92.63; K<-95; Bcurrent<-1
r<-0.005; sigma<-0.387
d1<-(log(Scurrent/K)+(r+sigma^2/2)*tau)/(sigma*sqrt(tau))
d2<-d1-sigma*sqrt(tau)
s<-pnorm(d1); sS<-s*Scurrent; bB<--K*exp(-r*tau)*pnorm(d2);
b<-bB/Bcurrent
c<-sS+bB
```

Figure 14.1 shows the values of $c(t)$ computed according to the Black–Scholes formula at the end of each weekday, which is also the total capital of the hedging strategy. The figure also shows the positive amount $s(t)S(t)$ of that capital which is invested in stock and the negative amount $b(t)B(t)$ that is invested in the bond (or the bank account). Of course, at time T we have $c(T)$ equal to the benefit of the option holder, which in this example is by chance equal to zero. Obviously, the figure cannot be plotted at time 0 since we do not know then what the future values of the stock will be, but, as time goes by, we can add new points to the figure until we reach the final time T.

Exercise 14.3 *Consider a European call option on the stock mentioned above.*

(a) *What is the composition of the hedging strategy one month after the option issuing date (i.e. at time $t = 1/12$ years) if the stock price at that time is $S(\frac{1}{12}$ years$) = €78.64$? What is the fair price $c(\frac{1}{12}$ years$)$ of the option at that time?*

(b) *If for this option the market price at time $t = 0$ were to be €6.500 instead of €6.156, what would the implied volatility be (see definition below)?*

(c) *What would be $c(0)$ and the composition of the hedging strategy at time 0 if the strike price is at the money ($K = €92.63$)?*

(d) *What if $K = €85$?*

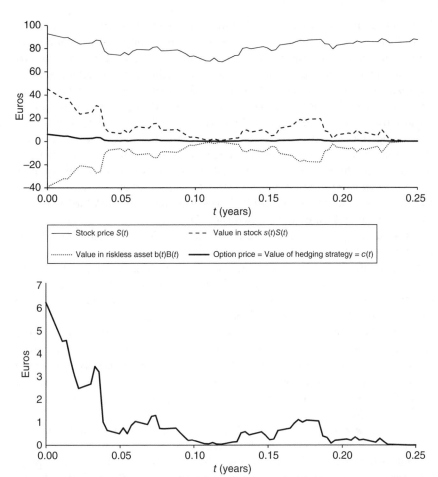

Figure 14.1 Daily (weekday closing) values in euros for the hedging strategy of a European call option on the Renault stock. In this example, the option was sold on 31 December 2015 ($t = 0$), had an expiration date of 31 March 2016 ($T = 0.25$ years), and had a strike price of €95. The riskless asset is assumed to have a return rate of $r = 0.5\% = 0.005$ per year and the volatility is assumed to be 0.387 per $\sqrt{\text{year}}$ (which is the historical volatility value estimated from the previous 4 years). On the top figure we find, from top to bottom, the evolution of: (a) The stock price $S(t)$ (thin solid line). At the issuing date of the option $t = 0$, only the value $S(0) = 92.63$ was known. The value at time T turned out to be $S(T) = €87.32$. (b) The value $s(t)S(t)$ of the stock component of the hedging strategy (dashed line). (c) The total value $c(t) = b(t)B(t) + s(t)S(t)$ of the hedging strategy (thick solid line), which is also the option price at time t. At time 0, only the value of $c(0) = €6.156$ was known. This function is repeated at the lower figure, where we have amplified the vertical axis in order to provide a better visualization of the changes $c(t)$ suffered over time. (d) The value $b(t)B(t)$ of the riskless asset component of the hedging strategy (dotted line).

Of course, the value of the volatility we have used is an estimate and so it has an estimation error (although quite likely a small error, since the confidence interval was small). Furthermore, in real markets μ and σ do not stay exactly constant for long periods of time as the Black–Scholes model assumes as an approximation. What would really matter is the volatility σ that will hold in the time period $[0, T]$, which we do not know. Actually, the options, like the stocks, are also traded in the markets and their prices are subject to the typical market bargaining. If we were to use the Black–Scholes formula in reverse, i.e. knowing the market price $c(0)$ and trying to match that price using the Black–Scholes formula with different values of σ, the value of σ that would work might not be the *historical volatility* that we had estimated and used above, but rather the value that the market 'expects' or 'believes' is going to be valid for the period $[0, T]$; that value is called the *implied volatility*.

For example, if the market value of the option was $c^*(0) = €6.50$, the implied volatility would be $\sigma^* = 0.4056/\sqrt{\text{year}}$ (check using this value in the Black–Scholes formula and see if it gives the option price of €6.50). Things are actually a little more complex. It would be expected that European call options with the same issuing and expiration dates should have the same implied volatility, independent of the strike price K, so if we plot the values of the implied volatility on the vertical axis against the strike price on the horizontal axis, we would expect to see a straight horizontal line. However, what is typically observed is a line with the shape $\smile$, with options quite in the money (values of K much lower than $S(0)$) or quite out of the money (values of K much higher than $S(0)$) having a higher implied volatility. This is known as the *smile effect*.

In the financial markets it is quite common to use sensitivity measures that give a quick way of determining how the variation of a parameter or a value will affect the option price. This is given by the derivative of the option price with respect to that parameter or value. Since they are traditionally denoted by Greek letters, they are called *Greeks*.

For example, the Greek *delta* is denoted by the Greek letter Δ and we will use an index c to indicate that it refers to a European call option; its value at time t measures the sensitivity of $c(t)$ to the stock price $x = S(t)$, so Δ_c is given at time t by using the expression of $c(t)$ from the Black–Scholes formula and taking the derivative with respect to $x = S(t)$, i.e. $\frac{\partial C(t,x)}{\partial x} = \Phi(d_1(t, x))$; particularizing $x = S(t)$, we get $\frac{\partial C(t,S(t))}{\partial S(t)} = \Phi(d_1(t))$. It means that, for a unit variation in the stock price, the option price will change by approximately Δ_c.

In the example above, at time 0, $\Delta_c = 0.489$. If the information on this value $\Delta_c = 0.489$ and on the current value $c(0) = €6.156$ of the option is given, one can immediately say, without having to use the Black–Scholes formula, that, if the stock price was €1 higher than it is, the option price would increase by approximately €0.489, becoming approximately equal to €6.156 + €0.489 =

€6.645. Similarly, if the stock price was €0.50 lower than it is, the option price would decrease approximately €0.50 × €0.489 ≈ €0.245, becoming approximately equal to €6.156 − €0.245 = €5.911. This is, of course an approximation. One can compute, using the Black–Scholes formula, the exact values of the option if the stock price was €1.00 higher or €0.50 smaller than it really is; the option prizes would be, respectively, €6.656 and €5.914.

The Greek *rho* is represented by the Greek letter ρ (or, indicating that our option is a European call option, by ρ_c) and measures the sensitivity of the option price to the riskless asset return rate r. It is therefore obtained by taking the derivative with respect to r in the expression of $c(t)$ given by the Black–Scholes formula. Here $r = 0.5\% = 0.005$. If r increases, say by 0.001 (i.e. becomes = 0.006 = 0.6%), then $c(t)$ would change approximately by $0.001 \times \rho_c$.

The Greek *theta* is represented by the Greek letter Θ (or, indicating that our option is a European call option, by Θ_c) and measures the sensitivity of the option price to the passing of time t. It is therefore obtained by taking the derivative with respect to t (or, equivalently, w.r.t. $-\tau = -(T - t)$) in the expression $c(t)$ given by the Black-Scholes formula.

The Greek *vega* is represented by the Greek letter v (or, indicating that our option is a European call option, by v_c), but is not named after it. It measures the sensitivity of the option price to the volatility σ and is obtained by taking the derivative with respect to σ in the expression of $c(t)$ given by the Black–Scholes formula.

Exercise 14.4 *Consider a European call option on a stock.*

(a) *Show that the expression for the Greek Δ_c at time t is indeed given by $\Phi(d_1(t))$. You can take the derivative using the expression (14.17) with $x = S(t)$ or directly (14.18) and simplify the expressions. Note that*
$$\Phi(z) = \int_{-\infty}^{z} \frac{1}{\sqrt{2\pi}} \exp\left(-\frac{u^2}{2}\right) du \text{ and so } \frac{d\Phi(z)}{dz} = \phi(z) = \frac{1}{\sqrt{2\pi}} \exp\left(-\frac{z^2}{2}\right).$$

(b) *Since Δ_c is always positive, this shows that the higher the stock price, the higher is the option price.*

(c) *Determine the value of Δ_c at time 0 for the example of the stock used in this section.*

(d) *Obtain expressions, in the simplest form, for the other Greeks and determine the concrete values of these Greeks at time 0 for the example given in this section.*

(e) *Can you establish the sign of at least some of the above Greeks? This indicates if the option price will increase (if the Greek is positive) or decrease (if the Greek is negative) when there is an increase in the corresponding parameter.*

There are also second-order Greeks like, for example, Γ_c, which is $\frac{\partial^2 C(t,x)}{\partial x^2}$, i.e. the derivative of first-order Greek Δ_c with respect to the stock price.

To conclude, let us just mention that, if a stock pays *dividends* before the expiration date T, at previously fixed dates and amounts, the pricing of the European call option needs to be adjusted. In fact, after dividend payment the stock price is reduced by the amount of the dividend and so we must use not $S(0)$ but rather an adjusted spot price that deducts from the nominal price the present value (at the riskless return rate) of the future dividends. So, just to give an example, if a stock with nominal spot price S at time 0 will distribute two dividends D_1 and D_2 at times t_1 and t_2 with $0 < t_1 < t_2 < T$, then the adjusted spot price at time 0 would be $S - D_1 e^{-rt_1} - D_2 e^{-rt_2}$. With such adjustment, we can use the treatment given in this section. The situation where the dividends occur in dates and/or amounts that depend on the market scenario is more complex and beyond the scope of this book.

14.4 The Black–Scholes formula via Girsanov's theorem

We will now look for a different method for arriving at the Black–Scholes formula, taking advantage of Girsanov's theorem. Although in this section we will deal again with the same European call option on a stock following the same Black–Scholes model and the same assumptions we made in the previous sections, this different method can be used when the asset, which may not even be a stock, follows some other model.

In order to be dealing with martingales with respect to the new riskless probability (based on Girsanov's theorem), we shall work with discounted prices with a discount rate r equal to the return rate of the riskless asset. We can assume that the riskless asset is a bond with fixed return rate r or a bank account with interest rate r. The value at time t of one unit of the riskless asset is $B(t) = B(0)e^{rt}$, where $B(0)$ is the value at time 0 of one unit of the bond or, in case of a bank account, the value at time 0 of one monetary unit, say €1 (in which case $B(0) = €1$). The discounted price of any asset at time t is just its nominal price (say in euros) divided by $B(t)$, for example if the riskless asset is a bond, this is just the number of units of the bond required to match at time t the price of the asset.

The discounted prices of the bond, of the stock, and of the European call option on that stock are, therefore, given by

$$\begin{aligned}
\tilde{B}(t) &= B(t)/B(t) \equiv 1 \\
\tilde{S}(t) &= S(t)/B(t) = S(t)\exp(-rt)/B(0) \\
\tilde{c}(t) &= c(t)/B(t) = c(t)\exp(-rt)/B(0).
\end{aligned} \qquad (14.21)$$

By Itô theorem,

$$d\tilde{S}(t) = \frac{1}{B(0)}\{-rS(t)\exp(-rt)dt + \exp(-rt)dS(t) + 0(dS(t))^2\},$$

and so

$$d\tilde{S}(t) = (\mu - r)\tilde{S}(t)dt + \sigma\tilde{S}(t)dW(t). \tag{14.22}$$

Therefore, $\tilde{S}(t)$ is a homogeneous diffusion process with drift and diffusion coefficients

$$\begin{aligned} f(x) &= (\mu - r)x \\ g^2(x) &= (\sigma x)^2. \end{aligned} \tag{14.23}$$

In order to work with a martingale, we want to apply *Girsanov's theorem* to change the drift coefficient of $\tilde{S}(t)$ to

$$f^*(x) = 0 \tag{14.24}$$

The function $\theta(x)$ appearing in (13.4) is, in this case, given by

$$\theta(x) \equiv \frac{\mu - r}{\sigma} \tag{14.25}$$

and is called the *market price of risk*. It is the advantage of the stock average return rate μ over the riskless return rate r measured in terms of the amount of the risk σ taken. In other words, it is the extra average rate premium the market 'pays' per unit of risk σ to the holder of the risky asset.

From (13.5), we get

$$M(t) = \exp\left(-\frac{\mu - r}{\sigma}W(t) - \frac{(\mu - r)^2}{2\sigma^2}t\right) \tag{14.26}$$

and one can easily check that this is a martingale with respect to the original probability P (the true probability of the market scenarios). By Girsanov's theorem, we construct a new probability P^* defined by its Radon–Nikodym derivative

$$\frac{dP^*}{dP} = M(T) = \exp\left(-\frac{\mu - r}{\sigma}W(T) - \frac{(\mu - r)^2}{2\sigma^2}T\right). \tag{14.27}$$

From (13.7), we obtain

$$W^*(t) = W(t) + \frac{\mu - r}{\sigma}t, \tag{14.28}$$

which is a Wiener process for the new probability P^*. From (13.8) and (14.24), we see that

$$d\tilde{S}(t) = \sigma\tilde{S}(t)dW^*(t), \tag{14.29}$$

which solution is

$$\tilde{S}(t) = \tilde{S}(0) + \int_0^t \sigma\tilde{S}(u)dW^*(u). \tag{14.30}$$

By the properties of the stochastic integral, $\tilde{S}(t)$ is a martingale for P^*. Since P^* transforms $\tilde{S}(t)$, the price of the stock discounted by the return rate r of the riskless asset, into a martingale, it is called an *equivalent martingale probability measure*. It is also called *risk neutral probability* because it distorts the probabilities of the market scenarios in such a way that the stock prices look neutral to the risk, i.e. as seen from (14.29), the discounted values of the stock have a zero (arithmetic) average return rate for P^*. This means that the actual (non-discounted) values of the stock have a return rate for P^* identical to the return rate r of the riskless asset, and so the market price of risk under P^* is zero. It is very easy to check that indeed the non-discounted values of the stock have a return rate for P^* equal to r; in fact, applying the Itô theorem to $S(t) = \tilde{S}(t)B(t)$ and using (14.29), one gets

$$dS(t) = d(\tilde{S}(t)B(t)) = rS(t)dt + \sigma S(t)dW^*(t). \tag{14.31}$$

If there is a *hedging strategy* $(b(t), s(t))$, i.e. a self-financing strategy that reproduces the value $c(t)$ of the option, then

$$\begin{aligned}
c(t) &= b(t)B(t) + s(t)S(t) \\
&= b(0)B(0) + s(0)S(0) + \int_0^t b(u)dB(u) + \int_0^t s(u)dS(u)
\end{aligned}$$

and so

$$\begin{aligned}
dc(t) &= b(t)dB(t) + s(t)dS(t) \\
&= b(t)rB(t)dt + s(t)(rS(t)dt + \sigma S(t)dW^*(t)) \\
&= r(b(t)B(t) + s(t)S(t)) + \sigma s(t)S(t)dW^*(t) \\
&= rc(t)dt + \sigma s(t)S(t)dW^*(t)
\end{aligned}$$

and

$$\begin{aligned}
d\tilde{c}(t) &= \frac{1}{B(0)}d(e^{-rt}c(t)) \\
&= \frac{1}{B(0)}(re^{-rt}c(t)dt + e^{-rt}dc(t) + 0(dc(t))^2) \\
&= \frac{1}{B(0)}(e^{-rt}\sigma s(t)S(t)dW^*(t)) \\
&= \sigma s(t)\tilde{S}(t)dW^*(t).
\end{aligned}$$

Then,

$$\tilde{c}(t) = \tilde{c}(0) + \int_0^t \sigma s(u)\tilde{S}(u)dW^*(u) \tag{14.32}$$

is also a martingale for P^* with respect to the natural filtration of the Wiener process.

Therefore, for $0 \leq t \leq T$,

$$\tilde{c}(t) = \mathbb{E}^{P^*}[\tilde{c}(T) \mid \mathcal{M}_t]. \tag{14.33}$$

This has a very nice interpretation: The discounted price of the option $\tilde{c}(t)$ at time $t \in [0, T]$ is just the conditional mathematical expectation, under the risk neutral probability P^*, of its value $\tilde{c}(T)$ at the expiration date T, where we condition with respect to the information available at time t. Fortunately, we know the expression for $\tilde{c}(T)$. Note that the information $\mathcal{M}_t$ given by the Wiener process is equivalent to that given by the stock price process and that, as seen in Section 14.2, the option price at time t only depends on $S(t)$ (not on past values). So, conditioning on $\mathcal{M}_t$ is equivalent to conditioning on the value of the stock at time t. Therefore, a direct 'translation' of (14.33), putting $c(t) = C(t, S(t))$, is

$$\frac{C(t, S(t))}{B(0)e^{rt}} = \mathbb{E}^{P^*}\left[\left(\frac{S(T) - K}{B(0)e^{rT}}\right)^+\Bigg| S(t)\right],$$

from which

$$C(t, x) = \mathbb{E}^{P^*}[(S(T)e^{-r(T-t)} - Ke^{-r(T-t)})^+ | S(t) = x]. \tag{14.34}$$

From (14.31),

$$d \ln S(t) = (r - \sigma^2/2)dt + \sigma dW^*(t) \tag{14.35}$$

and so

$$\ln S(T) = \ln S(t) + \int_t^T (r - \sigma^2/2)ds + \int_t^T \sigma dW^*(s)$$
$$= \ln S(t) + (r - \sigma^2/2)(T - t) + \sigma(W^*(T) - W^*(t)).$$

So, conditional on $S(t) = x$ and with respect to the probability P^*, the r.v. $Z = \ln S(T)$ has distribution

$$\mathcal{N}(\ln x + (r - \sigma^2/2)(T - t), \sigma^2(T - t)),$$

the r.v.

$$U = \frac{Z - \ln x - (r - \sigma^2/2)(T - t)}{\sigma\sqrt{T - t}}$$

is standard Gaussian and we have

$$(S(T)e^{-r(T-t)} - Ke^{-r(T-t)})^+$$
$$= (xe^{-\sigma^2(T-t)/2+\sigma\sqrt{T-t}\,U} - Ke^{-r(T-t)})^+.$$

So, from (14.34), we see that $C(t, x)$ is the mathematical expectation of

$$(xe^{-\sigma^2(T-t)/2+\sigma\sqrt{T-t}\,U} - Ke^{r(T-t)})^+,$$

with U being a standard Gaussian r.v. Therefore, using the expressions of $d_1(t, x)$ and $d_2(t.x)$ defined in (14.17), we have

$$C(t, x) = \int_{-\infty}^{+\infty} (xe^{-\sigma^2(T-t)/2+\sigma\sqrt{T-t}\,u} - Ke^{-r(T-t)})^+ \frac{1}{\sqrt{2\pi}}e^{-u^2/2}du$$

$$= \int_{-d_2(t,x)}^{+\infty} (xe^{-\sigma^2(T-t)/2+\sigma\sqrt{T-t}\,u} - Ke^{-r(T-t)})\frac{1}{\sqrt{2\pi}}e^{-u^2/2}du$$

$$= x\int_{-d_2(t,x)}^{+\infty} \frac{1}{\sqrt{2\pi}}e^{-(u-\sigma\sqrt{T-t})^2/2}du - Ke^{-r(T-t)}\int_{-d_2(t,x)}^{+\infty} \frac{1}{\sqrt{2\pi}}e^{-u^2/2}du$$

$$= x \int_{-d_1(t,x)}^{+\infty} \frac{1}{\sqrt{2\pi}} e^{-v^2/2} dv - K e^{-r(T-t)} \int_{-d_2(t,x)}^{+\infty} \frac{1}{\sqrt{2\pi}} e^{-u^2/2} du$$

$$= x(1 - \Phi(-d_1(t,x))) - K e^{-r(T-t)}(1 - \Phi(-d_2(t,x)))$$

$$= x\Phi(d_1(t,x)) - K e^{-r(T-t)}\Phi(d_2(t,x)).$$

We have obtained again, using a different route, the Black–Scholes formula (14.17).

The previous deduction assumed the existence of a hedging strategy. From the expression of $C(t,x)$ one can obtain, by the same methods that we used in Section 14.2 for the same purpose, the hedging strategy expressions for $b(t)$ and $s(t)$, and show that it is indeed a hedging strategy, since its value reproduces at any time $t \in [0, T]$ the option price $c(t)$.

Exercise 14.5 *For the self-financing strategy, the capital invested in the riskless asset at time t is $b(t)B(t) = -K e^{-r(T-t)}\Phi(d_2(t))$, which is negative.*

Of course, at time t, the amount $K e^{-r(T-t)}$ is the present value of the strike price K, i.e. its value discounted by the interest rate r of the riskless asset. Put another way, having an amount of $K e^{-r(T-t)}$ at time t is equivalent to having an amount K at time T, since one can, without taking any risk, just invest that amount $K e^{-r(T-t)}$ in the riskless asset and simply wait until time T to get the amount K.

Show that, conditional on $S(t)$, $\Phi(d_2(t))$ is the P^*-probability that at time T the holder of the option will exercise it, i.e. the P^*-probability that $S(T) > K$.

So, $K e^{-r(T-t)}\Phi(d_2(t))$ is the risk neutral probability that the holder will exercise the option times the present value of the amount the holder will spend by exercising it (i.e. by buying one unit of stock at the strike price K). When the holder does not exercise the option, it does not spend anything. So, we can interpret $b(t)B(t) = -K e^{-r(T-t)}\Phi(d_2(t))$ as a loan (notice the negative sign) in the amount of the present value (value at time t) of the (neutral market) P^*-expected value of the holder's expense at time T.

The alternative method via Girsanov's theorem is advantageous in more complex models. In our case, we really did not need to use the alternative method since the risky asset follows the simple Black–Scholes model and we could determine an explicit expression for $C(t,x)$ by the method of Section 14.2. However, for more complex models, we often cannot obtain an explicit expression that way and the proof of existence and the determination of the hedging strategy may have to recur to new methods, namely the Yor's representation theorem, which is beyond the scope of this book. Even though we may not get an explicit expression for the option price for these more complex models, still its discounted price at time $t \in [0, T]$ is just the conditional mathematical expectation for the risk neutral probability P^* of its value at the expiration date T, where the conditioning is with respect to the information available at time t. Therefore, we can always have approximate values of the option price by approximating the conditional mathematical expectation through the use of Monte Carlo simulations or other techniques.

One way to approximate the conditional mathematical expectation is to discretize the interval $[0, T]$ and, at each time step, use the distribution with respect to P^* of the change incurred by the discounted option price, using the martingale property. It may not be possible to determine this distribution explicitly, but, for small time steps, it is approximately Gaussian (because $\Delta W^*(t)$ is Gaussian) and can therefore be approximated by an appropriate binomial distribution. For that, one needs also to discretize the state space, since the binomial distribution is discrete. This is the so-called binomial model or Cox–Ross–Rubinstein (CRR) model and, since it does not require a very fine discretization to give already a good approximation, it provides fast computing of the option price and of the hedging strategy.

In next section we will illustrate the use of this approximation modelling technique for the case of our European call option on a stock following a Black–Scholes model. In this simple case, we did not need this approximating method because we have the exact formula for the option price, the Black–Scholes formula. There are still good reasons for presenting this method in next section using this simple case as illustration: one gets a better grasp of what is happening by handling this simple case and the application of the method in more complex cases is done in a similar manner (with the obvious adaptations).

14.5 Binomial model

The *binomial model* or *Cox–Ingersol–Ross (CRR) model* consists of a discretization in time and in state space for the underlying asset process, working with the risk neutral probability P^* and approximating it, at each step, by a binomial r.v. As we have said, in this section we are going to illustrate the use of the technique when the underlying asset is a stock with price $S(t)$ following the Black–Scholes model and the derivative process is the price of a European call option $c(t)$ on that stock. For illustration of numerical computations, we will use the same numerical values as in Section 14.3.

We will approximate $S(t)$ by a discrete-time process $\overline{S}(t)$ with discrete state space such that, at each time step, the process moves to one of the two neighbour states. The time interval $[0, T]$ is discretized into n steps of equal duration $\Delta t = T/n$. In the numerical illustrations we will use $n = 3$ so that you can easily follow the procedure, but in real applications one should use a higher value of n.

From (14.31), we see that $S(t)$ is, for P^*, a diffusion process with drift coefficient rx and diffusion coefficient $\sigma^2 x^2$. Let $Z(t) = \ln S(t)$. From

(14.35), we see that $Z(t)$ is a homogeneous diffusion process with drift coefficient $r - \sigma^2/2$ and diffusion coefficient σ^2. The increment of $Z(t)$ (lets us call it log-increment) in the time interval from t to $t + \Delta t$ is $\Delta Z(t) = \ln S(t + \Delta t) - \ln S(t) = (r - \sigma^2/2)\Delta t + \sigma(W^*(t + \Delta t) - W^*(t))$, which has, for P^*, a Gaussian distribution with mean $(r - \sigma^2/2)\Delta t$ and variance $\sigma^2 \Delta t$.[7] Then, if $S(t)$ has the value S_t at time t, in the time step from t to $t + \Delta t$, it will move to a new value $S(t + \Delta t) = S_t e^{\Delta Z(t)}$. In each time step, we approximate the P^*-distribution of the log-increment $\Delta Z(t)$ by a Bernoulli distribution that can assume two possible values, Δz or $-\Delta z$, with probabilities p^* and $q^* = 1 - p^*$ appropriately chosen, where Δz should also be appropriately chosen.

We thus obtain an approximating discrete-time process $\overline{S}(t)$ ($t = 0, \Delta t$, $2\Delta t, \dots, n\Delta t = T$). If the approximating process is at state S_t a time t, in the time step from t to $t + \Delta t$, it will either move up to $S_t u$ or down to $S_t d$, where $u = e^{\Delta z}$ and $d = e^{-\Delta z} = 1/u$, with probabilities p^* and $q^* = 1 - p^*$, respectively. Of course, we want process $\overline{S}(t)$ to approximate the process $S(t)$ as $n \to +\infty$ (which implies $\Delta t \to 0$). This requires that $\Delta z = \sigma \sqrt{\Delta t}$ so that the variance of the Bernoulli distribution is proportional to $\sigma^2 \Delta t$. We are using, for similar reasons, the same discretization used in Section 4.2 to approximate a Wiener process $\sigma W(t)$ by a random walk. Ideally, we would like the p^* be such that the Bernoulli distribution would have the same mean and variance as $\Delta Z(t)$; this cannot be achieved exactly, but we can get a good approximation, with errors of lower order than Δt.

To do this, suppose that, at time t, $S(t)$ and $\overline{S}(t)$ have the value S_t; we will try to exactly match the expected value of $\overline{S}(t + \Delta t)$ under its binomial distribution with the expected value of $S(t + \Delta t) = S_t e^{\Delta Z(t)}$ under its Gaussian P^*-distribution. We have

$$\mathbb{E}^{Binomial}[\overline{S}(t + \Delta t) \mid \overline{S}(t) = S_t] = p^* S_t u + q^* S_t d = (p^* u + (1 - p^*)d)S_t$$

$$\mathbb{E}^{P^*}[S(t + \Delta t)|S(t) = S_t] = \mathbb{E}^{P^*}[S(t)e^{\Delta Z(t)}|S(t) = S_t] = S_t \mathbb{E}^{P^*}[e^{\Delta Z(t)}] = S_t e^{r\Delta t}.$$

We get an exact match when

$$p^* = \frac{e^{r\Delta t} - d}{u - d} = \frac{e^{r\Delta t} - e^{-\sigma\sqrt{\Delta t}}}{e^{\sigma\sqrt{\Delta t}} - e^{-\sigma\sqrt{\Delta t}}}. \tag{14.36}$$

With this value of p^* we see that the mean and variance of the Bernoulli approximation will be very close to the mean and variance of the log-increment under the P^*-distribution, with error terms of lower order than Δt.

---

7 For underlying assets following a different model, this distribution of $\Delta Z(t)$ may not be Gaussian, but, for small Δt, it is approximately Gaussian with mean equal to the drift coefficient of $Z(t)$ multiplied by Δt and variance equal to the diffusion coefficient of $Z(t)$ multiplied by Δt. In some of these other models, it may happen that $\ln S(t)$ is not Gaussian but some other transform $h(S(t))$ is Gaussian, in which case we may use $Z(t) = h(S(t))$ instead of using $Z(t) = \ln S(t)$.

Of course, the cumulative approximate log-increment over several steps, being in our example the sum of Bernoulli r.v. that are independent and identically distributed, will have a binomial distribution, which approaches the correct normal distribution of the true cumulative log-increment. The binomial distribution of $\ln \overline{S}(t)$ therefore approaches, as $n \to +\infty$, the Gaussian P^*-distribution of $\ln S(t)$.

Figure 14.2 illustrates one of the several possible future trajectories for the binomial model with $n = 3$, i.e. a trajectory of $\overline{S}(t)$ that approximates the corresponding $S(t)$. It also shows (see the nodes) all the possible points where the trajectories of $\overline{S}(t)$ may pass. We use as starting point the stock price $S = S(0) = €92,63$ of the Renault stock at Euronext Paris at time $t = 0$ corresponding to 31 December 2015. We assume the stock will follow from $t = 0$ to $T = 0.25$ years a Black–Scholes model with volatility $\sigma = 0.387$ per $\sqrt{\text{year}}$. Each time step has duration $\Delta t = T/n = 0.0833$ years (1 month) and we assume the return rate of the riskless asset to be $r = 0.5\% = 0.005$ per year. We use the risk neutral probability P^* and approximate it by the binomial model. Then, $\overline{S}(t)$ will, on the

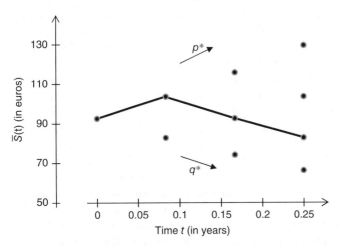

Figure 14.2 The future evolution from time $t = 0$ to $T = 0.25$ years of a stock price $S(t)$ governed by a Black–Scholes model with volatility $\sigma = 0.387$ per $\sqrt{\text{year}}$ and initial value $S(0) = S = €92.63$ using the binomial model approximation $\overline{S}(t)$. The risk neutral probability, corresponding to a return rate of the riskless asset of $t = 0.5\% = 0.005$ per year is assumed and a binomial approximation for its cumulative log-increments is used. One trajectory is depicted and the nodes indicate the points where the trajectories of the binomial approximation may pass. We have used $n = 3$ (in practice, one should use higher numbers). Therefore, the time steps have duration $\Delta t = T/n = 0.0833$ years $= 1$ month. The log-increments are $\pm \Delta z$, with $\Delta z = \sigma\sqrt{\Delta t} = 0.1117$, so that $u = e^{\Delta z} = 1.1182$ and $d = e^{-\Delta z} = 1/u = 0.8943$. The probability that, at a given time step, the trajectory moves up is $p^* = \frac{e^{r\Delta t}-d}{u-d} = 0.4740$ and the probability that it moves down is $q^* = 1 - p^* = 0.5260$.

time step from 0 to Δt, either move up to $Se^{\Delta z} = Su$ or down to $Se^{-\Delta z} = Sd$, where $u = e^{\sigma \sqrt{\Delta t}} = 1.1182$ and $d = e^{-\sigma \sqrt{\Delta t}} = 1/u = 0.8943$, with probabilities $p^* = \frac{e^{r\Delta t} - d}{u - d} = 0.4740$ and $q^* = 1 - p^* = 0.5260$, respectively. For the example of the trajectory shown in the figure, it has moved up to Su. Then, in the time step Δt to $2\Delta t$, it can move up to $Suu = Su^2$ or down to $Sud = S$ with probabilities p^* and q^*, respectively. In the example given in the figure it has moved down and so, at time $2\Delta t$, is at state S. Finally, in the last time step it can again move up to Su or down to Sd with probabilities p^* and q^*, respectively. In the example given in the figure, it has moved down.

Minor discrepancies in the numbers computed in the numerical illustrations below are not important. They are shown with just a few significant digits by rounding off the true computations, which were done with many more significant digits. So, if you do the computations based on the low number of significant digits shown, the result might be slightly different from the one we present based on higher precision computations.

Figure 14.3 shows, on top, the complete binomial tree with all the values of $\bar{S}(t)$ shown at all the nodes and the probabilities of the transitions written next to the corresponding transition arrows. This is very easy to obtain, one just needs to compute the values at the nodes one time step at a time starting at time zero. Each time one moves up, the value of the stock is multiplied by u, and every time one moves down, it is multiplied by d.

Figure 14.3 shows, in the middle, the tree of the European call option price $c(t)$ at each node, of course using the binomial model approximation. Here, we have used, as in Section 14.3, a strike price of $K = €95$. One first starts by computing the values of the nodes at the expiration date T, which are given by $c(T) = (S(T) - K)^+$. For example, for the upper node at time $T = 0.25$ years, we have $c(T) = (129.51 - 95)^+ = €34.51$ (in the figure we show one more significant digit for $c(t)$) and for the lower node we have $c(T) = (66.25 - 95)^+ = €0$. Then one moves back in time, one time step at a time, using the martingale property of the discounted option prices for the risk neutral probability P^*. Remember that, at time t, the discounted option price $\tilde{c}(t)$ is the conditional expectation, using probability P^*, of $\tilde{c}(T)$, conditioned on the value of $S(t)$. Of course, here we will use the binomial approximation instead of P^*.

Let us show a concrete computation of $c(t)$ at some node of the penultimate time $t = 2\Delta t = 0.1667$ years; we have chosen for illustration the middle node at that time, i.e. the node for which $S(0.1667) = €92.63$. Conditioned on being at that node at time $t = 0.1667$ years (i.e. conditioned on having $S(0.1667) = €92.63$), we know that, for the binomial model, at next time $t = T = 0.25$ years, it can only have two possible values, $S(0.25) = €103.58$ (corresponding to a move up in the time step from 0.1167 years to 0.25 years, having probability $p^* = 0.4740$ of occurring) or $S(0.25) = €82.84$ (corresponding to a move down in the time step from 0.1167 years to 0.25 years, having probability $q^* = 0.5260$ of occurring). In the first case, the discounted option price is

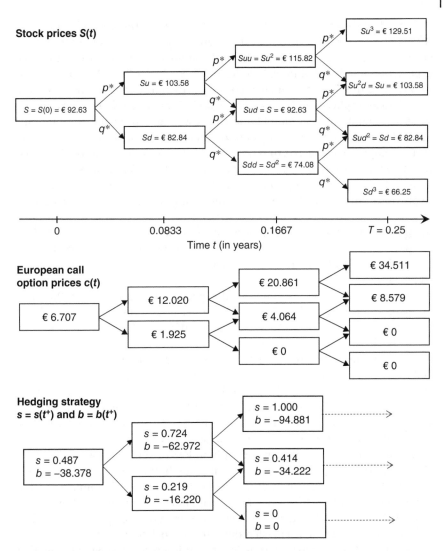

Figure 14.3 Using the same conditions and parameters as the example in the text and in Figure 14.2, we now show: (a) On top, the binomial tree with the stock prices $S(t)$ in euros at every node. Note that the transition from one time t to the next time $t + \Delta t$ occurs by a movement up or a movement down (the arrows show the possible movements), with probabilities $p^* = 0.4740$ and $q^* = 1 - p^* = 0.5260$, respectively (these are the probabilities under the binomial model, which approximates the risk neutral probability P^*). (b) In the middle, the binomial tree (with the approximations given by the binomial model) of the European call option prices $c(t)$ in euros at every node. These are also the values of the hedging strategy at the same nodes. (c) At the bottom, and assuming $B(0) = €1$, the binomial tree (with the approximations given by the binomial model) of the number of units of stock $s(t^+)$ and the number of units of the riskless asset $b(t^+)$ at the nodes. In this approximation, it is assumed that changes in numbers (i.e. the transactions involving the hedging strategy) occur immediately after the time t of a node (i.e. occur at time t^+) and numbers are kept unchanged till the next node; since at T^+ the option is already liquidated, no transactions occur at time T.

$\tilde{c}(T) = e^{-rT}c(T) = 0.99875 \times 8.579 = €8.568$ and in the second case it is $\tilde{c}(T) = e^{-rT}c(T) = €0$. Therefore, the conditional expectation of this $\tilde{c}(T)$ given $S(0.1167) = €92.63$ under P^* can be approximated by that conditional expectation under the binomial model and is $\approx 8.568 \times 0.4740 + 0 \times 0.5260 = €4.061$, and this is approximately $\tilde{c}(0.1167)$; therefore, at the node we have chosen to exemplify, $c(0.1167) = e^{0.1167r}\tilde{c}(0.1167) \approx €4.064$. Collecting together these computations and simplifying the resulting expression, the binomial approximation for the middle node of time $t = 0.1167$ years can be more easily computed using $c(0.1667) \approx e^{-r\Delta t}(8.579p^* + 0q^*) = €4.064$. A similar technique will give us the approximate values of $c(0.1167)$ at the other nodes.

Now that we have the option price at time $t = 0.1167$ years, we move back to previous time $t = 0.083$ years. An important property of conditional expectations is that you can condition in two steps, i.e. for $0 < t < u < T$, since the natural filtration of the Wiener process satisfies the property $\mathcal{M}_t \subset \mathcal{M}_u \subset \mathcal{M}_T$, we have $\tilde{c}(t) = \mathbb{E}^{P^*}[\tilde{c}(T) \mid \mathcal{M}_t] = \mathbb{E}^{P^*}[\mathbb{E}^{P^*}[\tilde{c}(T) \mid \mathcal{M}_u] \mid \mathcal{M}_t] = \mathbb{E}^{P^*}[\tilde{c}(u) \mid \mathcal{M}_t]$. So, putting $t = 0.0833$ years and $u = 0.1667$ years, you can obtain the values of $\tilde{c}(0.0833)$ at the different nodes starting from the now already known values of $\tilde{c}(0.1667)$ and taking the conditional expectation with respect to the value of $S(0.0833)$. For example, to obtain the upper node value at time $t = 0.0833$ years, we have $c(0.0833) \approx e^{-r\Delta t}(20.861p^* + 4.064q^*) = €12.020$.

Finally, there is one node remaining, the only node at time $t = 0$. Using the same argument as before, we get $c(0) \approx e^{-r\Delta t}(12.020p^* + 1.925q^*) = €6.707$, and that is the price of the option at time 0 computed with the binomial model approximation with $n = 3$. The true value, computed from the Black–Scholes formula is, as we have seen in Section 14.3, $c(0) = €6.156$. Being an illustration, we have used an extremely low value of n and the approximation would be much better if a larger more adequate value of n were used.

For simplicity of presentation and without loss of generality, we will assume $B(0) = €1$. The hedging strategy is shown at the bottom of Figure 14.3, which displays, for the node points, approximations based on the binomial model of the number $s(t)$ of units of stock and the number $b(t)$ of units of the riskless asset. This gives us information on how many units of stock and of the riskless asset we need to buy or sell at each time step, i.e. between a time t on the grid and the time $t + \Delta t$, in order to hedge the option price. We will assume that such transactions are made immediately after the beginning of the time step, more precisely at time t^+ (for practical terms, you can think of doing it, say, within one second after the interval starts). Note that, given the value $S(t)$ of the stock at time t, there are, under the binomial model, only two possible values at time $t + \Delta t$, obtained by multiplying the value at time t by u or by d, with probabilities p^* and q^*, respectively. Again, we should start at the expiration date T and move backward in time, time step by time step. Note that at time T there are no transactions (at time T^+, the option has already been liquidated, either exercising it or not). So, in our $n = 3$ example, the last transaction is at time 0.1667^+ years.

Let us see how to compute the hedging strategy at the middle node at that time, corresponding to a stock price of $S(0.1667) = €92.63$. We want to determine $b = b(0.1667^+)$ and $s = s(0.1667^+)$. Since there are no transactions after 0.1667^+, we have $b(T) = b$ and $s(T) = s$, and so $c(T) = b(T)B(T) + s(T)S(T) = bB(0)e^{rT} + sS(T) = be^{rT} + sS(T)$ (since we have assumed $B(0) = 1$). The two possible values of $S(T)$ in the binomial model are €103.58 and €82.84, which correspond to $c(0.25) = €8.579$ and $c(0, 25) = €0$, respectively. So, we have two equations, one for each possible value:

$$8.579 = e^{0.005 \times 0.25}b + 103.58s$$

$$0 = e^{0.005 \times 0.25}b + 82.84s.$$

Solving the system, we get $s(0.1667^+) \approx 0.414$ and $b(0.1667^+) \approx -34.222$. This gives at that time an amount invested in stock of $0.414 \times 92.63 = €38.31$ and an amount invested in bonds of $-34.222 \times B(0)e^{0.1667r} = -€34.25$. These numbers of units $s = 0.414$ and $b = -34.222$ are to be kept up to time T, but of course, since the value of each unit varies, the amount invested in stocks and the riskless asset also changes.

We now move to time $t = 0.0833$ years and choose to show, as an example, what happens at the upper node at that time, i.e. at the node corresponding to $S(0.0833) = 103.58$, for which we get a similar system of equations for $b = b(0.0833^+)$ and $s = s(0.833^+)$:

$$20.861 = e^{0.005 \times 0.1667}b + 115.82s$$

$$4.064 = e^{0.005 \times 0.1667}b + 92.63s.$$

That gives $s(0.0833^+) \approx 0.724$ and $b(0.0833^+) \approx -62.972$, values that are to be kept up to time 0.01667 years.

Finally, we go to time $t = 0$ and obtain in a similar fashion that $s(0^+) \approx 0.487$ and $b(0^+) \approx -38,378$, values that are to be kept up to time 0.0833 years. We can compare these values with the exact values given by the Black–Scholes formula: $s(0) = 0.489$ and $b(0) = -39.148$.

The method works very well (using of course a reasonable value of n) to determine the option price and the hedging strategy at time 0, and it gives these items for other times t, but only for the times and stock prices available in the nodes, which is quite limiting. There is, however, no impediment, since, even if t is not on the grid and/or the stock price S at time t is not a node, you can get an approximation by looking at a node having very close values of t and S. If there is no such node, since what matters is the time to expiration, you can start from scratch, resetting the initial conditions (i.e. the old time t becomes time 0 with a corresponding stock value S) and changing accordingly the expiration date to $T - t$, and then redoing the whole exercise with these new initial conditions and a new grid.

You can try the binomial model with different European call options, possibly with a larger n. As already mentioned, good approximations require larger values of n than the one we chose for illustration and so a lot of computations are required. You can write a computer code to do this or use the software already available (which includes other underlying asset models), for example:

- http://comisef.eu/?q=implementing_binomial_trees (R code)
- http://www.pricing-option.com/tree.aspx (free online computation)
- http://www.fintools.com/resources/online-calculators/options-calcs/binomial/ (idem)
- http://www.ivolatility.com/calc/ (idem)
- https://www.youtube.com/watch?v=cuviNtMMQsA (tutorial video on how to implement it in Excel)
- https://www.youtube.com/watch?v=fysNDIXSB2k (idem).

Since the binomial model deals with approximations, there might be slight differences among implementations due to the use of slightly different approximation criteria in one or more steps of the algorithm.

Of course, in the case of European call options on stock following the Black–Scholes model, there is no advantage in using the binomial model approximation, since it is easier and better to use the exact Black–Scholes formula. The advantage of the binomial model approximation comes in other situations for which we do not have explicit expressions for pricing the option. This includes the case of American put options, as we will see in Section 14.7.

14.6 European put options

A *European put option* with *expiration date* $T > 0$ (a deterministic time instant) and *strike price* $K > 0$ on a stock gives its holder the right, but not the obligation, of selling a unit of the stock at time K by the strike price K. If the market price $S(T)$ at time T is lower than K, the holder of the option will exercise it and sell one unit of the stock at the price K, having a benefit of $K - S(T) > 0$. In case $S(T)$ is higher than K, the holder will obviously not exercise it and will have a zero benefit. Therefore, a European put option gives the holder at time T the benefit

$$(K - S(T))^+ := \max(K - S(T), 0).$$

The put option is *in the money*, *at the money* or *out of the money*, according to whether the current value of the stock (its *spot price*) is, respectively, smaller than, equal to or larger than the strike price K, meaning that, if the expiration date was right now instead of at time T, the holder of the put option, should (s)he decide to exercise it, would make money, would be even or would loose money, respectively.

The treatment of European put options could be done in a similar fashion as for call options, with the obvious adjustments, and we would reach expressions similar to the Black–Scholes formula. But there is a connection, called *put-call duality* or *put-call parity*, between put options and call options that allows us to obtain immediately the expressions for the European put option from the analogous ones for European call options.

Let $p(t)$ be the price of a European put option at time $t \in [0, T]$ and let $(b_p(t), s_p(t))$ ($t \in [0, T]$) be the hedging strategy, i.e. a self-financing trading strategy that reproduces the price $p(t)$ of the option at time T (which is equal to the holder's benefit at that time) and so, by the no arbitrage principle, reproduces the option price ar any time $t \in [0, T]$).

The put-call parity is based on the comparison of two portfolios. Consider a portfolio made up of one unit of the European put option and minus one unit (short position) of a European call option on the same stock with the same expiration date T and equal strike price K. Consider another portfolio with minus one unit of the stock (short position) and $\frac{K}{B(0)\exp(rT)}$ units of the riskless asset (bond with fixed return rate r or bank deposit with fixed interest rate r). The value of the first portfolio at time T is $p(T) - c(T) = (K - S(T))^+ - (S(T) - K)^+ = -S(T) + K$ (you can check by looking separately at the different sign possibilities of $K - S(T)$). The second portfolio has at time T the value $-S(T) + \frac{K}{B(0)\exp(rT)}B(T) = -S(T) + K$ (since $B(T) = B(0)e^{rT}$). So, at time T, the two portfolios have exactly the same value for all market scenarios. By the no arbitrage principle, the two portfolios must have the same value at any time

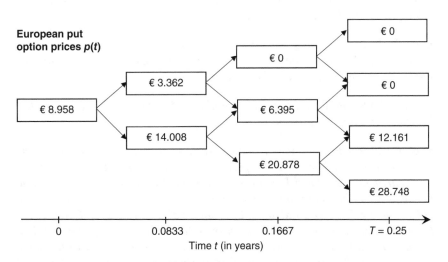

Figure 14.4 Using the same example of the Renault stock and the same expiration date $T = 0.25$ years and strike price $K = €95$ as before, the figure shows, for $n = 3$, the binomial tree with the approximations given by the binomial model for the European put option prices $p(t)$ in euros. These are also the values of the hedging strategy at the same nodes.

$t \in [0, T]$. Therefore,

$$p(t) - c(t) = -S(t) + \frac{K}{B(0)\exp(rT)}B(t)$$
$$= -S(t) + Ke^{-r(T-t)},$$

from which we get

$$p(t) = c(t) - S(t) + \frac{K}{B(0)\exp(rT)}B(t) = c(t) - S(t) + Ke^{-r(T-t)}. \quad (14.37)$$

Since we can easily compute $c(t)$ by the Black–Scholes formula, using (14.37) one immediately obtains $p(t)$.

Let $(b(t), s(t))$ and $(b_p(t), s_p(t))$ be the hedging strategies of, respectively, the European call option and the European put option. Then, $b(t)B(t) + s(t)S(t) = c(t)$ and $b_p(t)B(t) + s_p(t)S(t) = p(t)$. Replacing in (14.37), one obtains

$$\left(b_p(t) - b(t) - \frac{K}{B(0)\exp(rT)}\right)B(t) + (s_p(t) - s(t) + 1)S(t) = 0.$$

This suggests that the hedging strategy for the European put option is given by

$$b_p(t) = b(t) + \frac{K}{B(0)\exp(rT)}$$
$$s_p(t) = s(t) - 1. \quad (14.38)$$

Exercise 14.6 *Verify that the strategy defined by (14.38) is indeed the hedging strategy of the European put option. Verify also that the position of the strategy is long on the riskless asset and short on the stock.*

Exercise 14.7 *Assuming the same numerical values as in the example of Section 14.3, including the expiration date and strike price, determine the European put option price and hedging strategy.*

Exercise 14.8 *Perform an exercise similar to Exercises 14.3a, c and d, but using now a European put option.*

Exercise 14.9 *By Girsanov's theorem, we can, as in Section 14.4, conclude that the discounted price $\tilde{p}(t) = p(t)/B(t)$ of the option at time $t \in [0, T]$ is a martingale for the risk neutral probability P^*, and so its value at time t is equal to the conditional expectation, for P^*, of the discounted price $\tilde{p}(T) = (K - S(T))^+/B(T)$ at time T, conditional on the value of $S(t)$. Using this fact, design a binomial model approximation for the European put option. For $n = 3$ and for the European put option on the same stock with the same expiration date $T = 0.25$ years and the same strike price $K = €95$, the binomial tree for the stock price is obviously the same as in the top part of Figure 14.3. Figure 14.4 shows the binomial tree for the European put option price; see if you can obtain these values (apart some slight differences due to rounding* *off in intermediate computations). Of course, since n = 3 is too small, the approximation obtained for p(0) is not very good, namely is €8.958, while the exact value obtained using (14.37) is €8.407.*

Exercise 14.10 *Determine expressions for the Greeks Δ_p, ρ_p, Θ_p, and v_p of a European put option and compute them for the particular example we have considered.*

14.7 American options

American options, call or put, are similar to the European call or put options, except for the fact that the holder does not have to wait for the expiration date T to liquidate them. The holder can exercise an American option at any time up to the expiration date T or decide not to exercise it. So, at any time $t \in [0, T]$, the holder has to decide if it advantageous to exercise it at that time or wait longer (of course, if the wait goes beyond T, that means the holder does not exercise the option at all). This is a much more difficult problem mathematically and usually there are no explicit expressions available. We will now look at American options on a stock paying no dividends.

The binomial model can handle this situation, although only approximately. For American options, the computations for the option price at the nodes of the binomial tree are similar, apart from a possible adjustment, to the ones we made for European options; see, for European call options, Section 14.5 (with results shown in the middle part of Figure 14.3) and, for European put options, Section 14.6 (with results shown in Figure 14.4). For a certain time t on the grid and a particular node at that time, if there is an adjustment at that node that means that the option should be exercised at that moment.

Since for American call options there is never an adjustment and one obtains exactly the same option price results as for the European call option, exercising the option before the expiration date T would result in a financial loss. So, American call options are treated exactly as the European call options, using exactly the same methods as used in Sections 14.2 to 14.5.[8]

Let us look then at the possible adjustments required for American put options when computing the binomial tree for option prices. One still proceeds by moving backwards in time, starting at time T and moving towards time 0. At time T, there is obviously no difference between European and American put options and so the option price is in both cases $(K - S(T))^+$. At other times t, at each node, we have to determine the value of the option under two possibilities: exercising the option at that time or not. Of course, the holder should exercise the option only when the value of exercising it at that

---

8 Be aware, though, that if the stock distributes dividends to its holders in the time interval [0, T], this is no longer true.

time is higher than the value of not exercising it at that time (which means keeping the option). In any case, if the holder exercises the option at time t, its value is the benefit $V_i = (K - S(t))^+$ of doing it, and this is called the *intrinsic value* of the option (at that time and for the stock price $S(t)$ of that node); note that this intrinsic value is zero if the stock price $S(t)$ is above the strike price K. If the holder does not exercise the option at time t and keeps it, then its value is computed in the same manner used in the European put option (for which an early exercise is not possible) and so is given by $V_n = e^{-r\Delta t}(p_u(t + \Delta t)p^* + p_d(t + \Delta t)q^*)$, where $p_u(t + \Delta t)$ and $p_d(t + \Delta t)$ are the option values at the neighbouring up and down nodes at time $t + \Delta t$.

Since now we have an American option, if the holder is at a given node situation, it should exercise the option at that time if, for that node, $V_i > V_n$ and should not exercise the option (should keep it) if $V_i < V_n$. So, under the principle of no arbitrage, the true value of the American put option at that time and node is always the bigger of these two numbers and that is the value that should be placed on the binomial tree.

One can write in bold the nodes for which $V_i > V_n$, which allows a quick identification of the favourable occasions to early exercise of the option. If, at a given time $t < T$, the stock price $S(t)$ happens to be one of the node values of the stock at that time, the holder should exercise the option if the node is in bold and should not exercise the option if it is not in bold. Of course, often $S(t)$ does not correspond to stock values at the nodes of the binomial tree at that time, but, if n is large, there will be a node with a very close value and the holder can follow its indication. At the expiration time T, the value of the option coincides with its intrinsic value and the holder should exercise the option if the stock price $S(T)$ is below K and should not exercise the option if $S(T)$ is above K.

Figure 14.5 illustrates for $n = 3$ the binomial trees of the stock price (on top, which is equal to the top part of Figure 14.3) and of the option prices (at the bottom) for an American put option satisfying the same conditions as the example in the previous section on European put options. Compare the bottom part of Figure 14.5 with Figure 14.4 containing the binomial tree of the European put option.

To exemplify, look at time 0.1667 years. At the top node, if the option is not exercised, the value is $e^{-r\Delta t}(0p^* + 0q^*) = €0$, and if its is exercised, the value (the intrinsic value) is also $(K - S(0.1667))^+ = (95 - 115.82)^+ = €0$, so the biggest of the two is €0. At the middle node, the value of the option if not exercised is $e^{-r\Delta t}(0p^* + 12.161q^*) = €6.395$, and the intrinsic value is $(95 - 92.63)^+ = €2.37$, so the bigger of the two is €6.395 and the option should not be exercised then. At the lower node, the value of the option if not exercised is $e^{-r\Delta t}(12.161p^* + 28.748q^*) = €20.878$, and the intrinsic value is $(95 - 74.08)^+ = €20.92$ (using more digits in the computations gives 20.918), so the bigger of the two is €20.918 and the option should be exercised then. In this example, there are no more nodes corresponding to early exercise of the option.

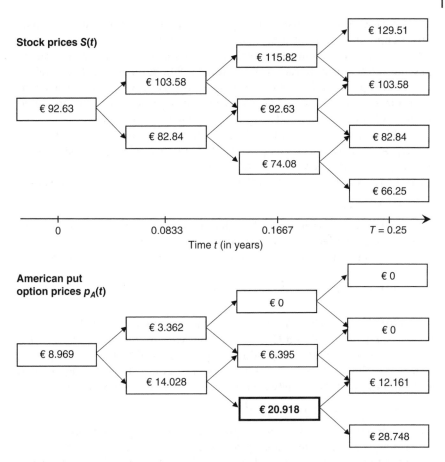

Figure 14.5 Using the same example of the Renault stock and the same expiration date $T = 0.25$ years and strike price $K = €95$ as before, the figure shows, for $n = 3$, the binomial tree with the approximations given by the binomial model for the stock prices (top) and for the American put option prices $p_A(t)$ (bottom) in euros. These are also the values of the hedging strategy at the same nodes. The nodes corresponding to an early (before time T) exercise of the option are in bold.

14.8 Other models

What we have done in previous sections can be considerably generalized. We can consider several stocks or other underlying assets (like exchange rates, interest rates, stock indices, other indices, commodities, etc.) with prices $S_1(t), S_2(t), \ldots, S_n(t)$. Such assets may have average return rates and volatilities with behaviour differing from the Black–Scholes model (e.g. exchange rates may follow a Vasicek or a CIR model, a stock may follow a Black–Scholes model but with μ and/or σ changing over time). The riskless asset can be any secure investment with fixed or variable return rate. Other more complex types of options, like, for example, barrier options, may be considered.

A basic study of this more general framework can be seen, for example, in Øksendal (2003) or in Karatzas and Shreve (1998), but many further developments have appeared in the specialized literature. Here, we just give you a hint on how this is achieved, based on essentially the same ideas that were presented in this book, but generalized to this more demanding framework.

Consider time t varying in $[0, T]$ ($T > 0$) and a m-dimensional standard Wiener process, $\mathbf{W}(t) = [W_1(t), \dots, W_m(t)]^T$, in a complete probability space $(\Omega, \mathcal{F}, P)$ and let $\mathcal{M}_t$ be its natural filtration.

For the riskless asset price we will use here $S_0(t)$ instead of $B(t)$; it satisfies

$$dS_0(t) = r(t, \omega)S_0(t)dt$$

with known initial condition $S_0(0)$ with a return rate r that now may vary, as long as it is a bounded function adapted to the filtration $\mathcal{M}_t$.

We now consider n risky assets with prices $S_i(t)$ ($i = 1, \dots, n$) satisfying

$$dS_i(t) = F_i(t, \omega)dt + \sum_{j=1}^{m} G_{ij}(t, \omega)dW_j(t)$$

with initial known values $S_i(0)$. Let $\mathbf{F}$ be the column vector of the F_i, $\mathbf{G}$ the $n \times m$-matrix of the G_{ij} and $\mathbf{G}_i$ be its ith line. We can then write $dS_i = F_i dt + \mathbf{G}_i d\mathbf{W}(t)$ ($i = 1, \dots, n$). Note that $\mathbf{F}$ and $\mathbf{G}_i$ play roles similar to the roles μx and σx played in the Black–Scholes model of a stock. Assume $\mathbf{F}$ and $\mathbf{G}$ to be adapted to the filtration $\mathcal{M}_t$ and satisfy conditions such that the column vector

$$\mathbf{S}(t) = [S_0(t), S_1(t), \dots, S_n(t)]^T$$

is an Itô process. Of course, under appropriate regularity conditions, this can be written as a *multidimensional stochastic differential equation* that generalizes the two equations we had before for $B(t) = S_0(t)$ and for the stock $S(t)$ (the Black–Scholes SDE model). Like before, the equation on the riskless asset $S_0(t) = S_0(0) \exp\left(\int_0^t r(u, \omega)du\right)$ does not have a term in $d\mathbf{W}(t)$.

$\mathbf{S}(t)$ is called a *market*. The *normalized market*

$$\tilde{\mathbf{S}}(t) = [\tilde{S}_0(t), \tilde{S}_1(t), \dots, \tilde{S}_n(t)]^T = [1, S_1(t)/S_0(t), \dots, S_n(t)/S_0(t)]^T$$

is obtained by discounting using as discounting process $S_0(t)$ and plays the role of our discounted prices in Section 14.4. We have

$$d\tilde{\mathbf{S}}(t) = \frac{1}{S_0(t)}(d\mathbf{S}(t) - r(t)\tilde{\mathbf{S}}(t)dt).$$

A *portfolio*, also called a *trading strategy*, $\mathbf{s}(t, \omega)$ ($t \in [0, T]$), which we abbreviate to $\mathbf{s}(t)$, is a line vector

$$\mathbf{s}(t) = [s_0(t), s_1(t), \dots, s_n(t)]$$

adapted to the filtration $\mathcal{M}_t$, which represents the number of units (it may be even negative or assume fractional values) of each of the $n + 1$ financial assets in the portfolio. The only difference is that now the portfolio can have more risky assets. Its value is

$$V(t) = V_s(t, \omega) = \mathbf{s}(t)\mathbf{S}(t) = \sum_{i=0}^{n} s_i(t)S_i(t).$$

The portfolio is *self-financing* if

$$V(t) = V(0) + \int_0^t \mathbf{s}(u)d\mathbf{S}(u),$$

or equivalently $dV(t) = \mathbf{s}(t)d\mathbf{S}(t)$, and if it also satisfies the technical conditions required for the defining expression to make sense. Only admissible portfolios (those having value bounded below to avoid unlimited debts) are considered.

Like before, an admissible portfolio is an *arbitrage* if it has zero value at time 0 and non-negative value at time T and if there is a positive probability of having a positive value at time T. This means that, with a null investment and without any risk, there is a positive probability of making a profit. Like before, the working hypothesis is that the market does not allow the existence of an arbitrage.

A probability measure P^* defined on $(\Omega, \mathcal{M}_T)$, equivalent (i.e. having the same sets of zero probability) to P, is an *equivalent martingale probability measure* if the normalized market $\tilde{\mathbf{S}}(t)$ is a martingale for the probability P^* (with respect to the filtration $\mathcal{M}_t$)

An important result (see, for example, Øksendal (2003)) is that, if there exists an equivalent martingale probability, then there is no arbitrage in the market. The reciprocal may fail and a stronger property than the no arbitrage is required to ensure that there is an equivalent martingale probability measure. A more general version of Girsanov's theorem gives sufficient conditions for that.

Although even further generalizations are possible, we mention here the case of a more general European option, a *European contingent claim*, which can be any $\mathcal{M}_T$-measurable lower bounded r.v. C, corresponding to the benefit of its holder at the expiration date T. The claim, being European, is to give the benefit only at time T. We have studied the case of European call options and European put options, but there are many other complex types of options that give different types of benefits at time T. This option is said to be an *attainable option* in the market $\mathbf{S}(t)$ if there at least an admissible self-financing portfolio $\mathbf{s}(t)$ having the same value C at time T, in which case it is called a *hedging portfolio*. Under the no arbitrage assumption, the value of that portfolio at any time $t \in [0, T]$ should be the fair price at that time for buying the right to hold the attainable claim. The market is said to be *complete* if all the above European contingent claims are attainable. It can be proved that, in order to know if an option is attainable, it is indifferent to work with the original market or with the normalized market. So, the original and the normalized market are either both complete or both incomplete.

As before, if some conditions are satisfied, one can use a generalized version of Girsanov's theorem to construct the martingale equivalent measure P^* such that the normalized market is a martingale for P^*. Similarly to the requirement (13.4) (in which we must put $f^*(x) = rx$ to ensure a zero drift for the discounted process and therefore its martingale property), we need to assume that there is an m-dimensional column vector $\theta(t, \omega)$ in $H^2[0, T]$ such that

$$\mathbf{G}(t)\theta(t) = \mathbf{F}(t) - r(t)[S_1(t), \dots, S_n(t)]^T$$

and assume that $\theta(t)$ verifies some additional appropriate requirement. Typically, it is assumed that it satisfies a Novikov condition $\mathbb{E}^P\left[\exp\left(\frac{1}{2}\int_0^T \theta^T(u)\theta(u)du\right)\right] < \infty$. Assume these assumptions are met.

It is then proved that the market is complete if and only if the matrix $\mathbf{G}(t)$ has characteristic m for almost all (t, ω). Note that this condition is the multidimensional equivalent to having, in our case of one risky asset, $G(t) = g(S(t)) \neq 0$ so that our θ function is uniquely defined by (13.4). That condition was met in our case, since our stock had $g(x) = \sigma x$ and $S(t)$ was always positive.

Of course, we can only have a complete market if $n \geq m$. The number of independent sources of randomness is the dimension m of the Wiener process. The fact that $\mathbf{G}(t)$ has characteristic m means that, among the n risky assets that can be used for hedging purposes, there are m non-redundant assets. There is redundancy among the risky assets if the value of one of them depends solely on the values of the others and on the value of the riskless asset. For example, a stock and a European call option on that stock form a redundant set of risky assets, since the value of the option is a function of the stock price and of the riskless asset. Also, a fixed set of 20 stocks and a stock market index based on the average of their market prices is necessarily redundant since the information contained in the index can be obtained from the 20 stocks.

We have dealt here with risky assets that had an average trend (average speed of change) and are subjected to market changes that are driven by the Wiener processes, which are very irregular but have the nice property of possessing independent increments (which guarantees the very useful Markov property for the assets) and of having continuous trajectories. Therefore, although random changes in the prices of the assets are quite frequent, typically very irregular (not smooth) and occasionally assuming cumulative large values in short time intervals, they are continuous, so there are no sudden discontinuous jumps. It is claimed that on top of that, 'catastrophes' may happen in the markets, although rarely (some claim that mid-size 'catastrophes' are less rare than ordinarily assumed), and, when that happens, sudden jumps may occur in the values of the assets. Note that 'catastrophes' can make prices jump down or jump up, so the informal term 'catastrophe' should not be considered necessarily in the negative sense. If jumps can occur, we need, besides the driving Wiener process for the effect of typical market random changes, also a driving jump process that describes the effect on the assets of occasional 'catastrophic' events. That leads to the use of jump-diffusion processes, whose treatment is much more complex and beyond the scope of this book. The reader interested in these processes can, for example, consult Cont and Tankov (2003), Shoutens (2003) or Øksendal and Sulem (2007). It might, however, be very difficult to apply such a complex model and to estimate its parameters, particularly if the 'catastrophic' events are rare and are likely to not occur or have a very small number of occurrences in the historical data available. If this is the case, as it typically is, the estimation of the jump component parameters would be very difficult or even impossible, even if we could identify the times and sizes of jump occurrences. The situation is even worse because it is often very difficult, looking at the discrete-time observations of the process that are available, to distinguish, with a reasonable success probability, large changes due solely to the Wiener processes from a large change due to jump processes. So, for most financial applications (and also, for the same reasons, for most biological applications), working with the models we have studied, where market (nature) variability is driven by the Wiener processes, is likely to be the best option. Above all, they are simpler to deal with.

15

Synthesis

We conclude this book with a brief synopsis of the main points it has covered, using some informality that skips the more technical issues.

Stochastic differential equations (SDEs) are useful tools in several scientific and technological fields in order to model dynamical phenomena that would usually be described by ordinary differential equations if it were not for the fact that their dynamics are affected by a perturbing noise. The perturbing noise, a stochastic process in continuous time, can, for the sake of mathematical convenience, be approximated by a white noise $\varepsilon(t) = \varepsilon(t, \omega)$, which is a generalized stationary Gaussian stochastic process whose values at different times are independent. It is the derivative (in the sense of generalized functions since the ordinary derivative does not exist) of the Wiener process (also known as Brownian motion) $W(t, \omega)$ ($t \geq 0$), usually abbreviated to $W(t)$. Reversely, we may say that $W(t)$ is the integral of the white noise, and so it represents the accumulated noise up to time t.

The Wiener process, which we assume defined on a complete probability space $(\Omega, \mathcal{F}, P)$, is a (proper) stochastic process $W(t)_{t \in [0, +\infty[}$ with values in $\mathbb{R}$ and continuous trajectories (a.s.) that has independent increments, verifies $W(0) = 0$ and the increments $W(t) - W(s)$ ($s < t$) have the normal distribution $\mathcal{N}(0, t - s)$ with mean 0 and variance $t - s$. It was initially used by Bachelier in 1900 to describe the price of stocks at the Paris stock market, by Einstein in 1905 to describe (using its multidimensional version) the Brownian motion of a particle suspended in a fluid, and later, in the Black–Scholes model, to describe the effect of market fluctuations on the return rate of a stock. The first SDE, the Ornstein–Uhlenbeck model, whose solution is the Ornstein–Uhlenbeck process, appeared precisely to improve upon the initial Einstein model for the Brownian motion of a particle. Wiener and Lévy set up the mathematical theory.

The Wiener process is a Markov process, i.e. knowing its present value, future values are independent of past values. It is even a strong Markov process, i.e. the Markov property also works even if one takes as present time a random

Markov time τ. A Markov time τ is such that we can determine, for any fixed t, whether τ is $\leq t$ or is $> t$ based solely on the values of the process up to and including time t. The transition pdf of the Wiener process for $s < t$, i.e. the pdf of $W(t)$ given that $W(s) = x$, is the pdf of a $\mathcal{N}(x, t - s)$ distribution. The Wiener process is also a martingale (with respect to its natural filtration), i.e. for $s < t$,

$$\mathbb{E}[W(t)|W(u) : 0 \leq u \leq s] = W(s).$$

The trajectories of the Wiener process are continuous, but very irregular, a.s. of unbounded variation and so having no derivative a.s.

We have studied SDEs of the form

$$\frac{dX(t)}{dt} = f(t, X(t)) + g(t, X(t))\varepsilon(t), \quad X(0) = X_0, \quad t \in [0, T],$$

with an initial condition X_0 being a r.v. independent of the Wiener process (it could even be a deterministic constant). It is usually written in the form

$$dX(t) = f(t, X(t))dt + g(t, X(t))dW(t), \quad X(0) = X_0, \quad t \in [0, T],$$

which is equivalent to its integral form (the one that gives true meaning to the solution)

$$X(t) = X_0 + \int_0^t f(s, X(s))ds + \int_0^t g(s, X(s))dW(s), \quad t \in [0, T],$$

where we have abbreviated, as usual, $X(t, \omega)$ to $X(t)$.

The first integral can be defined for each ω (i.e. for each trajectory) as a Riemann or Lebesgue integral, but the second integral cannot be defined as a Riemann–Stieltjes integral because the integrator $W(t)$ has unbounded variation a.s. Except for very simple integrand functions, this usually means that the limits (e.g. mean square limits) of the Riemann–Stieltjes approximating sums will depend on the choice of the tag points where the integrand is computed.

We need therefore to define such stochastic integrals. We did consider integrals of the type $\int_0^t G(s, \omega)dW(s, \omega)$, abbreviated to $\int_0^t G(s)dW(s)$, with integrand functions $G(s, \omega)$ that are non-anticipative. By that we mean that G is jointly measurable in the two independent variables and unable to 'anticipate' or 'guess' the future (meaning that $G(s, \omega)$ is independent of future values of the perturbing noise, which is equivalent to being independent of future increments of the Wiener process). Although the definition becomes a bit more complicated for more general integrand functions $G(s, \omega)$, in the case they are non-anticipative and mean square continuous, the stochastic Itô integral can be defined by

$$\int_0^t G(s)dW(s) = \int_0^t G(s, \omega)dW(t, \omega)$$
$$= \text{l.i.m.}_{n \to +\infty} \sum_{k=1}^n G(t_{n,k-1})(W(t_{n,k}) - W(t_{n,k-1})).$$

Here $0 = t_{n,0} < t_{n,1} < \cdots < t_{n,n} = t$ $(n = 1, 2, \ldots)$ are partitions of the integration interval $[0, t]$ with mesh converging to zero as $n \to +\infty$ and l.i.m.

represents the mean square limit. Notice that the tag point at which we have computed the integrand function G in each subinterval of the partition is always the initial point of the interval. This is the non-anticipative choice and translates the idea of 'no clairvoyance' with respect to the future, the reason behind the nice probabilistic properties of the Itô integral. This integral, however, does not follow ordinary calculus rules.

Other choices or combinations of choices of the tag points lead to other stochastic integrals, among which the Stratonovich integral plays a proeminent role. The Stratonovich integral is represented by $(S) \int_0^t G(s)dW(s)$ or $\int_0^t G(s) \circ dW(s)$ and it does follow ordinary calculus rules, but usually fails to have the nice probabilistic properties of the Itô integral.

We have worked with the Itô integral and have extended its definition to integrand functions in the Hilbert space $H^2[0, t]$ of non-anticipative functions such that $(\| G \|_{2*})^2 = \int_0^t \mathbb{E}[G^2]ds < +\infty; \| \cdot \|_{2*}$ is the norm of that space. The integral is a r.v. and is in the Hilbert space L^2 of r.v. with finite norm (in L^2, the norm of a r.v. X is $\mathbb{E}[X^2])^{1/2}$). We have verified that the integral has zero mathematical expectation and preserves the norm, i.e. $\mathbb{E}\left[\left(\int_0^t G(s)dW(s) \right)^2 \right] = \int_0^t \mathbb{E}[G^2(s)]ds$.

We have also verified that, as a function of the upper limit of integration t, the Itô integral is a continuous function and a martingale. In the particular case of a deterministic integrand G, the integral $\int_0^t G(s)dW(s)$ has a $\mathcal{N}(0, \int_0^t G^2(s)ds)$ distribution.

The Itô integral was then extended to functions in the class $M^2[0, t]$, i.e. non-anticipative functions such that $\int_0^t G(s)ds < +\infty$ a.s.; however, for these functions, there is no guarantee that the integral has a mean or that it is a martingale as a function of t. We have defined an Itô process as being a process of the form $X(t) = X_0 + \int_0^t F(s, \omega)ds + \int_0^t G(s, \omega)dW(s, \omega)$ (or, in differential notation, $dX(t) = F(t)dt + G(t)dW(t)$ with initial condition $X(0) = X_0$), with F being in $M^1[0, t]$ (non-anticipative such that $\int_0^t |F(s)|ds < +\infty$ a.s.) and $G \in M^2[0, t]$. The Itô integral follows special calculus rules. The rules are condensed in the Itô theorem/formula, which is a chain rule that says that, if $h(t, x)$ is a real-valued function of class $C^{1,2}$ and $X(t)$ an Itô process as defined above, then $Y(t) = h(t, X(t))$ is also an Itô process and

$$dY(t) = \frac{\partial h(t, X(t))}{\partial t}dt + \frac{\partial h(t, X(t))}{\partial x}dX(t) + \frac{1}{2}\frac{\partial^2 h(t, X(t))}{\partial x^2}(dX(t))^2,$$

where $(dX(t))^2$ is an extra term (compared to ordinary calculus) which should be computed using the mnemonic rules $(dt)^2 = 0$, $dtdW(t) = 0$ and $(dW(t))^2 = dt$ (this last one, due to the irregularity of the Wiener trajectories, is the one that breaks the ordinary rules).

Having defined the stochastic integral, the integral form of the SDE has now a meaning and we can look for conditions for the existence and uniqueness of the solution. We have presented an existence and uniqueness theorem for the solution of the SDE $dX(t) = f(t, X(t))dt + g(t, X(t))dW(t)$ ($t \in [0, T]$) with $X(0) = X_0$ independent of the Wiener process such that $\mathbb{E}[X_0^2] < +\infty$. It assumes that $f(t, x)$ and $g(t, x)$ satisfy a Lipschitz condition and a restriction on growth condition on the variable x. Under these assumptions, a continuous solution exists and is unique; it is shown also that the solution is in $H^2[0, T]$, is mean square continuous, has finite mean and variance (for which bounds were found), and is a Markov process. If, furthermore, f and g are continuous in t, the solution is also a diffusion process with drift coefficient (the speed at which the mean changes, also called infinitesimal mean) $a(s, x) = f(s, x)$ and diffusion coefficient (the speed at which the variance changes, also called infinitesimal variance) $b(t, x) = g^2(t, x)$, i.e. is a Markov process with a.s. continuous trajectories that verifies (5.1), (5.2), and (5.3). In this case, it verifies the Kolmogorov equations (see Section 5.2).

Autonomous SDEs, i.e. SDEs for which $f(t, x) \equiv f(x)$ and $g(t, x) \equiv g(x)$ do not depend directly on time t, are quite prominent in applications. In this case, the Lipschitz condition suffices since it implies automatically the restriction on growth condition, and the solution is a homogeneous diffusion process with drift coefficient $a(x) = f(x)$ and diffusion coefficient $b(x) = g^2(x)$. If, furthermore, the initial condition is known as having the deterministic value $X_0 \equiv x$, we say the solution is an Itô diffusion. In this case, the backward Kolmogorov equation for functionals $u(t, x) = \mathbb{E}_x[h(X(t))] := \mathbb{E}[h(X(t))|X(0) = x]$ with $h(x)$ a bonded continuous Borel-measurable function, takes the form $\frac{\partial u}{\partial t} = Du$, with $\lim_{u \downarrow 0} u(t, x) = h(x)$. Here $D = a(x)\frac{\partial}{\partial x} + \frac{1}{2}b(x)\frac{\partial^2}{\partial x^2}$ is the diffusion operator.

In the unidimensional case (we cannot extrapolate to higher dimensions), for autonomous SDEs, it suffices that f and g are of class C^1 to ensure the existence of a unique solution up to a possible explosion time. So, in this case, if we can show that the explosion time is a.s. $+\infty$, we can conclude that there exists a unique solution for all $t \geq 0$.

If one uses Stratonovich integrals instead of Itô integrals in the integral form of the SDE:

$$X(t) = X_0 + \int_0^t f(s, X(s))ds + (S) \int_0^t g(s, X(s))dW(s),$$

we have a Stratonovich SDE and represent it by

$$(S) \quad dX(t) = f(t, X(t))dt + g(t, X(t))dW(t), \quad X(0) = X_0.$$

The Stratonovich integral can be defined by

$$(S) \int_0^t g(s, X(s))dW(s) = \int_0^t g(s, X(s)) \circ dW(s)$$
$$= \text{l.i.m.} \sum_{k=1}^n g\left(t_{n,k-1}, \frac{X(t_{n,k-1}) + X(t_{n,k})}{2}\right)(W(t_{n,k}) - W(t_{n,k-1})).$$

One sees that the mentioned Stratonovich SDE is equivalent (meaning that it has the same solution) to the Itô SDE

$$dX(t) = \left(f(t, X(t)) + \frac{1}{2} \frac{\partial g(t, X(t))}{\partial x} g(t, Xt)) \right) dt + g(t, X(t)) dW(t).$$

This formula, used in the reverse way, allows an Itô SDE to be reduced to a Stratonovich SDE that, although it does look different, has, because of the different calculus rules, the same solution. This is very useful since to solve the Stratonovich equation we can use the calculus rules we are used to instead of the Itô calculus rules, which we are not used to. At the end, the solution we find is also the solution of the original Itô SDE.

A trivial generalization of the above ideas and methods to the multidimensional case allows us to deal with systems of SDEs.

In Chapter 8 we have looked in some detail at the Black–Scholes model (also known as the stochastic Malthusian model when applied to population growth) $dX(t) = RX(t)dt + \sigma X(t)dW(t)$, with deterministic initial condition $X(0) = x_0$. It is used to model the dynamics of a stock price in the stock market and the dynamics of the growth of a population with no resource limitations living in a randomly varying environment. Here, chance ω represents a market scenario or a state of nature chosen at random in the set Ω of all possible market scenarios or states of nature. The model can be written in the form $dX(t)/dt = (R + \sigma \varepsilon(t))X(t)$, which translates the idea that the return rate of the stock, or (*per capita*) growth rate of the population, R, instead of being constant as in a deterministic situation, is perturbed by a white noise. The parameter R can be interpreted now as an average rate of return or of growth. The parameter σ measures the intensity of the effect of environmental fluctuations on the return or growth rate; it is known in the financial literature as the volatility of the stock. The solution $X(t) = x_0 \exp((R - \sigma^2/2)t + \sigma W(t))$ could be easily obtained, either by direct resolution using Itô calculus (by applying the Itô formula to $\ln X(t)$) or by the trick of converting it to an equivalent Straonovich SDE and solving it with ordinary calculus rules.

SDEs with identical forms usually give different solutions under the Itô and the Stratonovich calculi and this gives rise to some controversy in the literature related to SDE application. The differences may even be of a qualitative nature. For example, in the stochastic Malthusian model, Stratonovich calculus ensures that the population will not become extinct if the average growth rate R is positive, while under Itô calculus the population becomes extinct if $R < \sigma^2/2$, including therefore positive values of R. The usual recipe, based on some limiting approximation theorems, is that the Stratonovich calculus solution gives a better approximation if the process intrinsically occurs in continuous time in nature with a perturbing noise that is coloured but which we approximate by a white noise in an appropriate manner. If, however, the process intrinsically occurs in discrete time with a perturbing discrete-time white noise (which is a proper stochastic process), i.e. if it is really governed by a stochastic difference equation and we approximate it by an SDE by making the time step converge to zero in an appropriate way, then the Itô calculus solution seems to be a better approximation. However, this hides the real issue and, besides, in real cases it is very difficult to see which situation better describes the phenomenon.

We have shown our resolution of the controversy in Chapter 9, first for this particular model (to allow for an easier description of the ideas behind the resolution) and then for quite general autonomous SDEs. We call the attention to the real meaning of parameter R, loosely taken in the literature as meaning the average growth rate and implicitly assuming the same type of average in both calculi. However, if one looks at the role of R in the dynamics of the population, it differs according to the calculus used. So, although the users of the two calculi may inadvertently use the same letter R, in reality R means two different things, namely two different types of average growth rate. Under Itô calculus, R is the arithmetic average growth rate. Under Stratonovich calculus, R is the geometric average growth rate. The two averages differ by exactly $\sigma^2/2$ and so, if we take that difference into account, we see that the two calculi are completely coincidental in their results; in particular, for both we conclude that extinction does not occur when the geometric average growth rate is positive.

In Section 11.1 the Ornstein–Uhlenbeck model used to improve the Einstein model of the Brownian motion of a particle, and a variant, the Vasicek model, which is quite used to model exchange rates and interest rates, were studied. In both cases we showed the existence of a stationary density, a kind of stochastic equilibrium. In these cases, the SDE solution $X(t)$ has a probability distribution that converges as $t \to +\infty$ to a limit distribution having the stationary density as its pdf. This fact was established for these models by recurring to the explicit solution $X(t)$.

However, as in ordinary differential equations, in most cases, although we know that the solution exists and is unique, we are unable to obtain an explicit expression for it or for its transient distribution (of course, we can use numerical methods and/or stochastic simulations to get approximations). Even in those cases, we can explicitly obtain the stationary density if it exists, as long as we are dealing with unidimensional autonomous SDEs that are regular diffusions (all states in the interior of the state space communicate) and some mild assumptions are satisfied. Let the drift and diffusion coefficients, $a(x)$ and $b(x)$, be continuous with $b(x) > 0$ in the interior of the state space. We presented in Chapter 11 methods for verifying the existence and to determine the stationary density without requiring solving explicitly the SDE. These methods are based on the classification of the two boundaries r_1 and r_2 of the state space, with $-\infty \le r_1 < r_2 \le +\infty$.

For this, one uses the scale density $s(\xi) = \exp\left(-\int_{\xi_0}^{\xi} \frac{2a(\eta)}{b(\eta)} d\eta\right)$ and the speed density $m(\xi) = \frac{1}{s(\xi)b(\xi)}$, where ξ_0 is a fixed arbitrary point in the interior of the state space (so these functions are defined up to a multiplicative constant which is irrelevant for the purposes in which we use them). The scale and speed functions are defined by $S(x) = \int_{x_0}^{x} s(\xi)d\xi$ and $M(x) = \int_{x_0}^{x} m(\xi)d\xi$, with x_0 a fixed arbitrary point in the interior of the space state (this gives another arbitrary constant, this time additive, which is again irrelevant). These define scale and speed measures by $S(]c,d]) = S(d) - S(c)$ and $M(]c,d]) = M(d) - M(c)$. If $h(x)$ is of class C^2, then $Dh(x) = \frac{1}{2} \frac{d}{dM(x)}\left(\frac{dh(x)}{dS(x)}\right)$.

If $r_1 < a < x < b < r_2$ and $X(0) = x$, we have presented (see Section 11.2) ordinary differential equations for certain functionals of the Markov time $T_{a,b} = \min\{T_a, T_b\}$ (time for first exit from the interval $]a,b[$, where $T_y = \inf\{t \geq 0 : X(t) = y\}$ is the first passage time by y). In particular, solving an appropriate equation, it is shown that $u(x) = P_x[T_b < T_a] := P[T_b < T_a | X(0) = x] = \frac{S(x)-S(a)}{S(b)-S(a)}$ (probability that the first exit from $]a,b[$ occurs at the point b). This implies that, for $r_1 < x < b < r_2$, the boundary r_1 is non-attractive (i.e. $P_x[T_{r_1^+} < T_b] = 0$) if and only if $S(]r_1, b]) = +\infty$. We obtain an analogous condition for the boundary r_2 to be non-attractive. Non-attractive boundaries cannot be reached in finite or infinite time (i.e. in the limit as $t \to +\infty$).

When both boundaries are non-attractive, the probability is 'pulled' away towards the interior of the state space when it gets close to a boundary and so it may happen that there is a tendency for the pdf to stabilize into a stationary distribution as $t \to +\infty$, reaching a kind of stochastic stable equilibrium or steady state (not for the solution, which keeps varying over time, but for its probability distribution). In fact, when both boundaries are non-attractive, this happens as long as the speed measure of the state space is finite, i.e. $M = \int_{r_1}^{r_2} m(\xi)d\xi < +\infty$, in which case the stationary density is given by $p(y) = m(y)/M$ ($r_1 < y < r_2$) (this expression comes out of the forward Kolmogorov equation when letting $t \to +\infty$). In this case, the process is also ergodic and therefore certain moments of the solution at the steady state can be approximated by the corresponding time-average sample moments obtained along a single trajectory of the process. This is extremely useful since, in many applications, particularly in finance and biology, we only have one trajectory, the observed trajectory for the concrete market scenario or state of nature that has occurred. In such applications we cannot go back in time and repeat the 'experiment' several times, with different randomly chosen market scenarios or states of nature, in order to estimate the desired moment by averaging over such 'experiments'.

In Chapter 12 we presented several applications in a diverse range of areas (exchange rates, interest rates, population growth under limited resources, fisheries, individual animal growth with an application in farming, time evolution of human mortality rates), applying the main theory and also the boundary classification methods of Chapter 11 to deal with interesting issues in finance, biology, and the sustainable optimization of resources exploitation (like in fishing and forestry). In this chapter we also dealt with, in a very introductory way, Monte Carlo simulations (required to handle many cases for which we cannot find explicit solutions) and with statistical issues of estimation and prediction required for real-life applicability.

Girsanov's theorem, presented in Chapter 13, is important for financial applications. It allows one to change the drift coefficient of an SDE $dX(t) = f(X(t))dt + g(X(t))dW(t)$ ($t \in [0, T]$) to a different coefficient of our choice f^*. This is particularly interesting when we change to a zero coefficient, since the solution in this case is just the initial condition plus a stochastic integral, and is therefore a martingale (assuming the integrand is in $H^2[0, T]$). The advantage of being a martingale is that its value at any time $t \in [0, T]$ can be obtained as the conditional mathematical expectation of its value at the terminal time T, conditional on the information available at time t. It so happens that, for important financial derivative products, like European options, we have the explicit expression of their value at time T and can therefore use this property to determine their fair market price at any time t.

To change the drift coefficient from f to f^* without changing the SDE solution, we need to change the Wiener process $W(t)$ to an appropriate process $W^*(t)$, which usually is no longer a Wiener process for the original probability P (the probability of occurrence of the market events). However, there is another probability P^* (which is quite different from P but is equivalent to P in the sense of having the same sets of zero probability) for which $W^*(t)$ is a Wiener process. Changing to the probability P^*, the solution of the initial SDE will follow a new SDE with a different drift coefficient and a new Wiener process. The reader may go back to Section 13.2 for the details of this procedure. Of course, the new probability P^* distorts the real market probabilities of the events. In the case where the drift is changed to 0 and we get a martingale, P^* is called an equivalent martingale probability measure.

In Chapter 14 we started by studying the famous (and widely used in the financial markets) Black–Scholes formula (due to the work of Black and the Nobel prize winners Scholes and Merton) that gives the fair price of a European call option with expiration time T on a stock following a Black–Scholes model. This is based on the idea that there is a self-financing trading strategy or portfolio that matches exactly the value of the option at time T, i.e. the benefit that the option holder will gain at that time. The strategy details the time evolution for that portfolio of the number of units (which can be positive or negative, integer or fractional) of the stock and of a riskless asset with fixed return rate r.

One assumes the principle of no arbitrage, which stands on the idea that markets are efficient and have equally informed operators. It states that it is not possible to have a riskless profit with a null investment (there are other equivalent definitions). If so, then the value of the portfolio should be equal to the price of the option at any time $t \in [0, T]$. For that reason, since it covers the risk of the seller of the option, the above-mentioned portfolio is called a hedging strategy.

Black, Merton, and Scholes developed the theory that allowed the determination of the value of the hedging strategy, which is also the fair price of the option, at any time t. To deduce it, we used the Feynman–Kac formula studied in Chaper 10. A numerical example was shown using real market data.

In Section 14.4 we reached the Black–Scholes formula using a different method, namely by showing that, using discounted prices (i.e. relative prices taking the price of the riskless asset as reference), one can, using Girsanov's theorem, change the drift of the discounted stock price to zero under a risk neutral probability P^*. Since the same happens to the discounted option price, this is therefore a martingale for the probability P^*. Since its expression is known at time T, the conditional expectation of that value, under the probability P^*, given the stock price at time t, gives the discounted option price at any time $t \in [0, T]$. The final result is obviously the same as in the previous method. The new probability P^* is known as *risk-neutral probability* because, under P^*, which distorts the probabilities of the market scenarios, the risky stock prices appear (just appear!) to have an average return rate exactly equal to the return rate of the riskless asset and so the market and the investors seem (just seem!) to be neutral with respect to risk.

The advantage of this new method based on Girsanov's theorem is that it can be easily approximated when we are unable to obtain an exact explicit expression whatever the method we use. One way to approximate it is to use the binomial model or Cox–Ross–Rubenstein (CRR) model, in which time is discretized into a sufficiently large number n of steps and P^* is discretized using a binomial distribution approximation (this means that the state space is also discretized). Although for European call options we had exact formulas, we have used the binomial model approximation with a toy $n = 3$ steps in Section 14.5 to illustrate how the method is applied.

In Section 14.6 we dealt with European put options on a stock by taking advantage of the put-call duality, which is based on the no arbitrage principle.

In Section 14.7 we dealt with American call options (which are treated the same way as European call options since the early exercise allowed in American options is never beneficial to the holder) and with American put options, where an early exercised is allowed and sometimes is beneficial to the holder. In this last case, no exact formulas to price the option are available and so the use of the binomial approximation is recommended and its application is shown with a real data example.

Finally, Section 14.8 takes a brief look at the immense and growing world of other financial applications.

Within this book we have covered the most relevant topics a beginner in the field should know and presented detailed applications to show the utility of SDEs and to help the reader, seeing them in real action, to better understand the concepts and methods. There is, however, a much bigger world that we cannot properly present in a book of this size, so there are other several important topics for which we could only give the readers some clues and direct the interested reader to other sources. We did not even give any clues about other more advanced areas such as Malliavin calculus, stochastic partial differential equations or control theory, which are also quite relevant and useful in applications. As for the applications, we have emphasized some biological and financial applications, but there are many other areas where SDEs have been successfully applied in modelling natural phenomena and possibly others in which SDEs could be but have not yet been applied.

If this introductory text has opened up for you the world of SDEs and their modelling applications and provided you with the basic instruments needed to travel further in this world, its mission has been accomplished.

References

Aït-Sahalia, Y. (2002) Maximum likelihood estimation of discretely sampled diffusions: a closed-form approximation approach. *Econometrica*, **70** (1), 223–262, doi:10.1111/1468-0262.00274.

Aït-Sahalia, Y. (2008) Closed-form likelihood expansions for multivariate diffusions. *Annals of Statistics*, **36** (2), 906–937, doi:10.1214/009053607000000622.

Allee, W.C., Emerson, A.E., Park, O., Park, T., and Schmidt, K.P. (1949) *Principles of Animal Ecology*, Saunders, Philadelphia.

Alvarez, L.H.R. and Shepp, L.A. (1998) Optimal harvesting in stochastically fluctuating populations. *Journal of Mathematical Biology*, **37** (2), 155–177, doi:10.1007/s002850050124.

Arnold, L. (1974) *Stochastic Differential Equations: Theory and Applications*, Wiley, New York.

Bachelier, L. (1900) Théorie de la speculation. *Annales Scientifiques de l'École Normal Supérieure 3ème Série*, **17**, 21–88.

Bandi, F.M. and Phillips, P.C.B. (2003) Fully nonparametric estimation of scalar diffusion models. *Econometrica*, **71** (1), 241–283, doi:10.1111/1468-0262.00395.

Basawa, I.V. and Rao, C.L.S.P. (1980) *Statistical Inference for Stochastic Processes*, Academic Press, London.

Beddington, J.R. and May, R.M. (1977) Harvesting natural populations in a randomly fluctuating environment. *Science*, **197** (4302), 463–463, doi:10.1126/science.197.4302.463.

Black, F. and Scholes, M. (1973) The pricing of options and corporate liabilities. *Journal of Political Economy*, **81** (3), 637–654, doi:10.1086/260062.

Bouleau, N. and Lépingle (1994) *Numerical Methods for Stochastic Processes*, Wiley, New York.

Braumann, C.A. (1985) Stochastic differential equation models of fisheries in an uncertain world: extinction probabilities, optimal fishing effort, and parameter estimation, in *Mathematics in Biology and Medicine* (eds V. Capasso, E. Grosso, and L.S. Paveri-Fontana), Springer, Berlin, pp. 201–206.

Introduction to Stochastic Differential Equations with Applications to Modelling in Biology and Finance, First Edition. Carlos A. Braumann.

© 2019 John Wiley & Sons Ltd. Published 2019 by John Wiley & Sons Ltd.

Companion Website: www.wiley.com/go/braumann/stochastic-differential-equations

Braumann, C.A. (1993) Model fitting and prediction in stochastic population growth models in random environments. *Bulletin of the International Statistical Institute*, **LV** (CP1), 163–164.

Braumann, C.A. (1999a) Comparison of geometric Brownian motions and applications to population growth and finance. *Bulletin of the International Statistical Institute*, **LVIII** (CP1), 125–126.

Braumann, C.A. (1999b) Variable effort fishing models in random environments. *Mathematical Biosciences*, **156** (1–2), 1–19, doi:10.1016/S0025-5564(98)10058-5.

Braumann, C.A. (2002) Variable effort harvesting models in random environments: generalization to density-dependent noise intensities. *Mathematical Biosciences*, **177 & 178**, 229–245, doi:10.1016/S0025-5564(01)00110-9.

Braumann, C.A. (2003) Modeling population growth in random environments: Ito or Stratonovich calculus? *Bulletin of the International Statistical Institute*, **LX** (CP1), 119–120.

Braumann, C.A. (2005) *Introdução às Equações Diferenciais Estocásticas e Aplicações*, Sociedade Portuguesa de Estatística, Lisboa.

Braumann, C.A. (2007a) Itô versus Stratonovich calculus in random population growth. *Mathematical Biosciences*, **206** (3), 81–107, doi:10.1016/j.mbs.2004.09.002.

Braumann, C.A. (2007b) Population growth in random environments: which stochastic calculus? *Bulletin of the International Statistical Institute*, **LXI**, 5802–5805.

Braumann, C.A. (2008) Growth and extinction of populations in randomly varying environments. *Computers and Mathematics with Applications*, **56** (3), 631–644, doi:10.1016/j.camwa.2008.01.006.

Braumann, C.A., Filipe, P.A., Carlos, C., and Roquete, C.J. (2009) Growth of individuals in randomly fluctuating environments, in *Proceedings of the International Conference in Computational and Mathematical Methods in Science e Engineering* (ed. J. Vigo-Aguiar), CMMSE, pp. 201–212.

Bravo, J.M.V. (2007) *Tábuas de Mortalidade Contemporâneas e Prospectivas: Modelos Estocásticos, Aplicações Actuariais e Cobertura do Risco de Longevidade*, Ph.D. thesis, Universidade de Évora, Évora.

Brites, N.M. (2017) *Stochastic Differential Equation Harvesting Models: Sustainable Policies and Profit Optimization*, Ph.D. thesis, Universidade de Évora, Évora.

Brites, N.M. and Braumann, C.A. (2017) Fisheries management in random environments: Comparison of harvesting policies for the logistic model. *Fisheries Research*, **195**, 238–246, doi:10.1016/j.fishres.2017.07.016.

Capocelli, R.M. and Ricciardi, L.M. (1974) A diffusion model for population growth in a random environment. *Theoretical Population Biology*, **5** (1), 28–41, doi:10.1016/0040-5809(74)90050-1.

Carlos, C. (2013) *Modelos de Crescimento Populacional em Ambiente Aleatório: Efeito de Incorreta Especificação do Modelo, Efeitos de Allee e Tempos de Extinção*, Ph.D. thesis, Universidade de Évora, Évora.

Carlos, C. and Braumann, C.A. (2005) Tempos de extinção para populações em ambiente aleatório, in *Estatística Jubilar* (eds C.A. Braumann, P. Infante, M.M. Oliveira, R. Alpizar-Jara, and F. Rosado), Edições SPE, Lisboa, pp. 133–142.

Carlos, C. and Braumann, C.A. (2006) Tempos de extinção para populações em ambiente aleatório e cálculos de Itô e Stratonovich, in *Ciência Estatística* (eds L.C. e Castro, E.G. Martins, C. Rocha, M.F. Oliveira, M.M. Leal, and F. Rosado), Edições SPE, Lisboa, pp. 229–238.

Carlos, C. and Braumann, C.A. (2014) Consequences of an incorrect model specification on population growth, in *New Advances in Statistical Modeling and Applications* (eds A. Pacheo, R. Santos, M.R. Oliveira, and C.D. Paulino), Springer, Heidelberg, pp. 105–113, doi:10.1007/978-3-319-05323-3_10.

Carlos, C. and Braumann, C.A. (2017) General population growth models with Allee effects in a random environment. *Ecological Complexity*, **30**, 26–33, doi:10.1016/j.ecocom.2016.09.003.

Carlos, C., Braumann, C.A., and Filipe, P.A. (2013) Models of individual growth in a random environment: study and application of first passage times, in *Advances in Regression, Survival Analysis, Extreme Values, Markov Processes and Other Statistical Applications* (eds J.L. daSilva, F. Caeiro, I. Natário, and C.A. Braumann), Springer, Berlin, pp. 103–111, doi:10.1007/978-3-642-34904-1_10.

Clark, C.W. (2010) *Mathematical Bioeconomics: The Optimal Management of Renewable Resources, 3rd edition*, Wiley, New York.

Clark, J.M.C. (1966) *The Representation of Nonlinear Stochastic Systems with Applications to Filtering*, Ph.D. thesis, Imperial College, London.

Cont, R. and Tankov, P. (2003) *Financial Modelling with Jum Processes*, Chapman & Hall/CRC, London/Boca Raton.

Cox, D.R. and Miller, H.D. (1965) *The Theory of Stochastic Processes*, Chapman & Hall/CRC, Boca Raton.

Dacunha-Castelle, D. and Florens-Zmirou, D. (1986) Estimators of coefficients of a diffusion from discrete observations. *Stochastics*, **19** (4), 263–284, doi:10.1080/17442508608833428.

Dennis, B. (1989) Allee effects: Population growth, critical density and the chance of extinction. *Natural Resource Modelling*, **3**, 481–539.

Dennis, B. (2002) Allee effects in stochastic populations. *Oikos*, **96** (3), 389–401, doi:10.1034/j.1600-0706.2002.960301.x.

Einstein, A. (1905) Über die von der molekularkinetischen theorie der wärme geforderte bewegung von in ruhenden flüssigkeiten suspendierten teilchen. *Annalen der Physik*, **322** (8), 549–560, doi:10.1002/andp.19053220806.

Einstein, A. (1956) *Investigations on the Theory of Brownian Movement*, Dover, New York.

Feldman, M.W. and Roughgarden, J. (1975) A population's stationary distribution and chance of extinction in a stochastic environment with remarks on the theory of species packing. *Theoretical Population Biology*, **7** (2), 197–207, doi:10.1016/0040-5809(75)90014-3.

Filipe, P.A. (2011) *Equações Diferenciais Estocásticas na Modelação do Crescimento Individual em Ambiente Aleatório*, Ph.D. thesis, Universidade de Évora, Évora.

Filipe, P.A., Braumann, C.A., Brites, N.M., and Roquete, C.J. (2010) Modelling animal growth in random environments: An application using nonparametric estimation. *Biometrical Journal*, **52** (5), 653–666, doi:10.1002/bimj.200900273.

Filipe, P.A., Braumann, C.A., and Carlos, C. (2015) Profit optimization for cattle growing in a randomly fluctuating environment. *Optimization: A Journal of Mathematical Programming and Operations Research*, **64** (6), 1393–1407, doi:10.1080/02331934.2014.974598.

Fleming, W.H. and Rishel, R.W. (1975) *Deterministic and Stochastic Optimal Control*, Springer, New York.

Florens-Zmirou, D. (1993) On estimating the diffusion coefficient from discrete observations. *Journal of Applied Probability*, **30** (4), 790–804, doi:10.2307/3214513.

Freedman, D. (1983) *Brownian Motion and Diffusion*, Springer, New York.

Garcia, O. (1983) A stochastic differential equation model for the height growth of forest stands. *Biometrics*, **39** (4), 1059–1072, doi:10.2307/2531339.

Gard, T.C. (1988) *Introduction to Stochastic Differential Equations*, Marcel Dekker, New York.

Gihman, I.I. and Skorohod, A.V. (1972) *Stochastic Differential Equations*, Springer, Berlin.

Gikhman, I.I. and Skorohod, A.V. (1969) *Introduction to the Theory of Random Processes*, W. B. Saunders, Philadelphia.

Gray, A.H. (1964) *Stability and Related Problems in Randomly Excited Systems*, Ph.D. thesis, California Institute of Technology Press.

Hanson, F.B. (2007) *Applied Stochastic Processes and Control for Jump Diffusions: Modeling, Analysis, and Computation*, Society for Industrial and Applied Mathematics.

Hanson, F.B. and Ryan, D. (1998) Optimal harvesting with both population and price dynamics. *Mathematical Biosciences*, **148** (2), 129–146, doi:10.1016/S0025-5564(97)10011-6.

Hening, A. and H. Nguyen, D. (2018) Persistence in stochastic Lotka–Volterra food chains with intraspecific competition. *Bulletin of Mathematical Biology*, **80** (10), 2527–2560, doi:0.1007/s11538-018-0468-5.

Iacus, S.M. (2008) *Simulation and Inference for Stochastic Differential Equations: With R Examples*, Springer.

Itô, K. (1951) *On Stochastic Differential Equations*, American Mathematical Society Memoirs No. 4.

Jiang, G.J. and Knight, J.L. (1997) A nonparametric approach to the estimation of diffusion processes, with an application to a short-term interest rate model. *Econometric Theory*, **13** (5), 615–645, doi:10.2307/3532619.

Kar, T.K. and Chakraborty, K. (2011) A bioeconomic assessment of the Bangladesh shrimp fishery. *World Journal of Modelling and Simulation*, **7** (1), 58–69.

Karatzas, I. and Shreve, S.E. (1991) *Brownian Motion and Stochastic Calculus*, Springer, New York.

Karatzas, I. and Shreve, S.E. (1998) *Methods of Mathematical Finance*, Springer, New York.

Karlin, S. and Taylor, H.M. (1981) *A Second Course in Stochastic Processes*, Academic Press, New York.

Kessler, M., Lindner, A., and Sørensen, M. (2012) *Statistical Methods for Stochastic Differential Equations*, Chapman and Hall/CRC, Boca Raton, London, New York.

Khashminsky, R.Z. (1969) *Ustoychivost' Sistem Diffentsual'nykh Uraveniy pri Sluchaynykh Vozmushcheniyskh* (in Russian), *Stability of Systems of Differential Equations in the Presence of Random Perturbations*, Nauka Press.

Kloeden, P.E. and Platen, E. (1992) *Numerical Solution of Stochastic Differential Equations*, Springer, Berlin.

Kloeden, P.E., Platen, E., and Schurz, H. (1994) *Numerical Solution of SDE through Computer Experiments*, Springer, Berlin.

Küchler, U. and Sørensen, M. (1997) *Exponential Families of Stochastic Processes*, Springer, New York.

Lagarto, S. (2014) *Modelos Estocásticos de Taxas de Mortalidade e Aplicações*, Ph.D. thesis, Universidade de Évora, Évora.

Lagarto, S. and Braumann, C.A. (2014) Modeling human population death rates: A bi-dimensional stochastic Gompertz model with correlated Wiener processes, in *New Advances in Statistical Modeling and Applications* (eds A. Pacheo, R. Santos, M.R. Oliveira, and C.D. Paulino), Springer, Heidelberg, pp. 95–103, doi:10.1007/978-3-319-05323-3_9.

Levins, R. (1969) The effect of random variations of different types on population growth. *Proceedings of the National Academy of Sciences USA*, **62** (4), 1062–1065.

Liu, M., Du, C., and Deng, M. (2018) Persistence and extinction of a modified Leslie–Gower Holling-type II stochastic predator–prey model with impulsive toxicant input in polluted environments. *Nonlinear Analysis: Hybrid Systems*, **27**, 177–190, doi:10.1016/j.nahs.2017.08.001.

Lungu, E. and Øksendal, B. (1997) Optimal harvesting from a population in a stochastic crowded environment. *Mathematical Biosciences*, **145** (1), 47–75, doi:10.1016/S0025-5564(97)00029-1.

Lv, Q. and Pitchford, J.W. (2007) Stochastic von Bertalanffy models, with applications to fish recruitment. *Journal of Theoretical Biology*, **244** (4), 640 – 655, doi:10.1016/j.jtbi.2006.09.009.

Malthus, T.R. (1798) *An Essay on the Principle of Population*, J. Johnson, in St. Paul's Church-yard, London. Library of Economics and Liberty [Online] available from www.econlib.org/library/Malthus/malPop.html.

May, R.M. (1973) Stability in randomly fluctuating versus deterministic environments. *American Naturalist*, **107** (957), 621–650, doi:10.1086/282863.

May, R.M. (1974) *Stability and Complexity in Model Ecosystems, 2nd edition*, Princeton University Press, New Jersey.

May, R.M. and MacArthur, R.H. (1972) Niche overlap as a function of environmental variability. *Proceedings of the National Academy of Sciences USA*, **69** (5), 1109–1113, doi:10.1073/pnas.69.5.1109.

McKean, H.P. (1969) *Stochastic Integrals*, Academic Press, New York.

Merton, R.C. (1971) Optimum consumption and portfolio rules in a continuous time model. *Journal of Economic Theory*, **3** (4), 373–413.

Merton, R.C. (1973) Theory of rational option pricing. *Bell Journal of Economics and Management Science*, **4** (1), 141–183.

Nicolau, J. (2003) Bias reduction in nonparametric diffusion coefficient estimation. *Econometric Theory*, **19** (5), 754–777, doi:10.2307/3533434.

Øksendal, B. (2003) *Stochastic Differential Equations. An Introduction with Applications, 6th edition*, Springer, New York.

Øksendal, B. and Sulem, A. (2007) *Applied Stochastic Control of Jump Diffusions, 2nd edition*, Springer, Berlin.

Pinheiro, S. (2016) Persistence and existence of stationary measures for a logistic growth model with predation. *Stochastic Models*, **32** (4), 513–538, doi:10.1080/15326349.2016.1174587.

Rudin, W. (1987) *Real and Complex Analysis, 3rd edition*, WCB/McGraw-Hill, New York.

Rudnicki, R. and Pichór, K. (2007) Influence of stochastic perturbation on prey–predator systems. *Mathematical Biosciences*, **206** (1), 108–119, doi:10.1016/j.mbs.2006.03.006.

Schuss, Z. (1980) *Theory and Applications of Stochastic Differential Equations*, Wiley, New York.

Shoutens, W. (2003) *Lévy Processes in Finance: Pricing Financial Derivatives*, Wiley, New York.

Sørensen, H. (2004) Parametric inference for diffusion processes observed at discrete points in time: a survey. *International Statistical Review*, **72** (3), 337–354, doi:10.1111/j.1751-5823.2004.tb00241.x.

Sørensen, M. (2009) Parametric inference for discretely sampled stochastic differential equations, in *Handbook of Financial Time Series* (eds T. Mikosch, J.P. Kreiß, R.A. Davis, and T.G. Andersen), Springer, Berlin, Heidelberg, pp. 531–553, doi:10.1007/978-3-540-71297-8_23.

Stratonovich, R.L. (1966) A new representation of stochastic integrals and equations. *SIAM Journal on Control*, **4** (2), 362–371, doi:10.1137/0304028.

Stroock, D.W. and Varadhan, S.R.S. (2006) *Multidimensional Diffusion Processes*, Springer, Berlin.

Suri, R. (2008) *Optimal Harvesting Strategies for Fisheries: A Differential Equations Approach*, Ph.D. thesis, Massey University, Albany New Zealand.

Turelli, M. (1978) A reexamination of stability in randomly varying environments with comments on the stochastic theory of limiting similarity. *Theoretical Population Biology*, **13** (2), 244–267, doi:10.1016/0040-5809(78)90045-X.

Uhlenbeck, G.E. and Ornstein, L.S. (1930) On the theory of Brownian motion. *Physical Review*, **36**, 823–841, doi:10.1103/PhysRev.36.823.

University of California, Berkeley USA and Max Planck Institute for Demographic Research, Germany (2018) Human Mortality Database, https://www.mortality.org/. Data downloaded on: 2018-08-26.

Wong, E. and Hajek, B. (1985) *Stochastic Processes in Engineering Systems*, Springer, New York.

Wong, E. and Zakai, M. (1965) The oscillations of stochastic integrals. *Probability Theory and Related Fields*, **4** (2), 103–112, doi:10.1007/BF00536744.

Wong, E. and Zakai, M. (1969) Riemann–Stieltjes approximation of stochastic integrals. *Probability Theory and Related Fields*, **12** (2), 87–97, doi:10.1007/BF00531642.

Index

Introduction to Stochastic Differential Equations with Applications to Modelling in Biology and Finance,
First Edition. Carlos A. Braumann.
© 2019 John Wiley & Sons Ltd. Published 2019 by John Wiley & Sons Ltd.
Companion Website: www.wiley.com/go/braumann/stochastic-differential-equations